CAMBRIDGE LIBRARY COLLECTION

Books of enduring scholarly value

Life Sciences

Until the nineteenth century, the various subjects now known as the life sciences were regarded either as arcane studies which had little impact on ordinary daily life, or as a genteel hobby for the leisured classes. The increasing academic rigour and systematisation brought to the study of botany, zoology and other disciplines, and their adoption in university curricula, are reflected in the books reissued in this series.

Narrative of a Voyage to the Pacific and Beering's Strait

Frederick William Beechey (1796–1856), naval officer and hydrographer, was born into a family of artists, joined the Navy at a young age and went on to travel the world to survey coastlines and oceans. He published several accounts of his expeditions to destinations including the Arctic and Africa. This two-volume work, first published in 1831, describes his voyage as commander of the *Blossom* in 1825–1828. The ship's mission was to support the exploration of the North-West Passage by travelling eastwards via the Bering Strait to meet the explorers Sir John Franklin and Sir Edward Parry who were travelling west from the North Atlantic. Volume 2 follows the expedition from California, where it had overwintered, via Hawai'i and China, back to the Bering Strait for a second summer. It includes a vocabulary of Eskimo words, notes on harbours and navigation, and a vivid description of the northern lights.

Cambridge University Press has long been a pioneer in the reissuing of out-of-print titles from its own backlist, producing digital reprints of books that are still sought after by scholars and students but could not be reprinted economically using traditional technology. The Cambridge Library Collection extends this activity to a wider range of books which are still of importance to researchers and professionals, either for the source material they contain, or as landmarks in the history of their academic discipline.

Drawing from the world-renowned collections in the Cambridge University Library, and guided by the advice of experts in each subject area, Cambridge University Press is using state-of-the-art scanning machines in its own Printing House to capture the content of each book selected for inclusion. The files are processed to give a consistently clear, crisp image, and the books finished to the high quality standard for which the Press is recognised around the world. The latest print-on-demand technology ensures that the books will remain available indefinitely, and that orders for single or multiple copies can quickly be supplied.

The Cambridge Library Collection will bring back to life books of enduring scholarly value (including out-of-copyright works originally issued by other publishers) across a wide range of disciplines in the humanities and social sciences and in science and technology.

Narrative of a Voyage to the Pacific and Beering's Strait

Performed in His Majesty's Ship Blossom, in the years 1825, 26, 27, 28

Volume 2

Frederick William Beechey

CAMBRIDGE UNIVERSITY PRESS

Cambridge, New York, Melbourne, Madrid, Cape Town,
Singapore, São Paolo, Delhi, Tokyo, Mexico City

Published in the United States of America by Cambridge University Press, New York

www.cambridge.org
Information on this title: www.cambridge.org/9781108031042

This edition first published 1831
This digitally printed version 2011

ISBN 978-1-108-03104-2 Paperback

NARRATIVE

OF A

VOYAGE TO THE PACIFIC

AND BEERING'S STRAIT,

TO CO-OPERATE WITH

THE POLAR EXPEDITIONS:

PERFORMED IN

HIS MAJESTY'S SHIP BLOSSOM,

UNDER THE COMMAND OF

CAPTAIN F. W. BEECHEY, R.N.

F.R.S. &c.

IN THE YEARS 1825, 26, 27, 28.

PUBLISHED BY AUTHORITY OF THE LORDS COMMISSIONERS OF THE ADMIRALTY.

IN TWO VOLUMES.

VOL. II.

LONDON:
HENRY COLBURN AND RICHARD BENTLEY,
NEW BURLINGTON STREET.

1831.

CONTENTS

OF THE SECOND VOLUME.

CHAPTER I.

Arrive at Sán Francisco—Description of the Harbour, Presisidio, and the Missions—Occupations—Dissatisfied State of the Garrison and the Priesthood—Contemplated Plan of settling the Indians in the Missions—Occupations of the converted Indians—Manner of making Converts—Expedition against the Tribe of Cosemenes—Official Despatch—Overland Journey to Monterey—Scarcity of Provisions at that place—Plan of the Voyage altered in consequence—Departure........................ Page 1

CHAPTER II.

Observations on the Country of California and its Trade—Climate—Meteorological Remarks—Short Account of the Wild Indians—Natural Productions—Monterey—Mission of San Carlos—Departure.................................... 66

CHAPTER III.

Passage to the Sandwich Islands—Woahoo—Historical Sketch of the Islands — Progress in Civilization — Sandal Wood—Resources of the Government—Slow Progress of Education—Efforts of the Missionaries—Unsuccessful Result of their Zeal—Sentiments of the King and Chiefs—Entertainment given by the King—Death of Krymakoo—Wailing Scene—Departure of Kahumana for Owyhee.................................... 88

CHAPTER IV.

Further Remarks on the Inhabitants—Treaty of Alliance—Climate—Medicinal Properties of the Ava—Supplies—Departure—Passage to China—Ladrone and Bashee Islands—Arrival at Macao—Transactions there—Departure—Botel Tobago Xima—Arrival at the Great Loo Choo 112

CHAPTER V.

Appearance of Loo Choo—Visits of the Natives—Deputation—Permission given to land—Excursions into the Country—Discover Money in circulation—Mandarin visits the Ship—Departure of a Junk with Tribute—Visit of the Mandarin returned—Further Intercourse—Transactions of the Ship—Departure—Observations upon the Religion, Manners, and Customs of the People; upon their Laws, Money, Weapons, and Punishments; their Manufactures and Trade—Remarks upon the Country, its Productions, and Climate—Directions for entering the Port—Historical Sketch of the Kingdom of Loo Choo 143

CHAPTER VI.

Passage from Loo Choo eastward—Arrive at Port Lloyd in the Yslas del Arzobispo—Description of those Islands—Passage to Kamtchatka—Arrival at Petropaulski—Notice of that Place—Departure—Pass Beering's Strait—Enter Kotzebue Sound—Prosecute the Voyage to the Northward Stopped by the Ice—Return to the Southward—Discover Port Clarence and Grantley Harbour—Description of these Harbours—Return to Kotzebue Sound—Ship strikes upon a Shoal................................ 227

CHAPTER VII.

Arrive at Chamisso Island—Find the Barge wrecked—Lieutenant Belcher's Proceedings—Conduct of the Natives—Approach of Winter—Final Departure from the Polar Sea—Observations upon the Probability of the North-West Passage from the Pacific—Remarks upon the Tribe inhabiting the North-West Coast of America—Return to California—Touch at San Blas, Valparaiso, Coquimbo, Rio Janeiro—Conclusion 273

APPENDIX.

Fossil Remains 331
Mexican Bees.. 357
Esquimaux Vocabulary............................... 366
Nautical Remarks.................................... 384
Geographical Positions.............................. 341
Aurora Borealis..................................... 345

Directions to the Binder for inserting the Plates and Charts.

Californians throwing the Lasso	..	*to face page* 63
View of Napa-Kiang (Loo Choo)	..	.. 144
Loo Choo Sepulchre	.. ..	.. 162
Natives of Loo Choo	.. ..	.. 167
Departure of a Loo Choo Junk with Tribute		.. 171
Kwan-yin (Loo Choo Deity) ..	..	.. 194
Fossil Remains, Plate 1	.. ..	.. 334
——————— Plate 2	.. ..	.. 335
——————— Plate 3	.. ..	.. 336
Section of a Mexican Bee-hive	..	.. 357

VOYAGE

TO THE

PACIFIC AND BEERING'S STRAIT.

CHAPTER I.

Arrive at Sán Francisco—Description of the Harbour, Presidio, and the Missions—Occupations—Dissatisfied State of the Garrison and the Priesthood—Contemplated Plan of settling the Indians in the Missions—Occupations of the converted Indians—Manner of making Converts—Expedition against the Tribe of Cosemenes—Official Despatch—Overland Journey to Monterey—Scarcity of Provisions at that Place—Plan of the Voyage altered in consequence—Departure.

Nov. 1826.

WHEN the day broke, we found ourselves about four miles from the land. It was a beautiful morning, with just sufficient freshness in the air to exhilarate without chilling. The tops of the mountains, the only part of the land visible, formed two ranges, between which our port was situated; though its entrance, as well as the valleys and the low lands, were still covered with the morning mist condensed around the bases of the mountains. We bore up for the opening between the ranges, anxious for the rising sun to withdraw the veil, that we might obtain a view of

CHAP. I. Nov. 1826.

the harbour, and form our judgment of the country in which we were about to pass the next few weeks. As we advanced, the beams of the rising sun gradually descended the hills, until the mist, dispelled from the land, rolled on before the refreshing sea wind, discovering cape after cape, and exhibiting a luxuriant country apparently abounding in wood and rivers. At length two low promontories, the southern one distinguished by a fort and a Mexican flag, marked the narrow entrance of the port.

We spread our sails with all the anxiety of persons who had long been secluded from civilized society, and deprived of wholesome aliment; but after the first effort of the breeze, it died away and left us becalmed in a heavy N. W. swell.

Off the harbour of Sán Francisco there is a bar which extends from the northern shore, gradually deepening its water until it approaches the peninsula on the opposite side*, where nine fathoms may be carried over it. Of this bar, however, we were ignorant, and naturally steered directly for the harbour, in doing which the depth of water gradually diminished to five fathoms. This would have been of no consequence, had it not been for a swell which rolled so heavily over the bank that it continually broke; and though our depth of water was never less than $4\frac{1}{2}$ fathoms, the ship on two or three occasions disturbed the sand with her keel. The tide was unfortunately against us, and the swell propelled the ship just sufficiently fast for her to steer without gaining any ground, so that we remained in this unpleasant situation several hours.

* The best part for crossing is with the island of Alcatrasses in one with the fort.

At length a breeze sprung up, and we entered the port, and dropped our anchor in the spot where Vancouver had moored his ship thirty-three years before. As we passed the entrance, a heavy sea rolling violently upon a reef of rocks on our left * bespoke the danger of approaching that side too close in light or baffling winds; while some scattered rocks with deep water round them skirting the shore on our right, marked that side also as dangerous; so that the entrance may be justly considered difficult. Beyond these rocks, however, near the fort, there is a bay in which, if necessary, ships may drop their anchor.

The fort, which we passed upon our right, mounts nine guns, and is built upon a promontory on the south side of the entrance, apparently so near to the precipice, that one side will, before long, be precipitated over it by the gradual breaking away of the rock. Its situation, nevertheless, is good, as regards the defence of the entrance; but it is commanded by a rising ground behind it. As we passed, a soldier protruded a speaking-trumpet through one of the embrasures, and hailed us with a stentorian voice, but we could not distinguish what was said. This custom of hailing vessels has arisen from there being no boat belonging to the garrison, and the inconvenience felt by the governor, in having to wait for a report of arrivals, until the masters of the vessels could send their boats on shore.

The port of Sán Francisco does not show itself to advantage until after the fort is passed, when it breaks upon the view, and forcibly impresses the spectator with the magnificence of the harbour. He then be-

* This reef lies three quarters of a mile from Punta Boneta.

CHAP. I. Nov. 1826.

holds a broad sheet of water, sufficiently extensive to contain all the British navy, with convenient coves, anchorage in every part, and, around, a country diversified with hill ánd dale, partly wooded, and partly disposed in pasture lands of the richest kind, abounding in herds of cattle. In short, the only objects wanting to complete the interest of the scene are some useful establishments and comfortable residences on the grassy borders of the harbour, the absence of which creates an involuntary regret, that so fine a country, abounding in all that is essential to man, should be allowed to remain in such a state of neglect. So poorly did the place appear to be peopled that a sickly column of smoke rising from within some dilapidated walls, misnamed the presidio or protection, was the only indication we had of the country being inhabited.

The harbour stretches to the S. E. to the distance of thirty miles, and affords a water communication between the missions of Sán José, Sánta Clára, and the presidio, which is built upon a peninsula about five miles in width. On the north the harbour is contracted to a strait, which communicates with a basin ten miles wide, with a channel across it sufficiently deep for frigates, though they cannot come near the land, on account of the mud. A creek on the N. W. side of this basin leads up to the new mission of Sán Francisco Solano; and a strait to the eastward, named Estrécho de Karquines, communicates with another basin into which three rivers discharge themselves, and bring down so large a body of water that the estrécho is from ten to eleven fathoms deep. These rivers are named Jesus Maria, El Sacraménto, and Sán Joachin: the first, I was informed, takes a northerly direction, passes at the back of Bodega, and extends beyond Cape Mendocino. El Sacraménto trends to the N. E., and is said to have its rise in the rocky mountains near the source of the Columbia. The other, Sán Joachin, stretches to the southward, through the country of the Bolbones, and is divided from the S. E. arm of the harbour by a range of mountains.

When Langsdorff was at this port, an expedition was undertaken by Don Louis Arguello and Padre Uria to make converts, and to inquire into the nature of the country in the vicinity of the Sierra nevada; and I learned from Don Louis, I believe a son of the commander, that they traced the Sacramento seventy or eighty leagues up, and that it was there very wide and deep, but that he had no boat to ascertain its depth. The Padre had it in contemplation to form a settlement in that direction, which he thought would become very rich in a short time by the number of Indians who would flock to it; but as it was never done, I presume he found material obstacles to his design.

As we opened out the several islands and stopping places in the harbour, we noticed seven American whalers at anchor at Sausalito, not one of which showed their colours; we passed them and anchored off a small bay named Yerba Buena, from the luxuriance of its vegetation, about a league distant from both the presidio and the mission of Sán Francisco. I immediately went on shore to pay my respects to Don Ignacio Martinez, a lieutenant in the Mexican army, acting governor in the absence of Don Louis, and to the priest, whose name was Tomaso, both of whom gave me a very hospitable and friendly reception, and offered their services in any way they might

CHAP. I. Nov. 1826.

be required. Our first inquiries naturally related to supplies, which we were disappointed to find not at all equal to what had been reported; in short, it seemed that with the exception of flour, fresh beef, vegetables, and salt, which might be procured through the missions, we should have to depend upon the American vessels for whatever else we might want, or upon what might chance to be in store at Monterey, a port of more importance than Sán Francisco, and from being the residence of a branch of a respectable firm in Lima, better supplied with the means of refitting vessels after a long sea voyage.

It was evident from this report that the supplies were likely to be very inadequate to our wants; but that no opportunity of obtaining them might be lost, I despatched Mr. Collie the surgeon, and Mr. Marsh the purser, overland to Monterey with Mr. Evans as interpreter, with orders to procure for the ship what medicines, provisions, and other stores were to be had, and to negotiate government bills, on which the exchange was far more favourable there than at the Sandwich Islands. The governor politely furnished a passport and a guard for this service; and our hospitable friend Tomaso, the padre of the mission, provided horses for them free of any charge. In the mean time we arranged with a relation of the governor for the daily supply of the ship's company, an arrangement which it afterwards appeared increased the jealousy that had long existed between the presidio and the missions, by transferring to the pocket of the commandant the profits that would otherwise have been reaped by the padre.

We were happy to find the country around our anchorage abounding in game of all kinds, so plentiful, indeed, as soon to lessen the desire of pursuit; still there were many inducements to both the officers and seamen to land and enjoy themselves; and as it was for the benefit of the service that they should recruit their health and strength as soon as possible, every facility was afforded them. Horses were fortunately very cheap, from nine shillings to seven pounds apiece, so that riding became a favourite amusement; and the Spaniards finding they could make a good market by letting out their stud, appeared with them every Sunday opposite the ship, ready saddled for the occasion, as this was a day on which I allowed every man to go out of the ship. Some of the officers purchased horses and tethered them near the place, but the Spaniards finding this to interfere with their market, contrived to let them loose on the Saturday night, in order that the officers might be compelled to hire others on the following day. The only obstacle to the enjoyment of this amusement was the scarcity of saddles and bridles, some of which cost ten times as much as a decent horse. The ingenuity of the seamen generally obviated these difficulties, while some borrowed or hired saddles of the natives: for my own part, I purchased a decent looking horse for about thirty-five shillings sterling, and on my departure presented it to a Spaniard, who had lent me the necessary accoutrements for it during my stay, which answered the purpose of both parties, as he was pleased with his present, and I had my ride for about a shilling a day: a useful hint to persons who may be similarly circumstanced.

Such of the seamen as would not venture on horseback made parties to visit the presidio and mission, where they found themselves welcome guests with the

CHAP. I. Nov. 1826.

Spanish soldiers. These two places were the only buildings within many miles of us, and they fortunately supplied just enough spirits to allow the people to enjoy themselves with their friends, without indulging in much excess—a very great advantage in a seaport.

The roads leading to these two great places of attraction in a short time became well beaten, and that to the mission very much improved, by having the boughs removed which before overhung it. It was at first in contemplation to hire a Spaniard to lop them; but our pioneers, who stopped at nothing, soon tore them all away, except one, a large stump, which resisted every attack, and unhorsed several of its assailants.

Martinez was always glad to see the officers at the presidio, and made them welcome to what he had. Indeed, nothing seemed to give him greater pleasure than our partaking of his family dinner; the greater part of which was dressed by his wife and daughters, who prided themselves on their proficiency in the art of cooking. It was not, however, entirely for the satisfaction of presenting us with a well-prepared repast that they were induced to indulge in this humble occupation: poor Martinez had a very numerous offspring to provide for out of his salary, which was then eleven years in arrears. He had a sorry prospect before him, as, a short time previous to our visit, the government, by way of paying up these arrears, sent a brig with a cargo of paper cigars to be issued to the troops in lieu of dollars; but, as Martinez justly observed, cigars would not satisfy the families of the soldiers, and the compromise was refused. The cargo was, however, landed at Monterey and placed under the charge of the governor, where all other tobacco is contraband; and as the Spaniards are fond of smoking, it stands a fair chance, in the course of time, of answering the intention of the government, particularly as the troops apply for these oftener than they otherwise would, under the impression of clearing off a score of wages that will never be settled in any other manner. Fortunately for Martinez and other veterans in this country, both vegetable and animal food are uncommonly cheap, and there are no fashions to create any expense of dress.

The governor's abode was in a corner of the presidio, and formed one end of a row, of which the other was occupied by a chapel; the opposite side was broken down, and little better than a heap of rubbish and bones, on which jackals, dogs, and vultures were constantly preying, the other two sides of the quadrangle contained storehouses, artificers' shops, and the gaol, all built in the humblest style with badly burnt bricks, and roofed with titles. The chapel and the governor's house were distinguished by being whitewashed.

Whether viewed at a distance or near, the establishment impresses a spectator with any other sentiment than that of its being a place of authority; and but for a tottering flag-staff, upon which was occasionally displayed the tri-coloured flag of Mexico, three rusty field pieces, and a half accoutred sentinel parading the gateway in charge of a few poor wretches heavily shackled, a visitor would be ignorant of the importance of the place. The neglect of the government to its establishments could not be more thoroughly evinced than in the dilapidated condition of

CHAP. I. Nov. 1826.

the building in question; and such was the dissatisfaction of the people that there was no inclination to improve their situation, or even to remedy many of the evils which they appeared to us to have the power to remove.

The plain upon which the presidio stands is well adapted to cultivation; but it is scarcely ever touched by the plough, and the garrison is entirely beholden to the missions for its resources. Each soldier has nominally about three pounds a month, out of which he is obliged to purchase his provision. If the governor were active, and the means were supplied, the country in the vicinity of the establishment might be made to yield enough wheat and vegetables for the troops, by which they would save that portion of their pay which now goes to the purchase of these necessary articles.

The garrison of Sán Francisco consists of seventy-six cavalry soldiers and a few artillerymen, distributed between the presidios and the missions, and consequently not more than half a dozen are at any time in one place.

They appeared to us to be very dissatisfied, owing not only to their pay being so many years in arrear, but to the duties which had been imposed both on the importation of foreign articles, and on those of the Mexican territory, amounting in the first instance to forty-two and a half per cent.; whereas, under the old government, two ships were annually sent from Acapulco with goods, which were sold duty free, and at their original cost in that country, and then, also, their pay being regularly discharged, they were able to purchase what they wanted. A further grievance has arisen by the refusal of the government to continue certain privileges which were enjoyed under the old system. At that time soldiers entered for a term of ten years, at the expiration of which they were allowed to retire to the Pueblós—villages erected for this purpose, and attached to the missions, where the men have a portion of ground allotted to them for the support of their families. This afforded a competency to many; and while it benefited them, it was of service to the government, as the country by that means became settled, and its security increased. But this privilege has latterly been withheld, and the applicants have been allowed only to possess the land and feed their cattle upon it, until it shall please the government to turn them off. The reason of this, I believe, was that Mexico was beginning to turn her attention to California, and was desirous of having settlers there from the southern districts, to whom it would be necessary to give lands; and until they could see what would be required for this purpose and for the government establishments, and had the limits of the property already allotted, defined, they did not wish to make any new grants. The real cause, however, was not explained to the soldiers; they merely heard that they would not have the land ceded to them for life as usual, and they were consequently much dissatisfied.

The same feeling of discontent that was experienced by the garrison, pervaded the missions, in consequence of some new regulations of the republican government, the first and most grievous of which was the discontinuance of a salary of 400 dollars per annum, heretofore allowed to each of the padres: the support the former government had given to the missions amounted, according to Langsdorff, to a million piastres a year. Another grievance was, the requisi-

CHAP. I. Nov. 1826.

tion of an oath of allegiance to the reigning authorities, which these holy men considered so egregious a violation of their former pledge to the king of Spain, that, until he renounced his sovereignty over the country, they could not conscientiously take it; and, much as they were attached to the place in which they had passed a large portion of their lives, and though by quitting it they would be reduced to the utmost penury—yet, so much did they regard this pledge, that they were prepared to leave the country, and to seek an asylum in any other that would afford it them. Indeed, the Prefect preferring his expulsion to renouncing his allegiance, had already received his dismissal, and was ready at the seaport of Monterey to embark in any vessel the government might appoint to receive him. A third grievance, and one which, when duly considered, was of some importance, not only to the missions but to the country in general, was an order to liberate all those converted Indians from the missions who bore good characters, and had been taught the art of agriculture, or were masters of a trade, and were capable of supporting themselves, giving them portions of land to cultivate, so arranged that they should be divided into parishes, with curates to superintend them, subservient to the clergy of the missions, who were to proceed to the conversion of the Indians as usual, and to train them for the domesticated state of society in contemplation.

This philanthropic system at first sight appeared to be a very excellent one, and every friend to the rights of man would naturally join in a wish for its prosperity; but the Mexican government could not have sufficiently considered the state of California, and the disposition of the Indians, or they would have known it could not possibly succeed without long previous training, and then it would require to be introduced by slow degrees.

The Indians whom this law emancipated were essential to the support of the missions, not only for conducting their agricultural concerns, but for keeping in subordination by force and example those whom disobedience and ignorance would exempt from the privilege ; and as a necessary consequence of this indulgence the missions would be ruined before the system could be brought into effect, even supposing the Indians capable of conducting their own affairs. So far from this being the case, however, they were known to possess neither the will, the steadiness, nor the patience to provide for themselves. Accustomed, many of them from their infancy, to as much restraint as children, and to execute, mechanically, what they were desired and no more, without even entertaining a thought for their future welfare, it was natural that such persons, when released from this discipline, should abandon themselves entirely to their favourite amusements, pastimes, and vices. Those also who had been converted in later life would return to their former habits, and having once again tasted the blessings of freedom, which confinement and discipline must have rendered doubly desirable, would forget all restraint, and then being joined by the wild discontented Indians, they would be more formidable enemies to the missions than before, inasmuch as they would be more enlightened. But I will not anticipate the result, which we had an opportunity of seeing on our return the following year ; and from which the reader will be able to judge how the system worked.

CHAP. I. Nov. 1826.

The padres, however, dreading the worst, were very discontented, and many would willingly have quitted the country for Manilla. The government appeared to be aware of this feeling, as they sent some young priests from Mexico to supplant those who were disaffected, and desired that they should be trained up in the mission, and should make themselves acquainted with the language and usages of the Indians, in order that they might not promote discontent by any sudden innovation.

The missions have hitherto been of the highest importance to California, and the government cannot be too careful to promote their welfare, as the prosperity of the country in a great measure is dependent upon them, and must continue to be so until settlers from the mother country can be induced to resort thither. As they are of such consequence, I shall enter somewhat minutely into a description of them. In Upper California there are twenty-one of these establishments, of which nine are attached to the presidios of Monterey and Sán Francisco, and contain about 7000 converts. They are in order as follow from north to south:—

Presidio	Mission	Established	Converts.
Sán Francisco.	Sán Francisco Solano, established in	1822,	about 1000
	Sán Raphael	1817	250
	Sán Francisco	1776	260
	Sán José	1797	1800
	Sánta Clara	1777	1500
Monterey.	Sánta Cruz	1797	300
	Sán Juan	1797	1100
	Sán Carlos	1770	200
	La Soledad	——	500
			6910

Sán Antonio	Buena Vistura	
Sán Miguel	Sán Fernando	
Sán Luis	Sán Gabriel	
De la Purissima	Sán Juan Capistram	
Sánta Ignes	Sán Luis Rey	3000
Sánta Barbara	Sán Tomaso	

I could not learn the number of Indians which are in each of the missions to the southward of Soledád, but they were stated collectively to amount to 20,000: on this head I must observe that the padres either would not say, or did not know exactly, how many there were, even in their own missions, much less the number contained in those to the southward: and the accounts were at all times so various that the above computation can be only an approximation. Almost all these establishments cultivate large portions of of land, and rear cattle, the hides and tallow of which alone form a small trade, of which the importance may be judged from the fact of a merchant at Monterey having paid 36,000 dollars in one year to a mission, which was not one of the largest, for its hide, tallow, and Indian labour. Though the system they pursue is not calculated to raise the colony to any great prosperity, yet the neglect of the missions would not long precede the ruin of the presidios, and of the whole of the district. Indeed, with the exception of two pueblos, containing about seven hundred persons, and a few farm houses widely scattered over the country, there are no other buildings to the northward of Monterey: thus, while the missions furnish the means of subsistence to the presidios, the body of men they contain keeps the wild Indians in check, and prevents their making incursions on the settlers.

CHAP. I. Nov. 1826.

Each mission has fifteen square miles of ground allotted to it. The buildings are variously laid out, and adapted in size to the number of Indians which they contain; some are inclosed by a high wall, as at Sán Carlos, while others consist merely of a few rows of huts, built with sun-burnt mud-bricks; many are whitewashed and tiled, and have a neat and comfortable appearance. It is not, however, every hut that has a white face to exhibit, as that in a great measure depends upon the industry and good conduct of the family who possess it, who are in such a case supplied with lime for the purpose. It is only the married persons and the officers of the establishment who are allowed these huts, the bachelors and spinsters having large places of their own, where they are separately incarcerated every night.

To each mission is attached a well-built church, better decorated in the interior than the external appearance of some would lead a stranger to suppose: they are well supplied with costly dresses for processions and feast days, to strike with admiration the senses of the gazing Indians, and on the whole are very respectable establishments. In some of these are a few tolerable pictures, among many bad ones; and those who have been able to obtain them are always provided with representations of hell and paradise: the former exhibiting in the most disgusting manner all the torments the imagination can fancy, for the purpose of striking terror into the simple Indians, who look upon the performance with fear and trembling. Such representations may perhaps be useful in exhibiting to the dull senses of the Indians what could not be conveyed in any other way, and so far they are desirable in the mission; but to an European the one is disgusting, and the other ludicrous. Each establishment is under the management of two priests if possible, who in Upper California belong to the mendicant order of Sán Francisco. They have under them a major-domo, and several subordinate officers, generally Spaniards, whose principal business is to overlook the labour of the Indians.

The object of the missions is to convert as many of the wild Indians as possible, and to train them up within the walls of the establishment in the exercise of a good life, and of some trade, so that they may in time be able to provide for themselves and become useful members of civilized society. As to the various methods employed for the purpose of bringing proselytes to the mission, there are several reports, of which some were not very creditable to the institution: nevertheless, on the whole I am of opinion that the priests are innocent, from a conviction that they are ignorant of the means employed by those who are under them. Whatever may be the system, and whether the Indians be really dragged from their homes and families by armed parties, as some assert, or not, and forced to exchange their life of freedom and wandering for one of confinement and restraint in the missions, the change according to our ideas of happiness would seem advantageous to them, as they lead a far better life in the missions than in their forests, where they are in a state of nudity, and are frequently obliged to depend solely upon wild acorns for their subsistence.

Immediately the Indians are brought to the mission they are placed under the tuition of some of the most enlightened of their countrymen, who teach them to repeat in Spanish the Lord's Prayer and certain passages in the Romish litany; and also, to cross themselves properly on entering the church. In a few days a willing Indian becomes a proficient in these mysteries, and suffers himself to be baptized, and duly initiated into the church. If, however, as it not unfrequently happens, any of the captured Indians show a repugnance to conversion, it is the practice to imprison them for a few days, and then to allow them to breathe a little fresh air in a walk round the mission, to observe the happy mode of life of their converted countrymen; after which they are again shut up, and thus continue to be incarcerated until they declare their readiness to renounce the religion of their forefathers.

I do not suppose that this apparently unjustifiable conduct would be pursued for any length of time; and I had never an opportunity of ascertaining the fact, as the Indians are so averse to confinement that they very soon become impressed with the manifestly superior and more comfortable mode of life of those who are at liberty, and in a very few days declare their readiness to have the new religion explained to them. A person acquainted with the language of the parties, of which there are sometimes several dialects in the same mission, is then selected to train them, and having duly prepared them takes his pupils to the padre to be baptized, and to receive the sacrament. Having become Christians they are put to trades, or if they have good voices they are taught music, and form part of the choir of the church. Thus there are in almost every mission weavers, tanners, shoemakers, bricklayers, carpenters, blacksmiths, and other artificers. Others again are taught husbandry, to rear cattle and horses; and some to cook for the mission: while the females card, clean, and spin wool, weave, and sew; and those who are married attend to their domestic concerns.

In requital of these benefits, the services of the Indian, for life, belong to the mission, and if any neophyte should repent of his apostacy from the religion of his ancestors and desert, an armed force is sent in pursuit of him, and drags him back to punishment apportioned to the degree of aggravation attached to his crime. It does not often happen that a voluntary convert succeeds in his attempt to escape, as the wild Indians have a great contempt and dislike for those who have entered the missions, and they will frequently not only refuse to re-admit them to their tribe, but will sometimes even discover their retreat to their pursuers. This animosity between the wild and converted Indians is of great importance to the missions, as it checks desertion, and is at the same time a powerful defence against the wild tribes, who consider their territory invaded, and have other just causes of complaint. The Indians, besides, from political motives, are, I fear, frequently encouraged in a contemptuous feeling towards their unconverted countrymen, by hearing them constantly held up to them in the degrading light of *béstias!* and in hearing the Spaniards distinguished by the appellation of *génte de razón*.

The produce of the land, and of the labour of the Indians, is appropriated to the support of the mission, and the overplus to amass a fund which is entirely at the disposal of the padres. In some of the establishments this must be very large, although the padres will not admit it, and always plead poverty. The government has lately demanded a part of this profit, but the priests who, it is said, think the Indians are

CHAP. I. Nov. 1826.

more entitled to it than the government, make small donations to them, and thus evade the tax by taking care there shall be no overplus. These donations in some of the missions are greater than in others, according as one establishment is more prosperous than another ; and on this also, in a great measure, depends the comforts of the dwellings, and the neatness, the cleanliness, and the clothiug of the people. In some of the missions much misery prevails, while in others there is a degree of cheerfulness and cleanliness which shows that many of the Indians require only care and proper management to make them as happy as their dull senses will admit of under a life of constraint.

The two missions of Sán Francisco and Sán José are examples of the contrast alluded to. The former in 1817 contained a thousand converts, who were housed in small huts around the mission ; but at present only two hundred and sixty remain—some have been sent, it is true, to the new mission of Sán Francisco Solano, but sickness and death have dealt with an unsparing hand among the others. The huts of the absentees, at the time of our visit, had all fallen to decay, and presented heaps of filth and rubbish ; while the remaining inmates of the mission were in as miserable a condition as it was possible to conceive, and were entirely regardless of their own comfort. Their hovels afforded scarcely any protection against the weather, and were black with smoke : some of the Indians were sleeping on the greasy floor ; others were grinding baked acorns to make into cakes, which constitute a large portion of their food. So little attention indeed had been paid even to health, that in one hut there was a quarter of beef suspended opposite a window in a very offensive and unwholesome state, but its owners were too indolent to throw it out. Sán José, on the other hand, was all neatness, cleanliness, and comfort; the Indians were amusing themselves between the hours of labour at their games; and the children, uniformly dressed in white bodices and scarlet petticoats, were playing at bat and ball. Part of this difference may arise from the habits of the people, who are of different tribes. Langsdorff observes, that the Indians of the mission of Sán José are the handsomest tribe in California, and in every way a finer race of men; and terms the neophytes of Sán Francisco pigmies compared with them. I cannot say that this remark occurred to me, and I think it probable that he may have been deceived by the apparently miserable condition of the people of Sán Francisco.

The children and adults of both sexes, in all the missions, are carefully locked up every night in separate apartments, and the keys are delivered into the possession of the padre; and as, in the daytime, their occupations lead to distinct places, unless they form a matrimonial alliance, they enjoy very little of each other's society. It, however, sometimes happens that they endeavour to evade the vigilance of their keepers, and are locked up with the opposite sex; but severe corporeal punishment, inflicted in the same manner as is practised in our schools, but with a whip instead of a rod, is sure to ensue if they are discovered. Though there may be occasional acts of tyranny, yet the general character of the padres is kind and benevolent, and in some of the missions, the converts are so much attached to them that I have heard them declare they would go with them, if they were obliged to quit the country. It is greatly to be regretted that with the

CHAP. I. Nov. 1826.

influence these men have over their pupils, and with the regard those pupils seem to have for their masters, that the priests do not interest themselves a little more in the education of their converts, the first step to which would be in making themselves acquainted with the Indian language. Many of the Indians surpass their pastors in this respect, and can speak the Spanish language, while scarcely one of the padres can make themselves understood by the Indians. They have besides, in general, a lamentable contempt for the intellect of these simple people, and think them incapable of improvement beyond a certain point. Notwithstanding this, the Indians are, in general, well clothed and fed; they have houses of their own, and if they are not confortable, it is, in a great measure, their own fault; their meals are given to them three times a day, and consist of thick gruel made of wheat, Indian corn, and sometimes acorns, to which at noon is generally added meat. Clothing of a better kind than that worn by the Indians is given to the officers of the missions, both as a reward for their services, and to create an emulation in others.

If it should happen that there is a scarcity of provisions, either through failure in the crop, or damage of that which is in store, as they have always two or three years in reserve, the Indians are sent off to the woods to provide for themselves, where, accustomed to hunt and fish, and game being very abundant, they find enough to subsist upon, and return to the mission, when they are required to reap the next year's harvest.

Having served ten years in the mission, an Indian may claim his liberty, provided any respectable settler will become surety for his future good conduct. A piece of ground is then allotted for his support, but he is never wholly free from the establishment, as part of his earnings must still be given to them. We heard of very few to whom this reward for servitude and good conduct had been granted; and it is not improbable that the padres are averse to it, as it deprives them of their best scholars. When these establishments were first founded, the Indians flocked to them in great numbers for the clothing with which the neophytes were supplied; but after they became acquainted with the nature of the institution, and felt themselves under restraint, many absconded. Even now, notwithstanding the difficulty of escaping, desertions are of frequent occurrence, owing probably, in some cases, to the fear of punishment—in others to the deserters having been originally inveigled into the mission by the converted Indians or the neophytes, as they are called by way of distinction to Los Gentíles, or the wild Indians—in other cases again to the fickleness of their own disposition.

Some of the converted Indians are occasionally stationed in places which are resorted to by the wild tribes for the purpose of offering them flattering accounts of the advantages of the mission, and of persuading them to abandon their barbarous life; while others obtain leave to go into the territory of the Gentiles to visit their friends, and are expected to bring back converts with them when they return. At a particular period of the year, also, when the Indians can be spared from the agricultural concerns of the establishment, many of them are permitted to take the launch of the mission, and make excursions to the Indian territory. All are anxious to go on such occasions, some to visit their friends, some to procure the

CHAP. I. Nov. 1826.

manufactures of their barbarous countrymen, which, by the by, are often better than their own; and some with the secret determination never to return. On these occasions the padres desire them to induce as many of their unconverted brethren as possible to accompany them back to the mission, of course, implying that this is to be done only by persuasion; but the boat being furnished with a cannon and musketry, and in every respect equipped for war, it too often happens that the neophytes, and the génte de razón, who superintend the direction of the boat, avail themselves of their superiority, with the desire of ingratiating themselves with their masters, and of receiving a reward. There are, besides, repeated acts of aggression which it is necessary to punish, all of which furnish proselytes. Women and children are generally the first objects of capture, as their husbands and parents sometimes voluntarily follow them into captivity. These misunderstanding and captivities keep up a perpetual enmity amongst the tribes, whose thirst for revenge is almost insatiable.

We had an opportunity of witnessing the tragical issue of one of these holyday excursions of the neophytes of the mission of Sán José. The launch was armed as usual, and placed under the superintendance of an alcalde of the mission, who, it appears from one statement (for there are several), converted the party of pleasure either into one of attack for the purpose of procuring proselytes, or of revenge upon a particular tribe for some aggression in which they were concerned. They proceeded up the Rio Sán Joachin until they came to the territory of a particular tribe named Cosemenes, when they disembarked with the gun, and encamped for the night near the village of *Los*

Gentiles, intending to make an attack upon them the next morning; but before they were prepared, the Gentiles, who had been apprised of their intention, and had collected a large body of friends, became the assailants, and pressed so hard upon the party that, notwithstanding they dealt death in every direction with their cannon and musketry, and were inspired with confidence by the contempt in which they held the valour and tactics of their unconverted countrymen, they were overpowered by numbers, and obliged to seek their safety in flight, and to leave the gun in the woods. Some regained the launch and were saved, and others found their way overland to the mission; but thirty-four of the party never returned to tell their tale.

There were other accounts of this unfortunate affair, one of which accused the padre of authorising the attack; and another stated that it was made in self-defence; but that which I have given appeared to be the most probable. That the reverend father should have sanctioned such a proceeding is a supposition so totally at variance with his character, that it will not obtain credit; and the other was in all probability the report of the alcalde to excuse his own conduct. They all agreed, however, in the fatal termination of their excursion, and the neophytes became so enraged at the news of the slaughter of their companions, that it was almost impossible to prevent them from proceeding forthwith to revenge their deaths. The padre was also greatly displeased at the result of the excursion, as the loss of so many Indians to the mission was of the greatest consequence, and the confidence with which the victory would inspire the Indians was equally alarming. He, therefore, joined with the

CHAP. I. Nov. 1826.

converted Indians in a determination to chastise and strike terror into the victorious tribe, and in concert with the governor planned an expedition against them. The mission furnished money, arms, Indians, and horses, and the presidio provided troops, headed by the alférez, Sanchez, a veteran who had been frequently engaged with the Indians, and was acquainted with every part of the country. The troops carried with them their armour and shields, as a defence against the arrows of the Indians: the armour consisted of a helmet and jerkin made of stout skins, quite impenetrable to an arrow, and the shield might almost vie with that of Ajax in the number of its folds.

The expedition set out on the 19th of November, and we heard nothing of it until the 27th; but two days after the troops had taken the field, some immense columns of smoke rising above the mountains in the direction of the Cosemenes, bespoke the conflagration of the village of the persecuted Gentiles. And on the day above-mentioned, the veteran Sanchez made a triumphant entry into the mission of San José, escorting forty miserable women and children, the gun that had been taken in the first battle, and other trophies of the field. This victory, so glorious, according to the ideas of the conqueror, was achieved with the loss of only one man on the part of the Christians, who was mortally wounded by the bursting of his own gun; but on the part of the enemy it was considerable, as Sanchez the morning after the battle counted forty-one men, women, and children, dead. It is remarkable that none of the prisoners were wounded, and it is greatly to be feared that the Christians, who could scarcely be prevented from revenging the death of their relations upon those who were brought to the mission, glutted their brutal passion on all the wounded who fell into their hands. The despatch which the alférez wrote to his commanding officer on the occasion of this successful termination of his expedition, will convey the best idea of what was executed, and their manner of conducting such an assault.

Translation—"Journal kept by citizen José Antonio Sanchez, ensign of cavalry of the presidio of Sán Francisco, during the enterprise against the Gentíles, called Cosemenes, for having put to death the neophytes of the mission of Sán José."—Written with gunpowder on the field of battle!

"On the morning of the 20th the troop commenced its march, and, after stopping to dine at Las Positas, reached the river Sán Joachin at eleven o'clock at night, when it halted. This day's march was performed without any accident, except that neighbour José Ancha was nearly losing his saddle. The next day the alférez determined to send forward the 'auxiliary neophytes' to construct balsas* for the troop to pass a river that was in advance of them. The troop followed, and all crossed in safety; but among the last of the horses that forded the river was one belonging to soldier Leandro Flores, who lost his bridle, threw his rider, and kicked him in the face and forehead; and as poor Flores could not swim, he was in a fair way of losing his life before he came within sight of the field of battle: assistance was speedily rendered, and he was saved. As Sanchez wished to surprise the enemy, he encamped until dusk, to avoid

* These are rafts made of rushes, and are the Indian substitute for canoes.

CHAP. I. Nov. 1826.

being seen by the wild Indians, who were travelling the country; several of whom were met and taken prisoners. At five they resumed their march; but neighbour Gexbano Chaboya being taken ill with a pain in his stomach, there was a temporary halt of the army: it however soon set forward again, and arrived at the river of Yachicumé at eleven at night, with only one accident, occasioned by the horse of neighbour Leandro Flores again throwing up his heels, and giving him a formidable fall.

"The troop lay in ambush until five o'clock the next evening, and then set out but here they were distressed by two horses running away; they were however both taken after a short march, which brought them to the river San Francisco, near the rancheria of their enemy the Cosemenes, and where the alférez commanded his troops to prepare for battle, by putting on their cuéros, or armour. The 23d the troop divided, and one division was sent round to intercept the Cosemenes, who had discovered the Christians, and were retreating; some of whom they made prisoners, and immediately the firing began. It had lasted about an hour, when the musket of soldier José Maria Garnez burst, and inflicted a mortal wound in his forehead; but this misfortune did not hinder the other soldiers from firing. The Gentiles also opened their fire of arrows, and the skirmishing became general. Towards noon a shout was heard in the north quarter, and twenty Gentiles were seen skirmishing with three Christians, two on foot and one on horseback, and presently another shout was heard, and the Christians were seen flying, and the Gentiles in pursuit of them, who had already captured the horse.

"It was now four o'clock, and the alférez, seeing that the Gentiles, who were in ambush, received little injury, disposed every thing for the retreat of the troops, and having burnt the rancheria, and seen some dead bodies, he retreated three quarters of a league, and encamped for the night. On the 24th the troops divided into two parties, one charged with booty and prisoners amounting to forty-four souls, mostly women.

"The other party went with the veteran Sanchez to the rancheria, to reconnoitre the dead bodies of which he counted forty-one men, women, and children. They met with an old woman there, the only one that was left alive, who was in so miserable a state that they showed their compassion by *taking no account of her*. The alférez then set out in search of the cannon that had been abandoned by the first expedition. The whole of the troop afterwards retreated, and arrived at the mission of Sán José on the night of the 27th."

This truly ludicrous account of an expedition of such trifling importance might appear to require an apology for its insertion, but it conveys so good an idea of the opposition to be expected by any power which might think proper to land upon the coast of California, that its omission might fairly be considered a neglect.

The prisoners they had captured were immediately enrolled in the list of the mission, except a nice little boy, whose mother was shot while running away with him in her arms, and he was sent to the presidio, and was, I heard, given to the alférez as a reward for his services. The poor little orphan had received a slight wound in his forehead; he wept bitterly at first, and refused to eat, but in time became reconciled to his fate.

CHAP. I. Nov. 1826.

Those who were taken to the mission were immediately converted, and were daily taught by the neophytes to repeat the Lord's prayer and certain hymns in the Spanish language. I happened to visit the mission about this time, and saw these unfortunate beings under tuition; they were clothed in blankets, and arranged in a row before a blind Indian, who understood their dialect, and was assisted by an alcalde to keep order. Their tutor began by desiring them to kneel, informing them that he was going to teach them the names of the persons composing the Trinity, and that they were to repeat in Spanish what he dictated.

The neophytes being thus arranged, the speaker began, "Santíssima Trinidáda, Dios, Jesu Cristo, Espíritu Santo"—pausing between each name, to listen if the simple Indians, who had never spoken a Spanish word before, pronounced it correctly, or any thing near the mark. After they had repeated these names satisfactorily, their blind tutor, after a pause added, "Santos"—and recapitulated the names of a great many saints, which finished the morning's tuition. I did not attend the next schooling to hear what was the ensuing task, but saw them arranged on their knees, repeating Spanish words as before.

They did not appear to me to pay much attention to what was going forward, and I observed to the padre that I thought their teachers had an arduous task; but he said they had never found any difficulty; that the Indians were accustomed to change their own gods, and that their conversion was in a measure habitual to them. I could not help smiling at this reason of the padre, but have no doubt it was very true; and that the party I saw would feel as little compunction at apostatizing again, whenever they should have an opportunity of returning to their own tribe.

The expenses of the late expedition fell heavy upon the mission, and I was glad to find that the padre thought it was paying very dear for so few converts, as in all probability it will lessen his desire to undertake another expedition; and the poor Indians will be spared the horrors of being butchered by their own countrymen, or dragged from their homes into perpetual captivity. He was also much concerned to think the Cosemenes had stood their ground so firmly, and he was under some little apprehension of an attack upon the mission. Impressed with this idea, and in order to defend himself the more effectually, he begged me to furnish him with a few fireworks, which he thought would strike terror into his enemies in case of necessity.

Morning and evening mass are daily performed in the missions, and high mass as it is appointed by the Romish Church, at which all the converted Indians are obliged to attend. The commemoration of the anniversary of the patroness saint took place during my visit at Sán José, and high mass was celebrated in the church. Before the prayers began, there was a procession of the young female Indians, with which I was highly pleased. They were neatly dressed in scarlet petticoats, and white bodices, and walked in a very orderly manner to the church, where they had places assigned to them apart from the males. After the bell had done tolling, several alguazils went round to the huts, to see if all the Indians were at church, and if they found any loitering within them, they exercised with tolerable freedom a long lash with a

CHAP. I. Nov. 1826.

broad thong at the end of it; a discipline which appeared the more tyrannical, as the church was not sufficiently capacious for all the attendants, and several sat upon the steps without; but the Indian women who had been captured in the affair with the Cosemenes were placed in a situation where they could see the costly images, and vessels of burning incense, and every thing that was going forward.

The congregation was arranged on both sides of the building, separated by a wide aisle passing along the centre, in which were stationed several alguazils with whips, canes, and goads, to preserve silence and maintain order, and, what seemed more difficult than either, to keep the congregation in their kneeling posture. The goads were better adapted to this purpose than the whips, as they would reach a long way, and inflict a sharp puncture without making any noise. The end of the church was occupied by a guard of soldiers under arms, with fixed bayonets; a precaution which I suppose experience had taught the necessity of observing. Above them there was a choir consisting of several Indian musicians, who performed very well indeed on various instruments, and sang the Te Deum in a very passable manner. The congregation was very attentive, but the gratification they appeared to derive from the music furnished another proof of the strong hold this portion of the ceremonies of the Romish church takes upon uninformed minds.

The worthy and benevolent priests of the mission devote almost the whole of their time to the duties of the establishment, and have a fatherly regard for those placed under them who are obedient and diligent; and too much praise cannot be bestowed upon them, considering that they have relinquished many of the enjoyments of life, and have embraced a voluntary exile in a distant and barbarous country. The only amusement which my hospitable host of the mission of Sán José indulged in, during my visit to that place, was during meal times, when he amused himself by throwing pancakes to the *muchachos*, a number of little Indian domestics, who stood gaping round the table. For this purpose, he had every day two piles of pancakes made of Indian corn ; and as soon as the ólla was removed, he would fix his eyes upon one of the boys, who immediately opened his mouth, and the padre, rolling up a cake, would say something ludicrous in allusion to the boy's appetite, or to the size of his mouth, and pitch the cake at him, which the imp would catch between his teeth, and devour with incredible rapidity, in order that he might be ready the sooner for another, as well as to please the padre, whose amusement consisted in a great measure in witnessing the sudden disappearance of the cake. In this manner the piles of cakes were gradually distributed among the boys, amidst much laughter, and occasional squabbling.

Nothing could exceed the kindness and consideration of these excellent men to their guests and to travellers, and they were seldom more pleased than when any one paid their mission a visit: we always fared well there, and even on fast days were provided with fish dressed in various ways, and preserves made with the fruit of the country. We had, however, occasionally some difficulty in maintaining our good temper, in consequence of the unpleasant remarks which the difference of our religion brought from the padres, who were very bigoted men, and invariably

CHAP. I. Nov. 1826.

introduced this subject. At other times they were very conversible, and some of them were ingenious and clever men; but they had been so long excluded from the civilized world, that their ideas and their politics, like the maps pinned against the walls, bore date of 1772, as near as I could read it for fly spots. Their geographical knowledge was equally backward, as my host at Sán José had never heard of the discoveries of Captain Cook; and because Otaheite was not placed upon his chart, he would scarcely credit its existence.

The Indians after their conversion are quiet and tractable, but extremely indolent, and given to intoxication, and other vices. Gambling in particular they indulge in to an unlimited extent: they pledge the very clothes on their backs, and not unfrequently have been known to play for each other's wives. They have several games of their own, besides some with cards, which have been taught them by the Spaniards. Those which are most common, and are derived from the wild Indians, are toussé, called by the Spaniards pares y nones, odd or even; escondido, or hunt the slipper; and takersia.

The first, though sometimes played as in England, generally consists in concealing a piece of wood in one hand, and holding out both for the guessing party to declare in which it is contained. The intense interest that is created by its performance has been amusingly described by Perouse. The second, escondido, needs no description; the last, takersia, requires some skill to play well, and consists in rolling a circular piece of wood with a hole in its centre along the ground, and throwing a spear through it as it rolls. If the spear pierces the hole, it counts ten towards the game; and if it arrests the wood in such a manner that it falls upon the spear, two is reckoned. It is a sport well calculated to improve the art of throwing the spear: but the game requires more practice to play it well than the Indians usually bestow upon it.

At some of the missions they pursue a custom said to be of great antiquity among the aborigines, and which appears to afford them much enjoyment. A mud house, or rather a large oven, called temeschal by the Spaniards, is built in a circular form, with a small entrance, and an aperture in the top for the smoke to escape through. Several persons enter this place quite naked and make a fire near the door, which they continue to feed with wood as long as they can bear the heat. In a short time they are thrown into a most profuse perspiration, they wring their hair, and scrape their skin with a sharp piece of wood or an iron hoop, in the same manner as coach horses are sometimes treated when they come in heated; and then plunge into a river or pond of cold water, which they always take care shall be near the temeschal.

A similar practice to this is mentioned by Shelekoff as being in use among the Konaghi, a tribe of Indians near Cook's River, who have a method of heating the oven with hot stones, by which they avoid the discomfort occasioned by the wood smoke; and, instead of scraping their skin with iron or bone, rub themselves with grass and twigs.

Formerly the missions had small villages attached to them, in which the Indians lived in a very filthy state; these have almost all disappeared since Vancouver's visit, and the converts are disposed of in huts as before described; and it is only when sickness prevails to a

CHAP. I. Nov. 1826.

great extent that it is necessary to erect these habitations, in order to separate the sick from those who are in health. Sickness in general prevails to an incredible extent in all the missions, and on comparing the census of the years 1786 and 1813, the proportion of deaths appears to be increasing. At the former period there had been only 7,701 Indians baptized, out of which 2,388 had died; but in 1813 there had been 37,437 deaths to only 57,328 baptisms.

The establishments are badly supplied with medicines, and the reverend fathers, their only medical advisers, are inconceivably ignorant of the use of them. In one mission there was a seaman who pretended to some skill in pharmacy, but he knew little or nothing of it, and perhaps often did more harm than good. The Indians are also extremely careless and obstinate, and prefer their own simples to any other remedies, which is not unfrequently the occasion of their disease having a fatal termination.

The Indians in general submit quietly to the discipline of the missions, yet insurrections have occasionally broken out, particularly in the early stage of the settlement, when father Tamoral and other priests suffered martyrdom.* In 1823, also, a priest was murdered in a general insurrection in the vicinity of Sán Luis Rey; and in 1827, the soldiers of the garrison were summoned to quell another riot in the same quarter.

The situations of the missions, particularly that of Sán José, are in general advantageously chosen. Each establishment has fifteen square miles of ground, of which part is cultivated, and the rest appropriated to

* Noticias de California, by Miguel Venegas.

the grazing and rearing of cattle; for in portioning out the ground, care has been taken to avoid that which is barren. The most productive farms are held by the missions of Sán José, Santa Clara, Sán Juan, and Sánta Cruz. That of Sán Francisco appears to be badly situated, in consequence of the cold fogs from the sea, which approach the mission through several deep valleys, and turn all the vegetation brown that is exposed to them, as is the case in Shetland with the tops of every tree that rises above the walls. Still, with care, more might be grown in this mission than it is at present made to produce. Sánta Cruz is rich in supplies, probably on account of the greater demand by merchant vessels, whalers in particular, who not unfrequently touch there the last thing on leaving the coast, and take on board what vegetables they require; the quantity of which is so considerable, that it not unfrequently happens that the missions are for a time completely drained. On this account it is advisable, on arriving at any of the ports, to take an early opportunity of ordering every thing that may be required.

A quantity of grain, such as wheat and Indian corn, is annually raised in all the missions, except Sán Francisco, which, notwithstanding it has a farm at Burri Burri, is sometimes obliged to have recourse to the other establishments. Barley and oats are said to be scarcely worth the cultivation, but beans, pease, and other leguminous vegetables are in abundance, and fruit is plentiful. The land requires no manure at present, and yields on an average twenty for one. Sán José reaps about 3,000 fanegas * of wheat annually.

Hides and tallow constitute the principal riches of the

* A fanega is one hundred pounds weight.

CHAP. I. Nov. 1826.

missions, and the staple commodity of the commerce of the country: a profitable revenue might also be derived from grain were the demand for it on the coast such as to encourage them to cultivate a larger quantity than is required by the Indians attached to the missions. Sán José, which possesses 15,000 head of cattle, cures about 2,000 hides annually, and as many bótas of tallow, which are either disposed of by contract to a mercantile establishment at Monterey, or to vessels in the harbour. The price of these hides may be judged by their finding a ready market on the Lima coast. Though there are a great many sheep in the country, as may be seen by the mission, Sán José alone possessing 3,000, yet there is no export of wool, in consequence of the consumption of that article in the manufacture of cloth for the missions.

Husbandry is still in a very backward state, and it is fortunate that the soil is so fertile, and that there are abundance of labourers to perform the work, or I verily believe the people would be contented to live upon acorns. Their ploughs appear to have descended from the patriarchal ages, and it is only a pity that a little of the skill and industry then employed upon them should not have devolved upon the present generation. It will scarcely be credited by agriculturists in other countries, that there were seventy ploughs and two hundred oxen at work upon a piece of light ground of ten acres; nor did the overseers appear to consider that number unnecessary, as the padre called our attention to this extraordinary advancement of the Indians in civilization, and pointed out the most able workmen as the ploughs passed us in succession. The greater part of these ploughs followed in the same furrow without making much impression, until they approached the padre, when the ploughman gave the necessary inclination of the hand, and the share got hold of the ground. It would have been good policy for the padre to have moved gradually along the field, by which he would have had it properly ploughed; but he seemed to be quite satisfied with the performance. Several of the missions, but particularly that of Sánta Barbara, make a wine resembling claret, though not near so palatable, and they also distil an ardent spirit resembling arrack.

In this part of California, besides the missions, there are several pueblos, or villages, occupied by Spaniards and their families, who have availed themselves of the privileges granted by the old government, and have relinquished the sword for the ploughshare. There are also a few settlers who are farmers, but, with these exceptions, the country is almost uninhabited. Perhaps I cannot convey a better idea of the deserted state of the country, or of the capability of its soil, than by inserting a short narrative which I have compiled from the journals of three of my officers who travelled over land from Sán Francisco to "the famous port of Monterey."

I have already stated that it was found expedient to make this journey to learn whether any supplies could be procured for the ship; and in consequence Mr. Collie, the surgeon, Mr. Marsh, the purser, and Mr. Evans, who was well acquainted with the Spanish language, were requested to proceed on this service. As it was of importance that no time should be lost in acquiring this information, they had very little time allowed them to prepare for so long, and, to seamen, so unusual a journey; but as the mode of travelling in that rude country admitted but few incumbrances, the

CHAP. I. Nov. 1826.

omission of these preparations was of less consequence.

In order to reach a tolerable halting place for the night, the first day's journey was necessarily long, and consequently by daylight on the 9th November the three officers were on their road to the mission; having found horses and an escort prepared in pursuance of previous arrangements.

Setting off at a round trot, they made the best of their way over three or four miles of ground so overgrown with dwarf oaks and other trees, that they were every moment in danger of being thrown from their horses, or having their eyes torn out by the branches as they passed. In half an hour, however, they reached the mission of Sán Francisco, and soon forgot the little annoyances they had hitherto met with in the hospitable welcome of the good priest, who regaled them with excellent pears and new milk. Nor was his conversation less palatable than his cheer; for, notwithstanding the introduction of half a dozen unnecessary *si senors* in each sentence, he contrived to amuse the vacant time with a flow of most genuine humour, for which Tomaso was always prepared, till the rattling accoutrements of a Californian dragoon announced the arrival of the passport from the governor. Intrusting their baggage to the care of two vaquéros (Indian cattle drivers) who were to accompany them, and receiving each a blessing from the padre, they set off with their escort about ten o'clock in the forenoon. The cavalcade consisted of three officers of the Blossom, the two vaquéros, and their champion the dragoon, preceded by nine or ten loose horses, driven on before as a relay, to be used when those they mounted should become fatigued. These

Rozinantes are not much inclined to deviate from the road, but if any thing should inspire them with a spirit of straying, the unerring lasso, the never-failing appendage to a Californian saddlebow, soon embraces their neck or their feet, and brings them back again to the right way.

I must not, however, permit the party to proceed farther without introducing to the notice of the reader the costume and equipment of this dragoon of California. As for his person, I do not find it described, but his dress consisted of a round blue cloth jacket with red cuffs and collar; blue velvet breeches, which being unbuttoned at the knees, gave greater display to a pair of white cotton stockings, cased more than half way in a pair of deer skin boots. A black hat, as broad in the brim as it was disproportionably low in the crown, kept in order, by its own weight, a profusion of dark hair, which met behind, and dangled half way down the back in the form of a thick queue. A long musket, with a fox skin bound round the lock, was balanced upon the pummel of the saddle; and our hero was further provided for defence against the Indians with a bull's hide shield, on which, notwithstanding the revolution of the colony, were emblazoned the royal arms of Spain, and by a double-fold deer skin cuirass as a covering for his body. Thus accoutred he bestrode a saddle, which retained him in his seat by a high pummel in front and a corresponding rise behind. His feet were armed at the heels with a tremendous pair of iron spurs, secured by a metal chain; and were thrust through an enormous pair of wooden box shaped stirrups. Such was the person into whose charge our shipmates were placed by the governor, with a passport which com-

CHAP. I. Nov. 1826.

manded him not to permit any person to interfere with the party either in its advance or on its return, and that it was to be escorted from place to place by a soldier.

Leaving the mission of Sán Francisco, the party receded from the only part of the country that is wooded for any considerable distance, and ascended a chain of hills about a thousand feet in height, where they had an extensive view, comprehending the sea, the Farallones rocks, and the distant Punta de los Reyes, a headland so named by the expedition under Sebastian Viscaino in 1602. The ridge which afforded this wide prospect was called Sierra de San Bruno, and for the most part was covered with a burnt-up grass, but such places as were bare presented to the eye of the geologist rocks of sandstone conglomerate, intersected by a few veins of jaspar. Winding through the Sierre de Sán Bruno, they crossed a river of that name, and opened out the broad arm of the sea which leads from the port to Sánta Clara, and is confined between the chain they were traversing and the Sierra de los Bolbones, distinguishable at a distance by a peaked mountain 3,783 feet high by trigonometrical measurement. Upon the summit of that part of the sierra bordering the arm of sea called Estrecho de Sán José, a thick wood, named Palos Colorados from its consisting principally of red cedar pine, stands conspicuous on the ridge. I mention this particularly, and wish to call attention to the circumstance, as the straggling trees at the south extreme of the wood are used as landmarks for avoiding a dangerous rock which we discovered in the harbour, and named after the Blossom.

About noon they reached a small cottage named

Burri Burri, about twelve miles from Sán Francisco; and being unused to travelling, especially upon Californian saddles, which are by no means constructed for comfort, they determined to rest, until the baggage that had been left in the rear should overtake them. The house in which they lodged was a small miserable mud cottage full of holes, which, however, afforded them some repose and some new milk. Its inhabitants had been engaged in tanning, in which process they used a liquid extracted from oak bark, contained in a hide suspended by the corners. They had also collected in great quantities a very useful root called in that country *amoles*, which seems to answer all the purposes of soap.

From Burri Burri, a continuation of the Sierra de San Bruno passes along the centre of the peninsula formed by the sea and the Estrecho de Sán José, and is separated from this arm of the harbour by a plain, upon which the travellers now descended from the mountains, and journied at a more easy and agreeable rate than they had done on the rugged paths among the hills. This plain near the sea is marshy, and having obtained the name of Las Salinas is probably overflowed occasionally by the sea. The number of wild geese which frequent it is quite extraordinary, and indeed would hardly be credited by any one who had not seen them covering whole acres of ground, or rising in myriads with a clang that may be heard at a very considerable distance. They are said to arrive in California in November, and to remain there until March. Their flesh in general is hard and fishy, but it was reported by padre Luis Gil, of the mission of Sánta Crux, that those which have yellow feet are exceptions to this, and are excellent eating. The

CHAP. I. Nov. 1826.

blackbirds are almost equally numerous, and in their distant flight resemble clouds. Among the marshes there were also a great many storks and cranes, which in Sán Francisco have the reputation of affording a most delicious repast.

Travelling onward, the hills on their right, known in that part as the Sierra del Sur, began to approach the road, which passing over a small eminence, opened out upon a wide country of meadow land, with clusters of fine oak free from underwood. It strongly resembled a nobleman's park: herds of cattle and horses were grazing upon the rich pasture, and numerous fallow-deer, startled at the approach of strangers bounded off to seek protection among the hills. The resemblance, however, could be traced no further. Instead of a noble mansion, in character with so fine a country, the party arrived at a miserable mud dwelling, before the door of which a number of half-naked Indians were basking in the sun. Several dead geese, deprived of their entrails, were fixed upon pegs around a large pole, for the purpose of decoying the living game into snares, which were placed for them in favourable situations. Heaps of bones also of various animals were lying about the place, and sadly disgraced the park-like scenery around. This spot is named Sán Matheo, and belongs to the mission of Sán Francisco.

Quitting this spot, they arrived at a farm-house about half way between Sán Francisco and Sánta Clara, called Las Pulgas (fleas); a name which afforded much mirth to our travellers, in which they were heartily joined by the inmates of the dwelling, who were very well aware that the name had not been bestowed without cause. It was a miserable habitation, with scarcely any furniture, surrounded by decaying hides and bones. Still, fatigue renders repose sweet upon whatsoever it can be indulged, and our party were glad enough to stretch themselves awhile upon a creaking couch, the only one in the hut, notwithstanding that the owner had a numerous family. Here, had there been accommodation, and had the place not acquired the reputation its name conveys, they would willingly have ended their day's journey; but the idea of las pulgas, sufficiently numerous in all the houses of California, determined them to proceed as soon as they conveniently could. The plain still continued animated with herds of cattle, horses, and sheep grazing; but the noble clusters of oak were now varied with shrubberies, which afforded a retreat to numerous coveys of Californian partridges, of which handsome species of game the first specimen was brought to England by the Blossom, and is now living in the gardens of the Zoological Society. They are excellent food; and the birds, in the country now under description, are so tame that they would often not start from a stone directed with Indian skill.

The sun went down before they reached Sánta Clara, which was to terminate that day's journey, and being unaccustomed to ride, the whole party were thoroughly fatigued. Indeed, so wearying was the journey even to the animals that bore them, that but for the relays of horses, which were now brought in with a lasso, they might have been compelled to pass the night upon the plain among the geese, the jackals, and the bears, which in the vicinity of Sánta Clara are by no means scarce. The pleasure of removing from a jaded horse to one that is fresh is not unknown probably to my readers, and our party rode

CHAP. I. Nov. 1826.

in comparative comfort the remainder of the journey, and reached the mission of Sánta Clara at eight o'clock.

Sánta Clara, distant by the road about forty miles from Sán Francisco, is situated in the extensive plain before described, which here, however, becomes more marshy than that part of the ground over which they had just travelled. It nevertheless continues to be occupied by herds of cattle, horses, sheep, and flocks of geese. Here, also, troops of jackals prowl about in the most daring manner, making the plain resound with their melancholy howlings; and indeed both wild and domesticated animals seem to lose their fear and become familiar with their tyrant man. The buildings of the establishment, which was founded in 1768, consists of a church, the dwelling-house of the priests, and five rows of buildings for the accommodation of 1,400 Indians, who since Vancouver's visit, have been thus provided with comparatively comfortable dwellings, instead of occupying straw huts, which were always wet and miserable. Attached to these are some excellent orchards, producing an abundance of apples and pears. Olives and grapes are also plentiful, and the padres are enabled to make from the latter about twenty barrels of wine annually. They besides grow a great quantity of wheat, beans, peas, and other vegetables. On the whole this is one of the best regulated and most cleanly missions in the country. Its herds of cattle amount to 10,000 in number, and of horses there are about 300.

When our travellers visited the mission it was governed by padres José and Machin, two priests of the mendicant order of Sán Francisco, to which class belong all the priests in Upper California. They appeared to lead a comfortable life, though not over well provided with its luxuries.

We will not, however, pry too narrowly into the internal arrangements of the good fathers' dwelling; let it suffice, that they gave our travellers a cordial welcome, and entertained them at their board in a most hospitable manner. After joining them in a dram of aquadente, they allowed their guests to retire to their sleeping apartment, where, stretched upon couches of bull-hide as tough and impenetrable as the cuirass of their friend the dragoon (who left them at this place), they soon fell asleep—thanks to excessive weariness—and slept as soundly as *las pulgas* would let them.

Having breakfasted the following morning with the padres, and being provided with fresh horses, a new escort and vaqueros, the party was about to start, but were delayed by the punishment of an Indian who had stolen a blanket, for which he received two dozen lashes with a leathern thong upon that part of the human frame, which, we learn from Hudibras, is the most susceptible of insult. Some other Indians were observed to be heavily shackled, but the causes of their punishment were not stated.

A beautiful avenue of trees, nearly three miles in length, leads from the mission to the pueblo of Sán José, the largest settlement of the kind in Upper California. It consists of mud houses miserably provided in every respect, and contains about 500 inhabitants—retired soldiers and their families, who under the old government were allowed the privilege of forming settlements of this nature, and had a quantity of ground allotted to them for the use of their cattle. They style themselves *Génte de Razón*,

CHAP. I. Nov. 1826.

to distinguish them from the Indians, whose intellectual qualities are frequent subjects of animadversion amongst these enlightened communities. They are governed by an alcalde, and have a chapel of their own, at which one of the priests of the mission occasionally officiates.

About eighteen miles from Sánta Clara, the party alighted upon the banks of a limpid stream, the first they had seen in their ride. It was too favourable a spot to be passed, and placing some milk and pears, which had been furnished by the hospitable priests at the mission, under the cool shade of an aliso-tree, they regaled themselves for a few minutes, and then resumed their journey. At the distance of eight leagues from Sánta Clara, they passed some remarkable hills near the coast named *El ójo del cóche;* and a few miles further on, they descended into the plain of *Las Llágas,* so called from a battle which took place between the first settlers and the Indians, in which many of the former were wounded. Stopping towards the extremity of this fertile plain at some cottages, named *Ranchas de las animas,* the only habitations they had seen since the morning, they dined upon some jerk beef, which, according to the old custom in this and other Spanish colonies, was served in silver dishes. Silver cups and spoons were also placed before our travellers, offering a singular incongruity with the humble wooden benches, that were substituted for chairs, and with the whole arrangement of the room, which, besides the board of smoking jerk beef, contained beds for the family, and a horse harnessed to a flour mill.

Leaving Lláno de las Llágas, they ascended a low range of hills, and arrived at a river appropriately named Rio de los Páxaros, from the number of wild ducks which occasionally resort thither. The banks of this river are thickly lined with wood, and being very steep in many places, the party wound, with some difficulty, round the trunks of the trees and over the inequalities of the ground ; but their Californian steeds, untrammelled with shoes, and accustomed to all kinds of ground, never once stumbled. They rode for some time along the banks of this river, which, though so much broken, were very agreeable, and crossing the stream a few miles lower down, they left it to make its way towards the sea in a south-west direction, and themselves entered upon the Lláno de Sán Juan, an extensive plain surrounded by mountains. It should have been told, before the party reached thus far, that as they were riding peaceably over the Lláno de las Animas, the clanking of their guide's huge broadsword, which had been substituted for the long musket of the soldier from the presidio, drew the attention of the party to his pursuit of a wild mountain-cat, which he endeavoured to ensnare with his lasso for the sake of its skin, which is said to be valuable. Two of these cats, which in species approach the ocelot, were shot by our sportsmen at Sán Francisco. Their skins were preserved to be brought to this country, but on opening the collection they were not found, and we have reason to suspect that a man who assisted the naturalist disposed of these, as well as many other specimens, to his own advantage.

Twilight approached as the party drew near to the mission of Sán Juan, where they alighted, after a ride of fifty-four miles, just as the bell tolled for vespers, and, stiff and tired, gladly availed themselves of the accommodation afforded by padre Arroyo, who in

CHAP. I. Nov. 1826.

hospitality and good humour endeavoured to exceed even the good father of Santa Clara. This worthy man was a native of Old Castile, and had resided in California since 1804, dividing his time between the duties of his holy avocation and various ingenious inventions. Supper was served in very acceptable time to the fatigued visiters, and the good-natured padre used every persuasion to induce them to do justice to his fare; treating them to several appropriate proverbs, such as "Un dia alégre vale cien ânos de pesadúmbre," (one day of mirth is worth a hundred years of grief,) and many more to the same purpose. Though so many summers had passed over his head in exile, his cheerfulness seemed in no way diminished, and he entertained his guests with a variety of anecdotes of the Indians and of their encounters with the bears too long to be repeated here. Nor was his patriotism more diminished than his cheerfulness, and on learning that one of the party had been at the siege of Cadiz, his enthusiasm broke forth in the celebrated Spanish patriotic song of "España de la guérra, &c." Having served them with what he termed the *viatico,* consisting of a plentiful supply of cold fricole beans, bread, and eggs, he led the party to their sleeping apartment amidst promises of horses for the morrow, and patriotic songs of his country adapted to the well-known air of Malbrook. Interrupting the good man's enthusiasm, they endeavoured to persuade the priest to allow them to proceed early in the morning, before the commencement of mass; this, however, was impossible, and he shut them into their apartment, repeating the proverb, "Oír mísa y dar cebàda no impede jornàda," (to hear mass and bestow alms will not retard your journey).

CHAP. I. Nov. 1826.

When the morning came, it was a holiday, and the vaquéros, not at all disposed to lose their recreation, had decamped with the saddles, and the party were obliged to pass the day at Sán Juan. After a small cup of chocolate, and a strip of dry bread, the only meal ever served in the missions until twelve o'clock, the party strolled over the grounds, and visited about thirty huts belonging to some newly converted Indians of the tribe of Tooleerayos *(bulrushes)*. Their tents were about thirty-five feet in circumference, constructed with pliable poles fixed in the ground and drawn together at the top, to the height of twelve or fifteen feet. They are then interwoven with small twigs and covered with bulrushes, having an aperture at the side to admit the inhabitants, and another at the top to let out the smoke. The exterior appearance of these wretched wigwams greatly resembles a bee-hive. In each dwelling were nine or ten Indians of both sexes, and of all ages, nearly in a state of nudity, huddled round a fire kindled in the centre of the apartment, a prey to vermin, and presenting a picture of misery and wretchedness seldom beheld in even the most savage state of society. They seemed to have lost all the dignity of their nature; even the black-birds *(oriolus niger)* had ceased to regard them as human beings, and were feeding in flocks among the wigwams. This was said to be the state in which the Indians naturally live, and the reader will not be surprised to hear that this party had voluntarily come from the mountains to be converted, and to join their civilized brethren at the mission. Happy would it be for these savages could they be once taught to make a proper use of that freedom which ought to follow their conversion to the pure religion of Christ, even under the restrained

CHAP. I. Nov. 1826.

form of Catholicism, that their minds might become by this means sufficiently improved to allow of their settling in independent Christian communities; but, judging from their present mental capacity, it must be long before so great and desirable a change can be effected. The experiment of liberating the Indians has been tried and has failed;* and appearances certainly justify the assertion that the Indian is happier under control than while indulging his free soul in the wilds of his native country.

What might seem a remarkable example of this was met with on turning from the dwelling of wretchedness just described to a scene of the greatest mirth and happiness amongst some converted Indians, who were passing their holiday in amusement. Some were playing at *takersia*, a game which, as already described, consists in trundling a hoop, or rather a piece of wood with a hole in it, and in endeavouring to pierce it with a short lance as it rolls. Another party were playing at a game resembling *hockey*, and in various parts of the plain adjoining the mission many others were engaged in pleasant recreations, passing their day in exercise, content, and enjoyment.

In the neighbouring meadows there were several large herds of cattle; and the geese settled there in flocks, as at the mission of Santa Clara. The rocks, where they protruded, were ascertained by Mr. Collie to be sandstone conglomerate with a calcareous basis.

The welcome peal of the mission bell assembled the party at dinner; but the padre, who for some time before had been earnestly engaged in endeavouring to

* The effect of emancipation on the Indians is spoken of more at large in an after part of this work.

convert one of his heretic guests, was unwilling to quit the train of theological disquisition which in his own opinion he had almost brought to successful issue, until reminded by his other visitors, who had not been accustomed to go so long without their breakfast, that they required something more substantial.

I will not attempt to stimulate the appetite of my reader by enumerating the various exquisite dishes which successively smoked on the board of the generous priest, suffice it that there were many good ones, as the padres in California are careful to have their table well supplied at all times of the year, and have an indulgence from the pope to eat meat even during the greater part of Lent, in consequence of the difficulty of procuring fish.

Having performed the honours of the table, padre Arroyo retired to indulge his usual siesta: this, however, caused but a brief suspension to the efforts he most industriously continued to make for the purpose of converting his heretical opponent to the true faith, reading him innumerable lectures in refutation of the Lutheran and Calvinistic doctrines, and in favour of the pope's supremacy, infallibility, and power of remitting offences.

It more than once occurred to the party—and I believe, not without good foundation for their opinion—that it was the hope of success in this conversion which occasioned all the little manœuvring to delay them, that I have before described. But having at length given his pupil over as irrevocably lost, he consented to their departure on the following morning. The padre appeared to be of an active mind, and had constructed a water clock which communicated with a

CHAP. I. Nov. 1826.

bell by his bedside, and which by being arranged at night could be made to give an alarm at any stated hour.

It was here that our travellers were surprised at the intelligence of the north-west passage having been effected by a Spaniard, and were not a little amused at the idea of having stumbled upon the long-sought north-west passage in an obscure mission of California.

The padre, however, was quite in earnest, and produced a work published by the Duke of Almodobar, Director of the Royal Academy in Spain, in which was transcribed at full length the fictitious voyage of Maldonado. It was in vain they endeavoured to persuade the padre that this voyage was not real, seeing that it bore even in its detail all the marks of truth, and that it emanated from such high authority. His credulity in this instance affords a curious proof of the very secluded manner in which these holy men pass their time, for it may be remembered, that it was in the very ports of California that both Vancouver and Quadra anchored, after having satisfactorily proved the voyage in question to have been a fabrication.

A still greater instance of the simplicity of the priest is related at his expense by persons in the mission. A youthful Indian couple who had conceived an affection for each other eloped one day, that they might enjoy each other's society without reserve in the wild and romantic scenery of the forests. Soldiers were immediately sent in pursuit, when, after a week's search, the fugitives were brought back; upon which padre Arroyo, to punish their misbehaviour, incarcerated them together, and kept them thus confined until he thought they had expiated their crime.

In addition to his other manifold accomplishments,

padre Arroyo was a grammarian, and said that he had written a vocabulary and grammar of the Indian languages, but he could not be prevailed upon to show them: such works, were they in existence, would, I believe, be the only ones of the kind; and it is a pity that they should not be given to the world as a matter of curiosity, though I cannot think they would be of much use to a traveller, as the languages of the tribes differ so materially, and in such short spaces, that in one mission there were eleven totally different dialects. I cannot omit to mention padre Arroyo's disquisition on the etymology of the name of the Peninsula of California. I shall observe first, that it was never known why Cortes gave to the bay* which he first discovered, a name which appears to be composed of the Latin words *calida* and *fornax*, signifying *heat* and *furnace*, and which was afterwards transferred to the peninsula. Miguel Venegas supposed it arose from some Indian words which Cortes misunderstood, and Burney, in his history of voyages in the Pacific,† observes, that some have conjectured the name to have been given on account of the heat of the weather, and says, it has been remarked that it was the only name given by Cortes which was immediately derived from the Latin language. Without entering into a discussion of the subject, which is not of any moment, I shall observe, that it was thought in Monterey to have arisen in consequence of a custom which prevails throughout California, of the Indians shutting themselves in ovens until they perspire profusely, as I have already described in speaking of the Temeschal. It is

* Bernal Diaz de Castillo, in his "Conquest of Mexico," calls California *a bay*.

† Vol. I. p. 178, 4to.

CHAP. I. Nov. 1826.

not improbable that the practice appeared so singular to Cortes that he applied the name of California to the country, as being one in which hot ovens were used for such singular purposes. Padre Arroyo, however, maintained that it was a corruption of *colofon*, which, in the Spanish language, signifies resin, in consequence of the pine trees which yield that material being so numerous. The first settlers, he said, at the sight of these trees would naturally exclaim, "Colofon," which, by its similarity to Californo, (in the Catalonian dialect, hot oven,) a more familiar expression, would soon become changed.

Our travellers, after taking leave of the hospitable and amusing priest the preceding evening, with the intention of proceeding early in the morning, experienced much delay in consequence of the refusal of the guard to start without hearing mass and receiving the benediction of the priest; but at length they quitted the plain of Sán Juan, and ascended with difficulty some steep hills commanding a view of the spacious bay of Monterey. Then winding among valleys, one of which was well wooded and watered, they entered an extensive plain called "Llano del Rey," which, until their arrival, was in the quiet possession of numerous herds of deer and jackals. This tract of land is bounded on the north, east, and south-east, by mountains which extend with a semicircular sweep from the sea at Santa Cruz, and unite with the coast line again at Point Pinos. It is covered with a rank grass, and has very few shrubs. In traversing this plain, before they could arrive at some ranchos, named Las Salinas, where they proposed to dine, the party had to wade through several deep ditches and the Rio del Rey, both of which were covered with wild ducks. The cottages called Las Salinas are on the farm of an old Scotchman, to whom the land was granted in consequence of some services which he rendered to the missions. They rested here, and to the provision they had brought with them very gladly added some pumpkins, procured from the Indians. Here, also, they were surprised with the novel occurrence of having water brought to them in baskets, which the Indians weave so close, that when wet they become excellent substitutes for bowls.

The remainder of the plain over which they passed toward Monterey was sandy, and covered with fragrant southernwood, broken here and there by dwarf oaks, and shrubs of the syngenesious class of plants. As they approached the town, pasture lands, covered with herds of cattle succeeded this wild scenery: and riding ownward, trees of luxuriant growth, houses scattered over the plain, the fort, and the shipping in the bay, announced the speedy termination of their journey. At five o'clock in the evening they alighted in the square at Monterey, and met a kind reception from Mr. Hartnell, a merchant belonging to the firm of Begg and Co. in Lima, who was residing there, and who pressed them to accept the use of his house while they remained in the town—an offer of which they thankfully availed themselves.

Gonzales, the governor to whom the party went to pay their respects, was an officer who had been raised by his own merit from the ranks to be captain of artillery and governor of Monterey: his family were residing with him, and having been educated in Mexico, complained bitterly of their banishment to this outlandish part of the world, where the population, the ladies in particular, were extremely ignorant,

CHAP. I. Nov. 1826.

and wretched companions for the *Mexicanus instruidas*. Besides, there were no balls or bull-fights in Monterey; and for all the news they heard of their own country, they might as well have been at Kamschatka. To compensate for these dreadful privations, the ladies generally amused themselves in the evening by smoking and playing cards, and relating the perils they encountered in the land journey from Mexico to the shores of the Pacific. Politeness and attention, however, were the characteristics of these good people, who offered our party every assistance in their power during their stay at Monterey.

Upon inquiry after the stores and medicines the ship stood in need of, the result was highly unfavourable; as there were no medicines to be had, and some stores which were essential to the ship could nowhere be procured. The exchange on bills was favourable, but there was no specie: Mr. Marsh therefore purchased what stores he could from the inhabitants and from the shipping in the roads, and arranged with a person who had come out from Ireland for the purpose of salting meat for the Lima market, to cure a quantity for the use of the ship, and to have it ready on her arrival at Monterey. They then hastened their departure, but the same difficulty arose about horses as before, and they were much inconvenienced in consequence, being obliged to alter a plan they had contemplated of returning by a different route. This, very unexpectedly to padre Arroyo, brought them again under his roof. The padre either did not like this second tax on his hospitality, or was put out of temper by the increase of a complaint to which he was subject, as he gave them a less cordial reception, and appeared very little disposed to conversation. It was imagined, however, that he still entertained hopes of the conversion of one of the party, and that with this view he again occasioned a delay in furnishing horses for the next day's journey; offering as excuses, that some of the horses of the mission were engaged by soldiers in pursuit of a Mexican exile, who had deserted; that others had been taken by the vaquéros to look after a male and female Indian, who had likewise absconded, and that the rest were gone to join the expedition against Los Gentiles, the Cosemenes. Vexed at this delay, the party endeavoured to hire horses at their own expense, but the price demanded was so exorbitant that they determined to wait the return of those that were said to be absent.

It is more than probable that some one of my readers may have been in the same predicament—in a strange town, in a strange country, with a beast fatigued to death, and an urgent necessity for proceeding; he will then easily remember the amiable and benevolent alacrity with which the inhabitants endeavoured to lighten his load of every stray crown they could obtain from him, on every pretence that ingenious cupidity can invent. So at least did the good people at Sán Juan, when padre Arroyo would no longer assist our poor companions. Private horses could be had, it was true, but the terms were either thirteen shillings sterling for the journey, or seventeen shillings sterling for the purchase of the horse, which in California is considered so exorbitant that our shipmates did not think proper to suffer the imposition, and awaited the horses belonging to the mission.

After a day's delay, during which they again heard many invectives against the new government of

CHAP. I. Nov. 1826.

Mexico, which had deprived the priesthood of their salaries, and obliged the missions to pay a tithe to the state, they resumed their journey, and arrived at Sán Francisco on the 17th of November.

In this route it will be seen that, with the exception of the missions and pueblos, the country is almost uninhabited; yet the productive nature of the soil, when it has been turned up by the missions, and the immense plains of meadow land over which our travellers passed, show with how little trouble it might be brought into high cultivation by any farmers who could be induced to settle there.

The unwelcome intelligence brought by this party of the nature of the supplies to be obtained at Monterey, obliged me to relinquish the plan I had contemplated of completing the survey of that part of the coast of California which had been left unfinished by Vancouver; and rendered it necessary that I should proceed direct either to Canton or to Lima, as the most likely places for us to meet with the medicines and stores of which we were in such imminent need. The western route of these two afforded the best opportunity of promoting the objects of the expedition, by bringing us into the vicinity of several groupes of islands of doubtful existence, at which, in the event of their being found, our time might be usefully employed until it should be necessary to proceed to Beering's Strait. An additional reason for this decision was, a request which I had made to the consul of the Sandwich Islands, if possible, to purchase provision for the ship at that place. I therefore determined, after taking on board the few stores that were purchased at Monterey, to proceed to the Sandwich Islands, searching in our way thither for some islands said to have been discovered by an American vessel, and from thence to prosecute the voyage to Canton.

While we remained in Sán Francisco refitting the ship, the boats were constantly employed sounding and surveying the harbour, in which duty we received every assistance from Martinez, the governor, who allowed us to enter the forts, and to take what angles and measures we pleased, requiring only in return for this indulgence a copy of the plan, when finished, for his own government: his proposal seemed so fair that I immediately acceded to it, and, on my return to the place the following year, fully complied with his request. It is impossible to pass unnoticed the difference between this liberal conduct of Martinez and that of the former Spanish authorities, who watched all Vancouver's actions with the greatest suspicion, and whose jealousy has been the subject of animadversion of almost every voyager who has touched at this port.

On the 12th of December a salute was fired from the battery; high mass was said in all the missions, and a grand entertainment, to which all the officers were invited, was given at the presidio, in honour of Santa Senora Guadaloupe. There was also to have been a fight between a bear and a bull, but for some reason not known to us—probably the trouble it required to bring the animal so far, as the bears do not come within many miles of the presidio—it did not take place; and we were all greatly disappointed, as we had offered to reward the soldiers for their trouble, and had heard so much of these exhibitions from every body, that our curiosity had been highly excited. This is a favourite amusement with the Californians, but it is of rare occurrence, as there is much trouble in getting a bear alive to the scene of combat, and there is also some risk and expense attending it. We were informed that when a fight is determined upon three or four horsemen are dispatched with lassos to the woods where the bears resort, and that when they come to an advantageous spot they kill a horse or a bullock as a bait, and hide themselves in the wood. Sometimes they have to wait a whole day or more before any of these animals appear, but when they come to partake of the food, the men seize a favourable opportunity, and rush upon them at different points with their lassos, and entangle one of them until he is thrown upon the ground, when they manage to suspend him between the horsemen, while a third person dismounts and ties his feet together; he is then extended upon a hide and dragged home; during which time it is necessary, they say, to keep him constantly wet to allay his thirst and rage, which amounts almost to madness—and woe to him who should be near if he were to break away from his fastenings. The entangling of the animal in the first instance appears to be by no means devoid of risk, as in case of the failure of a lasso it is only by speed that a rider can save himself and his horse. The bear being caught, two or three men are dispatched for a wild bull, which they lasso in an equally dexterous manner, catching him either by the horns or by whichsoever leg they please, in order to trip him up and retain him between them.

It is necessary to begin the fight as soon as the animals are brought in, as the bear cannot be tempted to eat, and is continually exhausting himself in struggling for his liberty. The two animals are then tied together by a long rope, and the battle begins, some-

Drawn by W^m. Smyth. Engraved by Edwd. Finden.

CALIFORNIANS THROWING THE LASSO.

London. Published by Henry Colburn & Richard Bentley, 1830.

times to the disadvantage of the bear, who is half dead with exhaustion, but in the end almost always proves fatal to the bull. It is remarkable that all the bears endeavour to seize the bull by the tongue, for which purpose they spring upon his head or neck and first grapple with his nose, until the pain compels the bull to roar, when his adversary instantly seizes his tongue, pierces it with his sharp talons, and is sure of victory. These battles were the everlasting topic of conversation with the Californians, who indeed have very little else to talk about, and they all agreed as to the manner of the fatal termination of the spectacle.

Subjoined is a spirited sketch of a Californian lassoing a bull, taken from life by Mr. Smyth, in which the method, as well as the costume of the natives is admirably delineated. The lasso, though now almost entirely confined to Spanish America, is of very great antiquity, and originally came from the east. It was used by a pastoral people who were of Persian descent, and of whom 8,000 accompanied the army of Xerxes.*

By Christmas-day we had all remained sufficiently long in the harbour to contemplate our departure without regret: the eye had become familiar to the picturesque scenery of the bay, the pleasure of the chase had lost its fascination, and the roads to the mission and presidio were grown tedious and insipid. There was no society to enliven the hours, no incidents to vary one day from the other, and to use the expression of Donna Gonzales, California appeared to be as much out of the world as Kamschatka.

On the 26th, being ready for sea, I was obliged to relinquish the survey of this magnificent port, which

* Rennell on the 20 Satrapies of Darius Hystaspes, p. 287.

CHAP. I. Dec. 1826.

possesses almost all the requisites for a great naval establishment, and is so advantageously situated with regard to North America and China, and the Pacific in general, that it will, no doubt, at some future time, be of great importance. We completed the examination of those parts of the harbour which were likely to be frequented by vessels for some years to come, in which it is proper to mention, in order to give as much publicity to the circumstance as possible, that we discovered a rock between Alcatrasses and Yerba Buena Islands, dangerous to both shipping and boats, in consequence of its rising suddenly from about seven fathoms, so near to the surface as to occasion strong overfalls with the tides. A shoal was also found to the eastward of the landing-place off the presidio, which ought to be avoided by boats sailing along shore. In my nautical remarks, I have given directions for avoiding both these dangers, which are the only hidden ones in that part of the harbour, which is at present frequented.

On the 28th we took leave of our hospitable and affable friends, Martinez and Padre Tomaso, full of gratitude for their kindness and attention to our wants; weighed anchor, and bade adieu to the Port of Sán Francisco, in which we had all received material benefit from the salubrity of its climate, the refreshing product of its soil, and the healthy exercise we had enjoyed there. In the ship's company, in particular, there was the most apparent amendment; some of them, from being so emaciated on their arrival that the surgeon could scarcely recognize them, were now restored to their former healthy appearance, and we had the satisfaction of sailing without a single case of sickness on board. We had to regret during our stay the loss of one of our best men, Joseph Bowers, a marine. He had accompanied one of the officers on a shooting excursion, and was led by his naturally ardent and bold disposition to plunge into a lake after some wild fowl that had been shot, forgetting that he could not swim. His eagerness led him beyond his depth, and in his attempt to regain his footing, he unfortunately perished before any aid could be brought. His body was interred at the burial ground near the presidio landing place, and was followed to the grave by all the officers. As the coffin was lowering into the ground, the good understanding that existed between the ship's company and the inhabitants was testified in the most gratifying manner, by the latter approaching and performing the last office for the deceased, by dropping the earth in upon his coffin. I cannot recollect ever having met with such conduct in any other foreign port, and the act, most certainly, did not lessen our regard for the inhabitants.

CHAPTER II.

Observations on the Country of California and its Trade—Climate—Meteorological Remarks—Short Account of the Wild Indians—Natural Productions—Monterey—Mission of San Carlos—Departure.

CHAP. II. Dec. 1826.

THE more we became acquainted with the beautiful country around Sán Francisco, the more we were convinced that it possessed every requisite to render it a valuable appendage to Mexico; and it was impossible to resist joining in the remark of Vancouver, "Why such an extent of territory should have been subjugated, and, after all the expense and labour bestowed upon its colonization, turned to no account whatever, is a mystery in the science of state policy not easily explained." Situated in the northern hemisphere, between the parallels of 22° and 39°, no fault can be found with its climate; its soil in general is fertile, it possesses forests of oak and pine convenient for building and contributing to the necessities of vessels, plains overrun with cattle, excellent ports, and navigable rivers to facilitate inland communication. Possessing all these advantages, an industrious population alone seems requisite to withdraw it from the obscurity in which it has so long slept under the indolence of the people and the jealous policy of the Spanish government. Indeed it struck us as lamentable to see such an extent of habitable country lying almost desolate and useless to mankind, whilst other nations are groaning under the burthen of their population.

It is evident, from the natural course of events, and from the rapidity with which observation has recently been extended to the hitherto most obscure parts of the globe, that this indifference cannot continue; for either it must disappear under the present authorities, or the country will fall into other hands, as from its situation with regard to other powers upon the new continent, and to the commerce of the Pacific, it is of too much importance to be permitted to remain long in its present neglected state. Already have the Russians encroached upon the territory by possessing themselves of the Farallones, and some islands off Sánta Barbara; and their new settlement at Rossi, a few miles to the northward of Bodega, is so near upon the boundary as to be the cause of much jealous feeling; —not without reason it would appear, as I am informed it is well fortified, and presents to California an example of what may be effected upon her shores in a short time by industry.

The tract situated between California and the eastern side of the continent of North America, having been only partially explored, has hitherto presented a formidable barrier to encroachment from that quarter; but settlements are already advancing far into the heart of the country, and parties of hunters have lately traversed the interior, and even penetrated to the shores of the Pacific;—not without the loss of lives from the attacks of the Indians, it is true, but with ease, compared with the labour and difficulty experienced by Lewis and Clarke, who had not the benefit

CHAP. II. Dec. 1826.

which more recent travellers have derived from the establishment of inland depôts by the American fur companies. One of these depôts, we were informed by a gentleman belonging to the establishment, whom we met at Monterey in 1827, is situated on the western side of the rocky mountains on a fork of the Columbia called Lewis River, near the source of a stream supposed to be the Colorado.

The trade of Upper California at present consists in the exportation of hides, tallow, manteca, horses to the Sandwich Islands, grain for the Russian establishments at Sitka and Kodiak, and in the disposal of provisions to whale-ships and other vessels which touch upon the coast,—perhaps a few furs and dollars are sent to China. The importations are dry goods, furniture, wearing-apparel, agricultural implements, deal-boards, and salt; and silks and fireworks from China for the decoration of the churches and celebration of the saints' days. In 1827 almost all these articles bore high prices: the former in consequence of the increased demand for them; and the latter, partly from the necessity of meeting the expenses of the purchase of a return cargo, and partly on account of the navigation act.

The missions and the inhabitants in general complained loudly of these prices, not considering that the fault was in a great measure their own, and that they were purchasing some articles which had been brought several thousand miles, when they might have procured them in their own country with moderate labour only. For example, they were actually living upon the sea-coast and amongst forests of pine, and yet were suffering themselves to buy salt and deal boards at exorbitant prices.

With a similar disregard for their interests, they were purchasing sea-otter skins at twenty dollars apiece, whilst the animals were swimming about unmolested in their own harbours; and this from the Russians, who are intruders upon their coast, and are depriving them of a lucrative trade: and again, they were paying two hundred dollars for carts of inferior workmanship, which, with the exception of the wheels, might have been equally well manufactured in their own country.

With this want of commercial enterprise, they are not much entitled to commiseration. With more justice might they have complained of the navigation laws, which, though no doubt beneficial to the inhabitants on the eastern coast of Mexico, where there are vessels belonging to the state in readiness to conduct the coasting trade, are extremely disadvantageous to the Californians, who having no vessels to employ in this service are often obliged to pay the duty on goods introduced in foreign bottoms. This duty for the encouragement of the coasting trade was made seventeen per cent. higher than that on cargoes brought in vessels of the state. Thus not only must the inhabitants purchase their goods on very disadvantageous terms, but, as a foreign vessel cannot break stowage without landing the whole of her cargo, they must in addition incur the expenses attending that, which will in general fall upon a few goods only, as the towns in California are not sufficiently populous, any one of them, to consume a whole cargo; and it is to be remembered, that no foreign vessel, after breaking stowage, can proceed to another port in the same dominion without being liable to seizure by the customs.

The imprudent nature of these laws, as regards Ca-

CHAP. II. Dec. 1826. lifornia, appears to have been considered by the authorities in that country, as they overlook the introduction of goods into the towns by indirect channels, except in cases of a gross and palpable nature. In this manner several American vessels have contrived to dispose of their cargoes, and the inhabitants have been supplied with goods of which they were much in need; but had the navigation laws been strictly attended to, the vessels must have returned unsuccessful, and the inhabitants have continued in want.

Far more liberal has been the hand of nature to this much neglected country, in bestowing upon it a climate remarkable for its salubrity. The Spanish settlers in California enjoy an almost uninterrupted state of good health. Many attain the age of eighty and ninety, and some have exceeded a hundred years. There have been periods, however, when the small pox and measles have affected the population, and particularly the Indians in the missions, who, unlike the Spaniards, appear to suffer severely from diseases of all kinds. The small pox many years ago prevailed to an alarming extent, and carried off several thousand Indians; but since the introduction of cattle into the country, and with them the cow pox, it has not reappeared. Vaccination was practised in California as early as 1806, and the virus from Europe has been recently introduced through the Russian establishment at Rossi. The measles have also at times seriously affected the Indians, and in 1806 proved fatal to thousands, while it is remarkable that none of the Spaniards affected with the disease died. Dysentery, the most prevalent complaint amongst the converted Indians, no doubt arises in a great measure from the coldness and dampness of their habitations, and becomes fatal through the want of proper medical assistance. They are happily free from the hooping cough.

This state of ill health does not extend to the uncivilized Indians; and, notwithstanding the mortality in the missions, the climate of California must be considered salubrious. Perouse, Vancouver, and Langsdorff were of the same opinion; and to judge of it by the general health of the Spanish residents, and by the benefit that our seamen derived from it during their short stay, it would certainly appear not to be surpassed. The summer and early part of the autumn are the least healthy parts of the year, in consequence of continued fogs, which occur at these periods.

It is, in all probability, in consequence of these fogs during the warmest part of the year that the coast of California has the reputation of being much colder than that of Chili in corresponding parallels of latitude. In the month of December the mean temperature of Sán Francisco was 53° 2′, the maximum 66°, and the minimum 46°. We nevertheless saw hoar frost upon the grass in the mornings, and in the following year observed snow lie several hours upon the ground. As the minimum of temperature was so many degrees above the freezing point, the former was in all probability occasioned by the radiation, which is very great in that country.

The winter of 1826 was said to be a very favourable season; we could not judge from our own experience, therefore, of what weather was usual on the coast at that period of the year. But there were very few days during our visit in which a vessel might not have approached the coast with safety. The strongest and most prevalent winds were from the north-west;

CHAP. II. Dec. 1826.

but these winds, though they blew directly upon the coast, were generally attended by clear weather, which would have enabled a vessel to find a port, had it been necessary. They were strongest about the full and change of the moon.

From the prevalence of the westerly swell off the harbour, and from the wind moderating as we approached the coast in both years, I am inclined to think that these winds do not usually blow home upon the shore.

There was a curious anomaly observed in the movements of the barometer and sympeisometer during our stay at Sán Francisco: the former rose with the winds which brought bad weather, and fell with those which restored serenity to the sky. The maximum height was 30·46, the minimum 29·98, and the mean 30·209.

The hygrometer on the whole indicated a dry atmosphere, and ranged from 0° to 20° of dryness on the thermometric scale, the mean degree of dryness being 6°, 6. The particulars of these observations are inserted in tables in the Appendix to the 4to edition.

The clear weather occasioned by the north-west wind was favourable for astronomical observations; but many were lost in consequence of a haze overhanging the land at night, and from the incovenience arising from a heavy deposition, which, besides occasioning much mirage, fell so profusely upon the glasses of the instruments that they were obliged to be repeatedly wiped, and sometimes at the most inconvenient moments.* Our observations, however, were very satisfactory, and are important, as the

* I found this in a great degree obviated by fixing a long paper tube to the field end of the telescope.

longitudes of the places between Nootka Sound and Sán Diego are dependent upon the situation of Sán Francisco and Monterey; Vancouver having, in his survey of the coast, rated his chronometers between the meridians of these places. My observatory was erected upon a small eminence near the anchorage at Yerba-Buena, from whence the observations were carefully reduced to the fort at the entrance of the harbour. The results have been published in the 4to edition, where will also be found some observations on the dip and variation of the needle, the tides, and other subjects.

I shall conclude this imperfect sketch of Upper California with a short description of the Indian mode of living, and of the natural productions of the country, derived principally from the information of the priests, and from the journals of the officers who went overland to Monterey. The Indians who enter the missions with which we became acquainted are divided in their wild state into distinct tribes, and are governed by a chief whose office is hereditary, but only in the male line, The widows and daughters, however, though not allowed to partake of this privilege, are exempted from labour, and are more respected than other women. Each tribe has a different dialect; and though their districts are small, the languages are sometimes so different, that the neighbouring tribes cannot understand each other. I have before observed, that in the mission of Sán Carlos there are eleven different dialects. Their villages consist of wigwams made with poles covered with bulrushes, and are generally placed in an open plain to avoid surprise. Like the Arabs and other wandering tribes, these people move about the country, and pitch

CHAP. II. Dec. 1826.

their tents whereever they find a convenient place, keeping, however, within their own district.

They cultivate no land, and subsist entirely by the chase, and upon the spontaneous produce of the earth. Acorns, of which there is a great abundance in the country, constitute their principal vegetable food. In the proper season they procure a supply of these, bake them, and then bruise them between two stones into a paste, which will keep until the following season. The paste before it is dried is subjected to several washings in a sieve, which they say deprives it of the bitter taste common to the acorn. We cannot but remark the great resemblance this custom bears to the method adopted by the South-sea Islanders to keep their bread fruit, nor ought we to fail to notice the manner in which Providence points out to different tribes the same wise means of preserving their food, and providing against a season of scarcity.

The country inhabited by the Indians abounds in game, and the rivers in fish; and those tribes which inhabit the sea-coast make use of muscles and other shell fish, of which the haliotis gigantea is the most abundant. In the chase they are very expert, and avail themselves of a variety of devices to ensnare and to decoy their game. The artifice of deceiving the deer by placing a head of the animal upon their shoulders is very successfully practised by them. To do this, they fit the head and horns of a deer upon the head of a huntsman, the rest of his body being painted to resemble the colour of a deer. Thus disguised, the Indian sallies forth, equipped with his bow and arrows, approaches the pasture of the deer, whose actions and voice he then endeavours to imitate, taking care to conceal his body as much as possible, for which purpose he generally selects places which are overgrown with long grass. This stratagem seldom fails to entice several of the herd within reach of his arrows, which are frequently sent with unerring aim to the heart of the animal, and he falls without alarming the herd; but if the aim should fail, or the arrow only wound its intended victim, the whole herd is immediately put to flight.

Their method of taking ducks and geese and other wildfowl is equally ingenious. They construct large nets with bulrushes, and repair to such rivers as are the resort of their game, where they fix a long pole upright on each bank, with one end of the net attached to the pole on the opposite side of the river to themselves. Several artificial ducks made of rushes are then set afloat upon the water between the poles as a decoy; and the Indians, who have a line fastened to one end of the net, and passed through a hole in the upper end of the pole that is near them, wait the arrival of their game in concealment. When the birds approach, they suddenly extend the net across the river by pulling upon the line, and intercept them in their flight, when they fall stunned into a large purse in the net, and are captured. They also spread nets across their rivers in the evening, in order that the birds may become entangled in them as they fly.

The occupation of the men consists principally in providing for their support, and in constructing the necessary implements for the chase and for their own defence. The women attend to their domestic concerns, and work a variety of baskets and ornamental parts of their dress, some of which are very ingenious, and all extremely laborious. Their closely wove

CHAP. II. Dec. 1826.

baskets are not only capable of containing water, but are used for cooking their meals. A number of small scarlet feathers of the oriolus phœniceus are wove in with the wood, and completely screen it from view on the outside; and to the rim are affixed small black crests of the Californian partridges, of which birds a hundred brace are required to decorate one basket:—they are otherwise ornamented with beads, and pieces of mother-of-pearl. They also embroider belts very beautifully with feathers of different colours, and they work with remarkable neatness, making use of the young quills of the porcupine, in a similar manner to the Canadian Indians; but here they manufacture a fine cloth for the ground, whereas the Canadians have only the bark of the birch-tree. They also manufacture caps and dresses for their chiefs, which are extremely beautiful; and they have a great many other feather ornaments, which it would be stepping beyond the limits of my work to describe.

The stature of the Indians which we saw in the missions was by no means diminutive. The Alchones are of good height, and the Tuluraios were thought to be, generally, above the standard of Englishmen. Their complexion is much darker than that of the South-sea Islanders, and their features far inferior in beauty. In their persons they are extremely dirty, particularly their heads, which are so thatched with wiry black hair that it is only by separating the locks with the hand that it can be got at for the purposes of cleanliness. Many are seen performing such acts of kindness upon their intimate friends; and, as the readiest means of disposing of what they find, consuming it, in the manner practised by the Tartars, who, according to Hakluyt—"cleanse one anothers' heades, and ever as thei take an animal do eate her, saeing, thus wille I doe to our enemies."*

Their bodies are in general very scantily clothed, and in summer many go entirely naked The women, however, wear a deer skin or some other covering about their loins: but skin dresses are not common among any of the tribes concerning whom we could procure any information. The women are fond of ornaments, and suspend beads and buttons about their persons, while to their ears they attach long wooden cylinders, variously carved, which serve the double purpose of ear-rings and needle-cases.

Tattooing is practised in these tribes by both sexes, both to ornament the person, and to distinguish one clan from the other. It is remarkable that the women mark their chins precisely in the same way as the Esquimaux.

The tribes are frequently at war with each other, often in consequence of trespasses upon their territory and property; and weak tribes are sometimes wholly annihilated, or obliged to associate themselves with those of their conquerors; but such is their warmth of passion and desire of revenge that very little humanity is in general shown to those who fall into their power. Their weapons consist only of bows and arrows: neither the tomahawk nor the spear is ever seen in their hands. Their bows are elegantly and ingeniously constructed, and if kept dry will discharge an arrow to a considerable distance. They resemble those of the Esquimaux, being strengthened by sinews at the back of the bow, but here one sinew, the size of the wood, occupies the whole extent of the back,

* Hakluyt's Selection of curious and rare Voyages, Supplement.

CHAP. II. Dec. 1826.

and embraces the ends, where they are turned back to receive the string; the sinew is fixed to the bow while wet, and as it becomes dry draws it back the reverse way to that in which it is intended to be used. The Indian manner of stringing these bows is precisely similar to that practised by the lovers of archery in England; but it requires greater skill and strength, in consequence of the increased curvature of the bow, and the resistance of the sinew.

The religion of all the tribes is idolatrous. The Olchone, who inhabit the seacoast between Sán Francisco and Monterey, worship the sun, and believe in the existence of a beneficent and an evil spirit, whom they occasionally attempt to propitiate. Their ideas of a future state are very confined: when a person dies they adorn the corpse with feathers, flowers, and beads, and place with it a bow and arrows; they then extend it upon a pile of wood, and burn it amidst the shouts of the spectators, who wish the soul a pleasant journey to its new abode, which they suppose to be a country in the direction of the setting sun. Like most other nations, these people have a tradition of the deluge; they believe also that their tribes originally came from the north.

The Indians in their wild state are said to be more healthy than those which have entered the missions. They have simple remedies, derived from certain medicinal herbs, with the property of which they have previously made themselves acquainted. Some of these roots are useful as emetics, and are administered in cases of sickness of the stomach: they also apply cataplasms to diseased parts of the body, and practise phlebotomy very generally, using the right arm for this purpose when the body is affected, and the left where the limbs. But the temiscal is the grand remedy for most of their diseases.

The very great care taken of all who are affected with any disease ought not to be allowed to escape a remark. When any of their relations are indisposed, the greatest attention is paid to their wants, and it was remarked by Padre Arroyo that filial affection is stronger in these tribes than in any civilized nation on the globe with which he was acquainted.

Our knowledge of the natural history of this country cannot be expected to be very extensive. In the woods not immediately bordering upon the missions, the black bear has his habitation, and when food is scarce it is dangerous to pass through them alone in the dusk of the evening; but when the acorns abound there is nothing to apprehend. It is said that the white bear also visits this district occasionally, from the northward. The lion *(felis concolor ?)* and the tiger *(felis onca ?)* are natives of these woods, but we never saw them; the inhabitants say they are small, and that the lion is less than the tiger, but more powerful. A large species of mountain cat *(gato del monte)* is common: a pole cat *(viverra putorius)* also is found in the woods: wolves and foxes are numerous, and the *cuiotas*, or jackalls, range about the plains at night, and prove very destructive to the sheep. The fallow-deer browses on the pasture land, not only in the interior, but also upon some of the islands and around the shores of the harbour: it is sought after for its skin, of which the Spaniards make boots, shoes, &c. The rein-deer also is found inland, particularly upon a large plain named Tulurayos, on account of the number of bulrushes growing there. In the months of May and June the Spaniards resort to this

CHAP. II. Dec. 1826.

plain with their lassos, and take as many of these animals as they can ensnare, for the sake of their fat, of which they will sometimes procure between four and five arobas from one animal.

The fields are burrowed by a small rat, resembling the *mus arvalis,* by a mountain rat of the *cricetus* species, and also by the ardillo, a species of *sciurus,* rather a pretty little animal, said to be good to eat: another of this species was seen among the branches of the trees. A small variety of *lepus cuniculus* is very common in the sand-hills near the presidio; hares are less common, and indeed it is doubtful whether any were seen by us. Raccoons are found in the mountains at a distance from the coast. The sea otter *(mustela lutris)* is not an unfrequent visiter in the harbour of Sán Francisco, but very few of them are taken, notwithstanding their fur is valuable. Judging from the accounts that have been published, these animals are becoming less numerous upon the coast: in 1786 it was stated that 50,000 of them might be collected annually, whereas at present the number is reduced to about 2,000. Porpoises and whales are numerous outside the harbour, and the common seal may occasionally be seen basking on the rocks of Yerbabuena, and other places.

The feathered tribe in Sán Francisco are very numerous, and have as yet been so little molested that there must be a rich harvest in store for the first naturalist who shall turn his attention to this place. We succeeded in killing a great many birds of different species, several of which were found to be quite new, and will be described in the natural history, which will shortly appear as a supplement to this voyage: but there are not many which delight, either by the brilliancy or beauty of their plumage, or by the melody of their note. The birds of prey are the black vulture *(vultus aura)*, sometimes large; several species of *falco*, one of which attacks the geese, and is in consequence called *mato gansas*, also a kite, and a sparrow hawk. The horned owl (a variety of the *strix virginiana?)* flies about after dark to the terror of the superstitious Indians, who imagine its screech forbodes evil. Several species of *oriolus* are met with in the plains, and one, the *oriolus phœniceus*, is seen in immense flocks. The natives say that this bird, which in its first year is of a greyish black colour, changes to deep black in the second, and ultimately becomes black with red shoulders; but Mr. Collie thinks there is some error in this. There is another oriolus which frequents moist and rushy places; crows in great numbers, some which are white, and smaller than those of England; and several species of finches, buntings, and sparrows, prove very destructive to the grain when sown. The magpie is also an inhabitant here, and a small blue jay frequents the woods. The California quail *(tetrao virginianus)*, wood pigeons with bronzed imbricated feathers on the back of the neck, plovers *(charadrius hiaticula?)*, snipes, several species of sanderlings *(tringa)*, razorbills *(hematopus)*, herons *(ardea)*, curlew *(scolopax linosa* aud *recurvirosta)*, and two species of *rallus*, afforded amusement to our sportsmen, as did also some of the many species of geese, ducks, widgeon, and teal, which frequent the lakes and plains. The two latter species and one of the *anas (erycthropus?)* were similar to those which had been seen in Kotzebue Sound; and the natives remark that they arrive from the north in the month of September, and depart again in May. The grey

CHAP. II. Dec. 1826.

geese are said to be good to eat, but we found them all fishy; not so the ducks, the greater part of which are very palatable: these birds, of which we procured about twenty species, and the mallard, are so common that several were frequently killed at one shot. It was observed that some kinds of ducks always preferred salt water to the lakes, particularly a species with a dark-coloured body and a white head, which we did not obtain. Among those which frequent the fresh water there were generally an abundance of water-hens. Pelicans *(pelicanus onocratulus)* may be seen morning and evening winging their long line of flight across the harbour, and settling upon the little island of Alcatrasses, which they have completely covered with their exuviæ, and rendered extremely offensive to persons passing near the place. Shags *(pelicanus graculus)* also abound in the harbour. I ought to have noticed in its proper place the humming bird, which, notwithstanding the high latitude of the country, is an inhabitant of the woods, and if we may rely upon Padre Tomaso, may be seen there all the year round. We noticed several of them fluttering about some gooseberry bushes near our anchorage, and shot one in full flesh: as this was in the middle of winter, the information of the padre was probably correct.

To this list of birds several were added the succeeding year at Mounterey, which, being found so near the place we are describing, may justly be classed with them: these consisted of the golden-winged woodpecker, a goat-sucker, several species of small birds unknown to us, and a golden-crested wren. At this place there were also several species of *picus*.

I shall pass rapidly over the reptiles, which are not numerous at Sán Francisco, and none were procured during our stay. The Spaniards assert that there is an adder in the wood which is venomous, and that there are rattlesnakes upon the island of Molate in the harbour; but we saw neither the one or the other, notwithstanding Mr. Elson and a boat's crew landed upon Molate, which is very small indeed.

Fish are not much sought after in California, in consequence of the productions of the land being so very abundant; several sorts, however, are brought to the tables of the missions. In the Bay of Monterey we noticed the scomber colias, and another kind of mackerel, the torpedo and another species of raia, achimara, and swarms of small fish resembling the sardinia. Muscles are found in considerable quantities upon the shores, and form a large portion of the food of the Indians bordering upon the coasts and rivers. At Monterey two species of *haliotis* of large size are also extremely abundant, and equally sought after by the Indians. They are found on the granite rocks forming the south-east part of the bay, which appears to be their northern limit. The natives make use of these shells for ornaments, and decorate their baskets with pieces of them. Besides these shell-fish, there were noticed a few *patella*, *limpet*, *turbo*, *cardium*, and *mya* shells, and among other *lepas*, a rare species of *l. anotifera* and a chiton *(tunicatus ?)*

The forests of this part of California furnish principally large trees of the pinus genus, of which the *p. rigida* and the red cedar are most abundant, and are of sufficient growth for the masts of vessels. Two kinds of oak arrive at large growth, but near the coast they do not appear to be very numerous. There is here a low tree with a smooth reddish-brown bark, bearing red berries, which from the hardness of its

CHAP. II. Dec. 1826.

wood, would serve the purpose of lignum vitæ: there are also some birch and plane trees; but there are very few trees bearing fruit which are indigenous; the cherry tree and gooseberry bush, however, appear to be so.

The shrubs covering the sand hills and moors are principally syngenesious, or of the order rhamnus, while those which prefer the more fertile and humid soils are a gaudy-flowered currant bush, and a species of honeysuckle; but the most remarkable shrub in this country is the yedra, a poisonous plant affecting only particular constitutions of the human body, by producing tumours and violent inflammation upon any part with which it comes in contact; and indeed even the exhalation from it borne upon the wind, is said to have an effect upon some people. It is a slender shrub, preferring cool and shady places to others, and bears a trefoil crenated leaf. Among other useful roots in this country there are two which are used by the natives for soap, *amole* and *jamate.*

From Sán Francisco we proceeded to Monterey to take in the stores that had been purchased at that place, and to procure some spars which grow more conveniently for embarkation there than at Sán Francisco. Though the distance between these two places is very little more than a hundred miles, our passage was prolonged to two days by light winds. On the last day of the year we passed Punto año nuevo, which with Punto Pinos forms the bay of Monterey. This is a spacious sandy bay about twenty miles across, and according to Perouse with anchorage near the shore in almost every part; but it is not advisable to enter it in any other place than that which is frequented as an anchorage, in consequence of a heavy swell which almost always rolls into it from the westward. The mission of Santa Cruz is situated at the north extremity of the bay near Punto año nuevo, and vessels occasionally anchor off there for fresh water and supplies of vegetables, neither of which are to be had in any quantity at Monterey. Care should be taken in landing at Santa Cruz, as the surf is very heavy, and the river of St. Lorenzo has a bar off it, which it is necessary to pass.

We dropped our anchor in Monterey Bay on the first of January, and with the permission of the governor, D. Miguel Gonzales, immediately commenced cutting the spars we required; for each of which we paid a small sum. Through the assistance of Mr. Hartnell, we procured several things from the missions which we should otherwise have sailed without, and our thanks are further due to him for his kindness and attention during our stay.

The anchorage of Monterey is about two miles south-east of point Pinos, in the south angle of the great bay just described. It is necessary to lie close to the shore, both on account of the depth of water, and in order to receive the protection of point Pinos, without which ships could not remain in the bay. It presents to the eye a very exposed anchorage, but no accidents have ever occurred to any vessel properly found in cables and anchors; in which respect it very much resembles the bay of Valparaiso, nearly in the same parallel in the southern hemisphere.

The village and presidio of Monterey are situated upon a plain between the anchorage and a range of hills covered with woods of pine and oak. The presidio is in better condition than that at Sán Fran-

CHAP. II. Jan. 1827.

cisco; still as a place of defence it is quite useless. The fort is not much better, and its strength may be judged of from its having been taken by a small party of seamen who landed from a Buenos Ayrean pirate in 1819, destroyed the greater part of the guns, and pillaged and burnt the town.

At the distance of a league to the southward of the presidio lies the mission of Sán Carlos, a small establishment containing 260 Indians. It is situated in a valley near the river St. Carmelo; a small stream emptying itself into a deep rocky bay. The shores of this bay, and indeed of the whole of the coast near Point Pinos, is armed with rocks of granite upon which the sea breaks furiously; and as there is no anchorage near them on account of the great depth of water, it is dangerous to approach the coast in light or variable winds. Fortunately some immense beds of sea weed *(fucus pyriformis)* lie off the coast, and are so impenetrable that they are said to have saved several vessels which were driven into them by the swell during calm and foggy weather. The ride from the presidio to San Carlos on a fine day is most agreeable. The scenery is just sufficiently picturesque to interest, while the hills are not so abrupt as to inconvenience a bold rider. The road leads principally through fine pasture lands, occasionally wooded with tall pine, oak, and birch trees; but without any underwood to give it a wildness, or to rob it of its park-like aspect. Before the valley of Sán Carmelo opens out, the traveller is apprized of his approach to the mission by three large crosses erected upon Mount Calvary; and further on by smaller ones placed at the side of the road, to each of which some history is attached. In the church is a drawing of the reception of La Perouse at the mission, executed on board the Astrolabe, by one of the officers of his squadron. I much wished to possess this valuable relic, with which however the padre was unwilling to part.

We found lying in the port of Monterey an American brig endeavouring to dispose of a cargo of dry goods, and to procure hides and tallow in return; and we opportunely received from her a supply of spirits, as the last cask was abroach. On the 4th a Russian brig, named the Baikal, belonging to the Russian American Fur Company, anchored in the bay. This vessel was employed upon the coast, trading between Sitka, Bodega, and several ports in California, either in carrying or arranging the supplies for the Russian settlements to the northward. She was commanded by an officer in the Russian navy, and had on board Mr. Klebnekoff, the agent. There are several of these vessels upon the coast carrying guns, and wearing pendants. On the 5th we took leave of our hospitable acquaintances, and put to sea on our passage to the Sandwich Islands.

CHAPTER III.

Passage to the Sandwich Islands—Woahoo—Historical Sketch of the Islands — Progress in Civilization — Sandal Wood — Resources of the Government—Slow Progress of Education—Efforts of the Missionaries—Unsuccessful Result of their Zeal—Sentiments of the King and Chiefs—Entertainment given by the King—Death of Krymakoo—Wailing Scene—Departure of Kahumana for Owyhee.

CHAP. III. Jan. 1827.

UPON leaving Monterey we steered to the southward with a fair wind, which carried us into the trades, and attended us the whole way to the Sandwich Islands. In our course we searched unsuccessfully for all the islands that were marked near our route, rounding to every night when near the position of any one, that it might not be passed unobserved, and making sail on a parallel of latitude during the day. In this manner we searched for Henderson's and Cooper's Islands, besides several others said to lie near them, and also for a group in the latitude of 16° N., and longitude between 130° and 133° W.; but we saw nothing of them, nor had we any of the usual indications of the vicinity of land; so that if any of these islands exist, they must be in some other parallel than that assigned to them in the American Geographical Table, published in 1825.*

* I have been recently informed that an island of moderate height has been seen by the Sultan American Whaler in latitude

On the 25th, after a pleasant passage of twenty days, we saw the Island of Owyhee; and the following day anchored in the harbour of Honoruru, the capital of the Sandwich Islands. We had the satisfaction to meet all our former acquaintances well, and to receive their congratulations on our return; we had also the pleasure to find Mr. Lay the naturalist ready to resume his occupations. During our absence, he had unfortunately been prevented pursuing his researches among the islands by a severe illness.

After the usual etiquette of salutes, I visited the king and Kahumana, who appeared very glad of our arrival, and being informed that the ship was to remain a few weeks in the harbour, they very kindly appropriated three houses to the use of the officers and myself, and seemed determined to show by other acts of attention that the regard they had always expressed for our nation was not merely an empty profession.

In my first visit to this place, I gave a sketch of the appearance of the town of Woahoo and of the inhabitants, with the advances which the country appeared to be making in civilization. It may not be superfluous here to insert a very concise account of the islands during the last few years, to enable my readers to judge more correctly of their progress, and to furnish information to such as may not have the history of them fresh in their memories.

At the time the Sandwich Islands were discovered by Captain Cook, Owyhee was under the sovereignty of Terreeoboo, or Teriopu, who died shortly after the departure of the discovery ships. Tamehameha, who

15° 30′ N., longitude between 130° and 134° W. And that another was landed upon in latitude 18° 22′ N., longitude 114° W.

CHAP. III. Jan. 1827.

afterwards became so celebrated, was the nephew of Terreeoboo. He is not mentioned in the official account of Cook's voyage, but in a narrative of the facts relating to the death of the great navigator, published by Mr. Samwell, the surgeon of the Discovery, Meah Meah, as he is called by that gentleman, is represented to have slept on board that ship, and to have had with him a magnificent feather cloak, with which he would not part, except for iron daggers, six of which he procured, and returned to the shore well pleased with his bargain. No doubt his intention was to wrest the sovereignty from the hands of the successor of Terreeoboo, an enterprize which he performed shortly afterwards by assembling his forces and defeating him in a pitched battle, in which he is said to have slain him with his own hands. After this victory, no other chief possessing sufficient power to oppose Tamehameha, we find that on the arrival of Vancouver in in 1792 he had acquired supreme authority both in Owyhee and Mowee. He soon afterwards attacked and conquered Woahoo, and, assisted by his valiant protegé Krymakoo, in 1817, became sovereign of all the Sandwich group.

Vancouver was very instrumental in establishing the power of this chief on a firm basis, by noticing Tamehameha in a manner which could not escape the observation of the other chiefs, and by building him a decked vessel, which gave him a decided superiority of force, and enabled him to keep them in subjection. In return for these important benefits, the grateful chief, in presence of Vancouver and the Eries of the group, made a formal cession of the islands to the king of Great Britain, and the natives have ever since considered themselves under the immediate protection of this country.

In the early stage of our intercourse with these islands, several acts, such as the death of Cook, the murder of Lieutenant Hergerst, and the treacherous seizure of an American vessel, rendered merchant vessels cautious of communicating with savages of apparently so ferocious a character; but when it was known that the perpetrators of these murders were punished by Tamehameha, and when his real character was made public by the voyages of Vancouver and other navigators, every vessel employed in the Pacific was desirous of visiting his dominions. In course of time a regular market was established for the sale of the productions of the islands; the natives were instructed to accept Spanish dollars and European clothing in exchange for their goods; and several foreigners, by the king's persuasion, were induced to settle upon the islands. The native chiefs, in imitation of their sovereign, began to dress in the European style. A fort was built for the protection of the principal town, and a number of the natives were instructed in the use of fire-arms. The harbour of Honoruru soon became crowded with ships of all nations, and latterly the place has assumed the appearance of an European colony.

The discovery of sandal wood in the mountains opened a profitable channel of commerce; and several adventurers, chiefly from the United States remained to collect it from the natives. They found a ready market for it in China; the goods of that country were brought in return to the Sandwich islands, and thus was laid the foundation of a trade which still continues. Tamehameha having purchased several vessels with this precious wood, attempted to conduct this trade with his own resources, and sent a schooner

CHAP. III. Jan. 1827.

bearing his flag to Canton; but, owing to the forms and impositions practised in China, and other circumstances which he could not control, the speculation failed, and this advantageous trade has since been carried on by the Americans.

In all these plans for the benefit of his country, for the introduction of civilization among his subjects, and for the establishment of his assumed authority, Tamehameha was greatly indebted to the advice and assistance of two respectable English seamen, Young and Davis, whom he persuaded to remain in the islands. Their services were not unrequited by the great chief, whose generous disposition and intimate knowledge of human nature induced him to bestow upon them both rank and fortune, by raising them to the station of chiefs, and giving them estates. They in turn proved grateful to their benefactor, and conducted themselves so properly that every visitor to the islands has spoken of them in the highest terms. Davis died in 1108, and was buried at Woahoo, where the place of his interment is marked by a humble tombstone: Young still survives, at the advanced age of eighty-two. Besides these advisers, Tamehameha had a faithful and wise counsellor in Krymakoo, afterwards better known by the appellation of Billy Pitt.

Tamehameha having seen his country emerge from barbarism under his well-directed efforts, and having conferred upon it other important benefits, died in May 1819, at the age of sixty-three. His biographer will do him injustice if he does not rank him, however limited his sphere, and limited his means, among those great men who, like our Alfred, and Peter the Great of Russia, have rescued their countries from barbarism, and who are justly esteemed the benefactors of mankind. His loss as a governor, and as a father to his people, was universally felt by his subjects. It is painful to relate that, though his death occurred so recently, several human victims were sacrificed to his manes by the priests in the morais; and, according to the custom of the islands, some who were warmly attached to him committed suicide, in order to accompany his corpse to the grave; while great numbers knocked out their front teeth, and otherwise mutilated and disfigured themselves.

Tamehameha was no sooner dead than his son Rio-Rio, who succeeded him, effected the most important change the country had yet experienced. Having held conferences with the chiefs, and obtained the sanction of Keopuolani, a powerful female of rank, he ordered all the morais to be destroyed, and declared the religion of the foreigners,—of the principles of which he was then very ignorant, should henceforth be the religion of the state. The burning of the idols and the abolition of the *taboo* immediately succeeded this destruction of the morais, and put an end to many cruel and degrading customs, both injurious to the interests of the country and oppressive to the people, especially to the females, who were thenceforth admitted to an equality with the men.

The prejudices of Tamehameha had always opposed this change in the religion of his subjects, not so much, I am informed, from his being bigoted to idolatry as from its being better adapted to his politics. The maxims of our religion he thought would tend to deprive him of that despotic power which he exercised over the lives and fortunes of his subjects. The terror inspired by human sacrifices, and the absolute command which the superstitions of his idolatrous

CHAP. III. Jan. 1827.

subjects gave him, suited the plan of his government better than any other religion, and he, consequently, opposed every attempt to propagate the gospel among his people.

Up to this period no missionaries had reached the Sandwich Islands, and for nearly a year there might be said to be no religion in the country; but at the expiration of that period (in 1820), several missionary gentlemen arrived from the United States, and immediately entered upon their vocations. Keopuolani became the first actual convert to the Christian religion, though in 1819 both Boki and Krymhakoo were baptized by the clergyman of Captain Freycinet's ship. Keopuolani being a chief of powerful influence, her example was followed by a great many persons, and the missionaries have since added daily to the number of their converts, and have been protected by the government, particularly by Kahumana and Kapeolani, two female chiefs next in rank to Keopuolani, and probably first in power in the islands.

Keopuolani died in 1823, after having received the sacrament. She was a grandchild of Terreeoboo, and a daughter of Kevalao, who was slain at Mowee. At the time of this victory, which added Mowee to the dominion of Tamehameha, Keopuolani was only thirteen years of age. She happened to be on the field at the moment of the defeat of her party, and became the prisoner of the conqueror, who, in order to secure his conquest by right as well as by victory, united her to himself in marriage. She had, however, afterwards, agreeably to the custom of the country, several husbands, of which one was Krymakoo, who also fell into the hands of the king at Mowee, and whose life was generously spared; and another, Hoapiri, who, though a plebeian, was admitted to the honour of being one of the favourites of the queen. This person is the reputed father of Kiukiuli the present king, while Tamehameha is said to have been the father of Rio-Rio. The queen, however, declared both her sons to be children of the illustrious chief, and they succeeded to the throne accordingly, in cases of this nature the declaration of the mother being held sufficient.

Rio-Rio is represented to have been far inferior in intellect to his predecessor, and his youth and inexperience encouraged the superior chiefs to plan means for recovering their independence. At the moment the order was given for the destruction of the idols, a chief named Kekoakalane treacherously seized the war god, and joined by a party of rebels fled with it to Owyhee, where he hoped to excite the inhabitants in his favour, and to establish himself as an independent chief; but he was closely pursued by the gallant Krymakoo, and slain at Lakelakee, and hence that place has become celebrated, as the spot on which the last struggle for idolatry occurred. Another insurrection soon afterwards occurred at Atooi, which was quieted by the courage and promptitude of Rio-Rio, who embarked with a few faithful followers in a canoe, and in a personal conference brought the rebels back to their duty. Atooi was the last of the Sandwich Islands that was reduced to subjection by Tamehameha, and its chiefs were constantly on the watch for opportunities of recovering their independence. Russia, or at least her subjects, taking advantage of the disaffected state of Atooi, landed some guns upon that island, and erected a fort, which was taken possession of by the natives. Krymakoo, however, with a body of followers from Woahoo, overthrew the rebels. The

CHAP. III. Jan. 1827.

chief being permitted to choose the manner of his death, desired that he might be carried to sea, and be drowned by having a weight fastened round his neck. In addition to this attempt of the Russians to separate Atooi from the kingdom, it was supposed that America was also desirous of forming a settlement upon one of the islands. Rio-Rio foreseeing that occasional rebellions might arise in his dominions, through the interference of foreign powers, determined on a voyage to England to have a personal interview with the king, under whose protection the islands had been placed by Tamehameha, and also, perhaps, from a desire to see the country which furnished articles so superior to the manufactures of his own dominions.

The death of Rio-Rio and his queen, it is well known, occurred in this visit to England. Their bodies were conveyed to the Sandwich Islands by Lord Byron, in H. M. Ship Blonde, and lodged in a house built for the purpose, where they still remain.* Lord Byron having given the chiefs, in Boki's words, "good advice," and having placed the crown upon the head of Kiukiuli, the brother of Rio-Rio, and seen the government confided to Krymakoo as regent, quitted the islands about ten months before our first arrival.

Previous to the death of Tamehameha, several European houses appeared in Woahoo. Vessels and warlike stores had been purchased with sandal wood. The navigation of the Pacific became more general in

* In 1827, some of the chiefs had been persuaded that it was improper to keep the bodies above ground, and these beautiful coffins covered with crimson velvet and silver were about to be lowered into the earth, as a commendable mortification of pride, when they were prevented by the timely arrival of a gentleman, from whom this account was derived.

consequence of the return of peace, and the islands were more frequently visited. The abolition of the taboo had already produced an entire change in the state of society, and frequent interviews with foreigners created amongst the inhabitants a desire for dress and for luxuries, which was increased by the visit of the chiefs to England. Thus improvement advanced, as might have been expected under such advantageous circumstances as those in which the Sandwich Islands were placed. At the period of our visit there were in Woahoo several respectable American merchants, in whose stores were to be found all the necessary articles of American manufacture, the productions of the China market, wines, and almost every article of sea store. There were also two hotels, at which a person might board respectably for a dollar a day; two billiard rooms, one of which was the property of Boki; and ten or a dozen public houses for retailing spirits. The houses of the chiefs were furnished with tables and chairs, and those belonging to Kahumana with silk and velvet sofas and cushions. Not contented with the comforts of life, they latterly sought its luxuries, and even indulged in its extravagances. Kahumana filled chests with the most costly silks of China, and actually expended four thousand dollars upon the cargo of one vessel. Boki paid three thousand dollars for a service of plate as a present for the king, notwithstanding he had other services in his possession; one of which was of expensively cut glass from Pellatt and Green in London.

This progress of luxury was attended by an equally remarkable change in the civil and political arrangements of the country. At the period of our visit the king was always attended by a guard under arms; a

CHAP. III. Jan. 1827.

sentinel presented his musket when an officer entered the threshold of the royal abode; soldiers paraded the ramparts of a fort mounting forty guns; and "all's well" was repeated throughout the town during the night. The harbour in the spring and autumn was crowded with foreign vessels, as many even as fifty having been seen there at one time; five thousand stand of arms were said to be distributed over the island; three hundred men were embodied and dressed in regimentals; and the Sandwich Island flag was daily displayed by five brigs and eight schooners. The islands had already received consuls from Great Britain and the United States; had concluded treaties of alliance with them; and we have just heard that their spirit of enterprise has induced them to fit out, and despatch an expedition to take possession of some of the islands of the New Hebrides.

This state of advancement, considering the remoteness of the situation of these islands, and the little intercourse they have hitherto held with the civilized world, could hardly have been anticipated; and we hope it may not prove too rapid to be advantageous to the country, which has now several expensive establishments to maintain, and extravagant ideas to satisfy, with means evidently diminishing, if not nearly exhausted. The treasures accumulated by Tamehameha, and the supply of that precious wood which has been so instrumental in bringing the islands into notice, have been drained to meet the expenses of ruinous purchases which have materially contributed to the apparent show of grandeur and prosperity above mentioned. The sandal wood, it is known, requires many years to arrive at a fit state for the market, and its cultivation not having been attended to, the wood is now becoming scarce, while the debt of the nation has considerably increased. During our visit, in order to avoid the expense attending the collection of this wood, it became necessary to levy a tax upon the people of a pecul, or 133lb. each, which they were required to bring from the mountains, under a penalty of four dollars, and to deposit with the authorities at Honoruru for the purpose of liquidating the debt of the nation. The greater part of the wood brought in was small and crooked, and only fit for the use of the Jos houses in China, where it is burned as incense, but the consumption of it there is diminishing in consequence of an order for its disuse in those places of worship. The odour of the sandal wood of the Sandwich Islands is very inferior to that of Malabar, Ceylon, and other parts of India. With the exception of the profits arising from the sale of sandal wood, of salt, and from the port dues, and from the advantage derived from merchant vessels visiting the islands for refreshments, there is no revenue of consequence; certainly none that is at all adequate to meet the expenses of the nation.

The chiefs, foreseeing the approaching crisis, are anxious to avail themselves of any prospect of an increase of revenue. Thus attempts have been made to manufacture sugar from the canes which grow very abundantly and in great luxuriance in the islands; and I sincerely hope that Mr. Marini, who has hitherto been of the greatest benefit to the government of Woahoo, may succeed in the mill which he was constructing for this purpose during our visit. But machines of this nature have already cost a very large sum, and have not hitherto succeeded, partly, perhaps, in consequence of the want of proper materials. A

CHAP. III. Jan. 1827.

cargo of this sugar it was hoped would be ready for exportation in 1827, which was then to be carried to the Californian market, where, as it has already been said, sugar attains a high price. But the Sandwich Islands, until much more advanced in the science of cultivation, will always have to compete with Manilla in the sale of this material. Tobacco, coffee, and spices have been introduced into the islands, and it is to be hoped they will succeed under the fostering hand of the indefatigable individual before mentioned. An attempt was made to encourage the planting of cotton, which was tolerably successful the first year, but for some reasons, which were ascribed to the rigid observance of the church duties, the labourers were prevented from gathering the crop, and it rotted in the pod. It is particularly unfortunate that the attempt to cultivate this plant, which would be of great advantage to the islands, should have failed both in the Society and Sandwich groups, as it will probably discourage the inhabitants from any further endeavour to produce it. Salt has been collected from some lakes near the town, and for some time past has produced a small revenue. Hereafter it is likely to be in greater request, for the purpose of curing meat for sea store, or for exportation to Kamschatka, where it is in great demand. Flax of a good quality grows upon Owyhee, and rope for the vessels of the country is made from a species of *urtica?* As yet, however, the sandal wood is the only material that has produced any revenue of consequence.

Soon after the Christian religion had been introduced into the Sandwich Islands, several of the chiefs were taught to read and write, and were so delighted at the idea of being able to communicate their thoughts to friends at a distance, without the necessity of disclosing them, and free from the risk of misinterpretation, that some of the scholars laboured at their task as if the prosperity of the islands depended upon penmanship alone. Education in other respects has made much slower progress than every well-wisher of the country could desire. A few individuals who have had the advantage of continued instruction have acquired a limited knowledge of the scriptures, but many remain ignorant even of the nature of the prayers they repeat; and in other subjects are entirely uninstructed.

The missionaries appear to be very anxious to diffuse a due knowledge of the tenets of the gospel among all the inhabitants, and have laboured much to accomplish their praiseworthy purpose: but the residents in Honoruru well know what little effect their exertions have produced, probably on account of the tutors having mistaken the means of diffusing education. In the Sandwich Islands, as in all other places, there is a mania for every thing new, and, with due reverence to the subject, this was very much the case with religion in Honoruru, where almost every person might be seen hastening to the school with a slate in his hand, in the hope of being able soon to transcribe some part of the *pala pala* (the scriptures). This feeling under judicious management might have produced the greatest blessings Woahoo could have enjoyed; and the gentlemen of the mission might have congratulated themselves on having bestowed upon the inhabitants very important benefits. But they were misled by the eagerness of their hopes, and their zeal carried them beyond the limits calculated to prove beneficial to the temporal interests of a people, still in the earliest stage of civilization. The apparent thirst after scriptural knowledge in Honoruru created a belief among the missionaries that this feeling was become general, and auxiliary schools were established in different parts of the island, at which we were informed every adult was required to attend several times a day.

While this demand upon their time was confined within reasonable limits, the chiefs, generally, were glad to find their subjects listen to instruction; but when men were obliged to quit their work, and to repair to the nearest auxiliary school so frequently during the day, so much mischief was produced by loss of labour, and such ruinous consequences threatened the country, that many of the chiefs became desirous of checking it. Kahumana and her party, however, persisted in considering it desirable, and in supporting the missionaries; while a powerful party, at the head of which were the king and the regent, exerted themselves to counteract their endeavours. Thus dissensions arose very prejudicial both to the cause of religion and to the interests of the country. The chiefs lost their influence, the subjects neglected their work, and hypocrisy on the one side, and intemperance on the other, became the prevailing errors of the time; the latter indulged in probably to a greater extent, with the view of bringing ridicule on the opposite party; a scheme in which it is said that Boki himself condescended to join.

At length the regent and other chiefs determined to break through this rigid discipline. The ten commandments had been recommended as the sole law of the land: this proposition was obstinately opposed; a meeting was called by the missionaries to justify their conduct, at which they lost ground by a proposal that the younger part of the community only should be obliged to attend the schools, and that the men should be permitted to continue at their daily labour. The king, whose riding, bathing, and other exercises had been restricted, now threw off all restraint, and appeared in public wearing the sword and feather belonging to the uniform presented to him from this country by Lord Byron, which his preceptor had forbid him to use, under the impression that it might excite his vanity. The boys, following the example of their youthful sovereign, resumed their games, which had been suppressed; and among other acts which, though apparently trifling, discovered to the common people a spirit of opposition, and an earnestness on the part of the chiefs to overthrow the system that had been brought into operation, Koañoa, who had long been enamoured of a female chief, Kenow, whom Kahumana intended for the king (although she was old enough to be his mother), being refused the marriage ceremony by the mission, carried off the object of his desire, and took her to his home.

This was the state in which we found Woahoo, and from it the missionaries might extract a useful lesson while imparting religious instruction to mankind, of the necessity of combining their temporal interests with those which relate to their prospects of futurity.

It was supposed, from the manner in which Kahumana persevered in her support of the missionaries, that she was actuated by a deeper policy than appeared. Her jealousy at the investment of the sovereign power in the king and Boki was well known; and it was surmised that she entertained hopes of creating a party which, in the event of the death of Pitt, then daily expected, would forward her ambitious

CHAP. III. Jan. 1827.

views. Whether this surmise was just I do not pretend to say; but she certainly did not succeed, that event having passed off during our stay without any movement in her favour.

Amidst this conflicting interest of parties, we were gratified to observe the greatest cordiality between the chiefs and the English and American residents, neither of whom took part in these state quarrels. To strengthen this feeling, a public dinner was given by the officers of the Blossom and myself to the king and all the royal family, the consuls, the chiefs, and the principal merchants resident in the place. On this occasion, the king was received with the honours due to his rank. He was dressed in full uniform, and altogether made a very elegant appearance. His behaviour at table was marked with the greatest propriety, and though he seemed fully aware of the superiority of Europeans, he appeared at the same time conscious that the attentions he received were no more than a just tribute to his rank. Boki, the regent, Koanoa, the colonel of the troops, and Manuia, the captain of the port, were dressed in the Windsor uniform; and Kahumana, and the two female chiefs next in rank, were arrayed in silk dresses, and had expended a profusion of lavender water upon their cambric handkerchiefs. Many loyal and patriotic toasts succeeded the dinner, some of which were proposed by Boki, in compliment to the king of England and the president of the United States, between both of whom and his royal protégé he expressed a hope that the warmest friendship would always subsist. The chiefs drank to the health of several persons who had shown them attention in London, and in compliment to the ladies of England proposed as a toast,

"The pretty girls of the Adelphi." Throughout the day the islanders acquitted themselves very creditably, and their conduct showed a close observance of European manners.

A few days afterwards the king gave an entertainment, at which his guests were seated at a long table spread in the European style, and furnished with some very good wines. Among other good things we had Leuhow, a dish of such delicious quality that excursions are occasionally made to the plantations for the pleasure of dining upon it; and, from this circumstance, a pic-nic and a Leuhow party have become nearly synonymous. The ingredients of the dish are generally the tops of the taro plant and mullet which have been fattened in ponds; these are wrapped in large leaves and baked in the ground, though sometimes fowls and pork are used. In order to amuse us, the king had also assembled several dancers and the best bards in the island; and we had the pleasure of witnessing some native performances, which were the more interesting, as these entertainments will shortly lose all their originality by the introduction of foreign customs. On the present occasion, indeed, it was difficult to procure performers of any celebrity, and both bards and dancers were sent for from a considerable distance; and even then only two of the latter were considered worth our notice. The performance opened with a song in honour of Tamehameha, to which succeeded an account of the visit of Rio Rio and his queen to England; their motives for undertaking the voyage were explained; their parting with their friends at Woahoo; their sea-sickness; their landing in England; the king's attempts to speak English; the beautiful women of this country; and

CHAP. III. Jan. 1827.

the sickness and death of the youthful royal pair, were described with much humour, good-nature, and feeling.

The natives were delighted with this performance, especially with that part which exhibited the sea-sickness, and the efforts of the king to speak English; but our slight acquaintance with the language did not enable us fully to appreciate the allusions. In the next performance, however, this defect was less felt. The song was executed by three celebrated bards, whose gray beards hung down upon their breasts: they were clothed in their rude native costume, and each had the under part of his right arm tattooed in straight lines from the wrist to the armpit. They accompanied themselves upon drums made of two gourds neatly joined together, and ornamented with black devices. Each bard had one of these instruments attached to his left wrist by a cord; the instrument was placed upon a cushion, and the performer throughout measured time by beating with his right hand upon the aperture of the gourd. The subject related to the illustrious Tamehameha, whose warlike exploits are the constant theme of these people. Occasionally the bards seemed to be inspired; they struck their left breasts violently with the palms of their hands, and performed a number of evolutions with their drums, all of which were executed simultaneously, and with ease, decision, and grace. On the whole it was an exhibition very creditable to the talents of the performers. To this succeeded several dances: the first, performed by a native of Atooi, was recommended principally by a display of muscular energy; the next was executed by a man who was esteemed the most accomplished actor of his time in

Woahoo, and the son of the most celebrated dancer the islands ever had. He wore an abundance of native cloth, variously stained, wrapped about his waist, and grass ornaments fixed upon his legs above the ancles. A garland of green leaves passed over his right shoulder and under his left arm, and a wreath of yellow blossoms, very commonly worn in the Sandwich Islands, was wound twice round his head. Unlike the former dance, the merit of this consisted in an exhibition of graceful action, and a repetition of elegant and unconstrained movements.

The dance of the females was spoiled by a mistaken refinement, which prevented their appearing, as formerly, with no other dress than a covering to the hips, and a simple garland of flowers upon the head; instead of this they were provided with frilled chemises, which so far from taking away the appearance of indecency, produced an opposite effect, and at once gave the performance a stamp of indelicacy. In this dance, which by the way is the only one the females of these islands have, they ranged themselves in a line, and began swinging the arms carelessly, but not ungracefully, from side to side; they then proceeded to the more active part of the dance, the principal art of which consisted in twisting the loins without moving the feet or the bust. After fatiguing themselves in accomplishing this to the satisfaction of the spectators, they jumped sidewise, still twisting their bodies, and accompanying their actions with a chorus, the words of which we supposed bore some allusion to the performance. We had afterwards a sham-fight with short spears, wherein very little skill was exhibited, and, compared with the dexterity of the warlike Tamehameha, who is said by Vancouver to have successfully

CHAP. III. Feb. 1827.

evaded six spears thrown at him at the same instant, the present representation was quite contemptible. These exercises are now seldom practised, and in a short time, no doubt, both they and the dances will cease to be exhibited.

On the 12th of February, we received the melancholy intelligence of the death of Krymakoo, who had long suffered under a dropsical complaint, for which he had undergone frequent operations. Only four days previously he went to bathe in the sea at Kairua, in Owyhee, and on coming out of the water he was taken ill, and died very soon afterwards. He was at an advanced age, and had been present at the death of our immortal countryman in Karakakoa Bay, and perfectly recollected that fatal transaction. Krymakoo, or, as he was more generally called, Pitt, from the circumstance of his being a contemporary prime minister with our great statesman, became a protégé of Tamehameha shortly after the departure of Cook's ships. He is first introduced to our notice by Vancouver, who particularly remarks his superior manners and conduct. His life was devoted to the advantage of his country, and to the support of his illustrious patron, in whose service he distinguished himself alike as a warrior and a counsellor. Intelligent, faithful, and brave, he was confided in and beloved by his king and his countrymen, and he was a chief in whom the foreign residents placed implicit reliance. His ardent spirit and anxiety for the welfare of his country led Tamehameha on one or two occasions of insurrection to suspect his fidelity, and in order to put it to the test he is said to have deprived him for the time of his estates; an act of injustice, calculated rather to increase than to allay any dissatisfaction that might have existed in his mind. Pitt, nevertheless, remained faithful, and fought by the side of his patron. After the death of Tamehameha, he enjoyed almost sovereign power, which he employed to the benefit and civilization of his countrymen. His command of temper was not less praiseworthy than his other virtues. On the occasion of some misunderstanding between the missionaries and the seamen of an American vessel, the crew went on shore with the view of burning Mr. Bingham's house, but mistaking the place, they set fire to one belonging to Pitt. The natives immediately flew to protect the property of their favourite chief, and a serious quarrel was about to take place, to the disadvantage of the Americans, when Pitt, who had escaped the flames, harangued the mob with the greatest composure, induced them to desist from acts of violence, and persuaded the crew, who by this time had discovered their mistake, to return to their vessel. It has been asserted of Pitt that he was extremely ambitious; but his ambition seems to have had no other object than the welfare of his country: had he aspired to the crown, there were many favourable opportunities of which he might have availed himself without much risk of failure, of which the death of Tamehameha, the revolt of Kekoakalane, the insurrection of Atooi and others are sufficient instances. He left one son, whom he was very anxious to have educated in England, and pressed his request so earnestly that I had consented to take him on board the Blossom, but the vessel which was sent to bring him from Owyhee returned hastily with the news of the death of the chief, which frustrated the plan. Immediately this event was known the flags of the forts and the shipping were lowered half-mast, and the shores of the bay resounded with the wailings of the inhabitants.

CHAP. III. Feb. 1827.

It had been supposed that the ambition and jealousy of Kahumana and the conflicting interests of the chiefs would have displayed themselves in insurrection on this occasion, and that the disaffected chiefs would have availed themselves of this moment to remove the supreme power from the hands of the young king; but whatever results this melancholy event might have produced had it occurred at an earlier date, nothing was now attempted. Boki, however, thought it prudent to assemble the troops in the fort, and the Blossom was put in readiness to preserve order, if necessary, and to receive the foreign residents, should their safety require it. Anxious to witness the effect of this occurrence upon the court, I immediately paid a visit of condolence to Kahumana, who was seated amidst a motley assemblage of attendants, looking very sorrowful. It appeared, however, from the following incident, that the sincerity of her grief was questionable. Happening to cast her eye upon a Bramah inkstand which I was conveying to the observatory, she seized it with both hands, and declared, her countenance brightening into a smile, how much she should like to have it. As it was the only one I possessed, I did not intend at first to gratify her majesty's wishes, but she fairly tore it from me: so that, making a virtue of necessity, I presented it to her. After bestowing some praise upon the invention, she passed it to Karui, a female chief next in rank to herself, and then dismissing her pleasant looks, she resumed her sorrow, and convinced every person present that she was quite an adept in this barbarous custom of her country.

Many of the court seemed to consider this moment one of apprehension, and every person who approached the queen's abode was at first supposed to be the bearer of the news of some insurrection or other convulsion of the state. As he entered the room, therefore, there was a dead silence; but when it was found that these visits were made merely to inquire after the health of the queen, the wailing, as if it had suffered by the disappointment, burst forth with redoubled energy. Kahumana herself evidently anticipated some disturbance, for she whispered to me to be upon my guard, as there was a probability that the people would be mischievous. Nothing, however, occurred to disturb the tranquillity of the town but the wailing around the royal abode.

It is unnecessary here to describe many instances of the extent to which this hypocritical affectation of grief was carried; suffice it to say, that several persons, as if determined to perpetuate the barbarous practice of self-mutilation, knocked out their front teeth with hammers.

The queen almost immediately after the death of her brother embarked for Owyhee in a native schooner, to the great satisfaction of the chiefs and the European residents in Woahoo. As it was probably the last time she would see us, she was complimented with a royal salute on leaving the harbour.

CHAPTER IV.

Further Remarks on the Inhabitants—Treaty of Alliance—Climate—Medicinal Properties of the Ava—Supplies—Departure—Passage to China—Ladrone and Bashee Islands—Arrival at Macao—Transactions there—Departure—Botel Tobago Xima—Arrival at the Great Loo Choo.

CHAP. IV. Feb. 1827.

On the return of the ship to the Sandwich Islands the chiefs were very anxious to learn where she had been, and to be informed whether in some of the countries she had visited, the produce of their dominions might not find a favourable market. Kahumana, in particular, was so much interested in these inquiries that she condescended to direct her attention to them, and laid aside a missionary book with which she had been instructing her mind while the back part of her body was undergoing the soothing operation of being pinched by one of her female attendants. The conversation happening to turn upon Bird Island, Boki, on hearing it was so near to the Sandwich group, meditated its addition to the dominions of the king, no doubt under the impression of its being similar to one of the Sandwich Islands, and was greatly disappointed when informed that the island was not worth his possession. The account of the high price of sugar in California quite put him in good humour with his sugar-mills, which for some time past had been a subject of annoyance to him, in consequence of the expense incurred by their continually breaking. All parties were evidently desirous to extend their commerce, and a spirit of enterprise appeared to have diffused itself amongst them, which it is to be hoped may continue.

During our absence two important political events had occurred—the negociation of a treaty of alliance between Captain Jones, of the United States' sloop Peacock, on the part of America, and Boki, the regent, on the part of the Sandwich Islands; by which the reception of the American vessels in the Sandwich Islands, on the footing of the most favoured nation, was guaranteed to America in the event of that nation being involved in hostilities with any other power. The other was the resignation of Pitt, who, being aware of his approaching dissolution, retired to Owyhee, and left his brother Boki to act as regent. Boki, who, it may be remembered, accompanied the late king Rio-Rio to England, appears to have derived much benefit from that visit, and on his return to the Sandwich Islands to have become very desirous of improving the condition of his countrymen. He was, however, a less active governor than Pitt, and less capable of effecting those changes which experience had nevertheless convinced him were necessary for their advancement.

The town of Honoruru had now a more cleanly and lively appearance than on our former visit, and the streets, occupied by happy little children who had resumed their games, wore a more cheerful aspect. There was an improvement also in the society of the place, arising apparently from the arrival of some Europeans, particularly of the consul's family, which

CHAP. IV. Feb. 1827.

was of very great advantage to the females of Woahoo, who seemed anxious to imitate their manners, and were so desirous of becoming acquainted with the method of arranging their different articles of dress, that it required an unusual share of good nature to avoid taking offence at the rude manner in which they gratified their curiosity. The females of Woahoo are shrewd observers of these matters, and on great occasions endeavour to imitate foreigners as nearly as they can ; but the powerful influence of fashion has not been yet able entirely to get the better of that other powerful principle, early habit, and the women of the Sandwich Islands in retirement still adhere to their old customs, affording as curious an instance as was ever beheld of barbarism walking hand in hand with civilization.

The lower class of the inhabitants of Woahoo have varied their dress very little from its original style; though in Honoruru some females may be seen clothed in the cotton of Europe, and even in the silks of China, with green and red shoes, and sometimes with parasols. They obtain these articles as presents from the crews of such ships as touch at the port. In every uncivilized country which has as much foreign intercourse as Woahoo, incongruities must be of frequent occurrence ; thus we were daily in the habit of seeing ladies disencumber themselves of their silks, slippers, and parasols, and swim off in fine style to different vessels, carrying their bundles on their heads, and resuming their finery when they got on board. Nor was it less amusing to observe them jump overboard soon after daylight, and continue sporting and swimming about the vessels in the harbour like so many nereids ; practices to which they adhere with as much fondness as ever. Many, however, now think it necessary to put on a bathing gown when they take this recreation.

The men make very tolerable seamen, and are particularly useful in boats. Accustomed from their infancy to the water, they are as much at home in that element as on land; and having frequently encountered gales of wind at sea in their open canoes, they have no apprehension of them on board a strongly built ship. They are active and honest, and many of them are taken on board merchant ships visiting the islands, as part of their crews.

In the course of time it is to be hoped that they will become sufficiently enlightened to navigate their own vessels, as they at present depend upon foreigners for the performance of that service. Their vessels are now generally chartered to Americans, who bear a certain proportion of the expenses of the voyage, and have carte blanche to proceed where they please, and to collect, sell, and purchase cargoes at their discretion, and as it may seem most advantageous for themselves and the owners, who divide the profits of the venture at the end of the voyage. Their occupation consists principally in trading with California and the islands of the Pacific, or in making sealing voyages; in which case the skins they obtain are carried to some foreign market, and the proceeds applied to the purchase of a new cargo adapted to the wants of the Sandwich Islanders; such as horses, or furniture, and other household materials. Upon the whole, these returns are said to be by no means equal to the risk and expenses of the voyage; and the ships, being built of slight materials, require constant repair, and soon wear out: so that their navy, at present, is of no great advantage to the state.

No duties have as yet been imposed on any goods, either imported or exported, and the only charges made by the government are the port dues, which are very prudently lighter on vessels touching at the islands for refreshments only, than upon those which bring cargoes of merchandize; the charge in the former case is six, and in the latter fifty, cents per ton for the outer anchorage, and ten and sixty cents per ton respectively for the inner anchorage.

The Sandwich Islanders will apparently make as good soldiers as they do sailors, and are so proud of the honour of being embodied in the corps of the state, that they cannot suffer a greater disgrace than to have the regimentals taken from them and to be turned out of the ranks. They were repeatedly drilled by our serjeant of marines, and though under the disadvantage of not understanding the language in which the word of command was given, they improved quite as much as men in general would have done who had been in the habit of seeing the exercise performed. The inhabitants appear disposed to learn any thing that does not require labour, and soldiering soon became so completely a mania, that the king had the choice of his subjects; and little boys were seen in all parts of the town tossing up a sugar cane, with a "shoulder ump!" and some of the troop, even after being dismissed, would rehearse the lesson of the day by themselves. The islanders have a good idea of acting in concert, derived from their early exercise of the *palalu*, so interestingly described by Vancouver, in which they were accustomed to form solid squares; and when engaged, presented a formidable phalanx, which it was not easy to force.

Among other services which we performed for the king was an inspection of his cannon in the forts, some of which were so corroded, that in all probability their discharge would have been productive of serious accidents to some of his subjects. We also furnished him with twenty tons of stones, which we had taken in at Chamisso Island as ballast, to be used in rebuilding the wall of his mud fort.

It is unnecessary to describe further the inhabitants of a country which has already been the subject of several volumes. Enough has been said to show that the people are fast imbibing foreign customs, and daily improving both in their manners and dress.

The harbour of Honoruru is the general rendezvous of all the whale ships employed in the North Pacific Ocean. In the spring time these vessels assemble here to the number of forty or fifty sail at a time, and take on board large supplies of vegetables and fruit, as sea stock, to enable them to remain upon their fishing ground until the autumn, when many of them return to the port. The fresh provision which they procure at these islands is of the greatest advantage to the crews of the whalers, who would otherwise be afflicted with scurvy; and the goods which they give in exchange are very acceptable to the inhabitants. A number of idle dissolute seamen however, discontented with their ships, generally remain behind, and live in the public houses until their money and clothes are expended, or attach themselves to females, and in either way become dependent upon the inhabitants for food. These characters do infinite mischief to the

CHAP. IV. Feb. 1827.

lower order of the natives, by encouraging them in intemperance, debauchery, idleness, and all kind of vice; nearly sufficient of themselves to counteract all the labours of the missionaries in the diffusion of morality and religion.

The harbour is formed by a coral reef, which extends along the coast from the Pearl River to Wytiete Bay, but connected with the shore at intervals, so as to impede the passage of vessels. The entrance is very narrow and intricate, and vessels are generally towed in early in the morning, before the breeze freshens. There is a rock nearly in mid-channel upon which the sea generally breaks. Sometimes indeed it breaks quite across the entrance, and renders it necessary at that time, in particular, to employ a pilot. The depth in the channel at high water is about eighteen feet; but as I did not make a plan of this port, in consequence of Lieutenant Malden of the Blonde having so recently executed all that was necessary in that respect, I cannot speak positively. In sailing along the reefs in boats it is necessary to keep at a considerable distance on account of the sudden rise of the sea, which is very apt to fill or upset them when it breaks; and boats should not at any time pull for the entrance until they have gained a proper station off it. I refer to the directions in my nautical remarks for finding this station, and also for further information regarding this port.

The climate of the Sandwich Islands is more refreshing than that of Otaheite, although the group is scarcely farther from the equator. I am not aware that any register has been kept for a whole year at Otaheite; but at Woahoo this has been done by the gentlemen attached to the missions, from which it appears that the mean temperature for 1821 was 75°, the maximum 88°, and the minimum 59°, and that the daily range on an average was about 13°. In the last fortnight of May 1826, we found the maximum 83°, and minimum 74°; and in the last fortnight of February 1827, maximum 80°, and minimum 58°.

The N. E. trade wind, in general, blows strong to the windward of the Sandwich Islands, though for many miles to leeward of them frequent calms and light baffling winds prevail, and impede the navigation between the islands. About the period of the rainy season these winds are interrupted by gales from north-west to south-west, but when they cease the trade resumes its usual course. The duration of this season at Woahoo is from February to May. In 1826 it was over on the 19th of May; and in 1827, it began on the 17th of February. At this period the rains are occasionally very heavy; in 1826 and 1830, I have been informed they were particularly so; at other times, however, the reverse takes place, and from August 1821, to the same month of the following year, it appears by the register of the missionaries that there were but forty days on which rain fell.

The windward sides of the islands are said to be much colder, and to be subject to more rain than those to leeward. They are also liable to fogs in the spring of the year, while those which are opposite are enjoying sunshine. The mountains, from their height, act upon the atmosphere as powerful condensers, and in particular times of the year are scarcely ever free from mists; these are occasionally detached by gusts of wind and carried over the leeward parts of the island, and it is not unusual in Honoruru to expe-

CHAP. IV. Feb. 1827.

rience a pretty sharp sprinkling of rain without perceiving any cloud from whence it proceeds.

Water-spouts not unfrequently visit these islands, one of which I was told burst over the harbour of Honoruru, discharging such a quantity of water that the sea rose three feet. I have repeatedly seen this phenomenon on a small scale carrying a column of dust along the plains near Honoruru, and whirling hats into the air; and I once saw a native boy greatly puzzled to escape from its influence.

I shall conclude these remarks with some observations on the use and effects of the ava, a root which was formerly in much use in the Pacific, taken from the Journal of the surgeon of the Blossom. The intoxicating property of the ava root, the cutaneous eruption which succeeds its use, and the renovating effect it has upon the constitution, have been noticed ever since the discovery of the Society Islands. Mr. Collie observes, that—" a course of it is most beneficial in renovating constitutions which have been worn out by hard living, long residence in warm climates, without, however, affections of the liver, and by protracted chronic diseases; more especially if the disorder be such as by the humoral pathologists would be attributed to an attenuated or acrid state of the blood." He had an opportunity of seeing " a gentleman, a foreigner, who had undergone a course of it to remove a cutaneous affection said to have been similar to St. Anthony's fire. It had affected at different times almost every part of the body, going from one place to another, but had been particularly obstinate in one leg. He took two doses a day of half a pint each, one before breakfast and one before dinner, by which his appetite was sharpened; and by the time he had finished his meal a most pleasing state of half intoxication had come on, so that he was just able to go to his couch, where he enjoyed a sound and refreshing sleep.

"About the second or third week, the eyes became suffused with blood, and the cuticle around them began to scale, when the whole surface of the body assumed the appearance above described. The first dose is continued for a week or so, according to the disease, and then gradually left off. The skin clears at the same time, and the whole system is highly benefited.

"I recommended the ava, and had an opportunity of seeing the first effects upon a man affected with chronic superficial ulceration, affecting the greater part of the toes, and the anterior part of the soles of the feet. The legs and feet were œdematous and swelled; the pain was very distressing, preventing any sound repose, and not permitting him even to lie down or bring them up, so as to be near a line horizontal with the body. The ulcers were covered with a tough, viscid, dark-coloured discharge that adhered to the surface, and entirely concealed it. His frame was emaciated, pulse quick and irritable, appetite gone, tongue dry and reddish; he had taken mercurial preparations at two previous periods, as he said, with considerable benefit; but for want of the medicines it was stopped, when the sores were nearly healed. He had been, and I believe still was, addicted to drinking spirituous liquors. The ava was given three times a day with the same immediate effects as before-mentioned, and at the end of ten days the ulcers were clean and healing. From the commencement of the course he had been able to lie down, allowing his feet

CHAP. IV. Feb. 1827.

to hang over the bed-side: he had slept soundly, and his appetite was good. Could he have procured and applied a suitable dressing for the ulcers, with appropriate support to the œdematous extremities, I have no hesitation in saying that the plan would have succeeded. Even with all these disadvantages, I am inclined to believe that a cure will be effected if he abstains from liquor."

In this account of the Sandwich Islands, I have avoided touching upon subjects connected with the mythology, traditions, and early manners and customs of the islanders, from a conviction that I could give but an imperfect sketch of them, and from a hope that they will hereafter be laid before the public by the author of Polynesian Researches, who from his intimate knowledge of the language, his long residence in the Pacific, and from the nature of his occupations, has greater opportunities of becoming acquainted with them than any other foreigner. My endeavour has been to give as faithful an account as I could of the government, and of the state of society in the islands at the time of our visit, and of the resources and commerce of the country. Had my occupations been less numerous, I might have done more justice to these subjects; but the determination of the position of the place, and attention to other observations, occupied my time so completely, that I had very little leisure for other pursuits.

The results of the observations that were made there will be given in the Appendix; and the natural history will form part of two volumes which will shortly appear before the public.

During the absence of the ship from the Sandwich Islands, Captain Charlton, the consul, had succeeded in procuring a supply of salt provision for her. This was the more opportune, as the meat which had been corned in California was found on examination to be so bad that it was necessary to throw the whole of it overboard. We at first imagined that this failure proceeded from our ignorance of the method of curing the meat, but that which had been prepared at Monterey, by a person brought up to the business, was found to be equally bad; and the failure, in all probability, arose from the heated and feverish state in which the animals were slaughtered. We frequently remonstrated with the governor of Sán Francisco against being obliged to kill the animals in this state, and begged he would have them penned up until the following day, as they were quite wild, had been harassed with lassos, and dragged many miles by tame bullocks. We did not however succeed, and if the animals were not slaughtered as they were delivered into our charge, they either made their escape, or, as was the case with several, broke their necks in their struggles for freedom. The present supply of provision was consequently of the greatest importance. In addition to this we procured a few other stores, but not sufficient for our purpose, and there were no medicines to be had, so that it was still necessary to proceed to China.

As soon as the ship was ready for sea, therefore, we endeavoured to sail, but the wind about this time blew from the south-west, and kept us imprisoned a fortnight; the harbour of Honoruru being so difficult of egress, that, unless the wind be fair, or there be a perfect calm, a vessel cannot proceed to sea. On the 4th of March, however, we took our leave of the authorities and residents of the place, from both of

CHAP. IV. March, 1827.

whom we had received the greatest attention, and put to sea on our way to Macao.

Upon leaving the Sandwich Islands I directed the course to the southward; and next day, having gained the latitude of 18° 32′ N., I stood to the westward, with the intention of pursuing the above-mentioned parallel as far as the Ladrone Islands. I did this with a view of keeping fairly within the limit of the trade wind, which, at the season of the year in which this passage was made, is frequently variable in a higher latitude, and even subject to interruptions from strong north-westerly winds. I was also desirous of ascertaining the position of an island bearing the name of Wake's Island, upon Arrowsmith's chart, situated directly in the route between the Sandwich Islands and China.

A fresh trade-wind attended us until the fifth day after our departure, when it was interrupted by a breeze from the southward. The serenity of the sky which accompanied the trade, now became obscured by heavy thunder clouds, which gathered around us until the night of the 6th, when they completely blackened the sky. We had lightning frequently during the day, which increased so much towards night, that from eight o'clock to daylight the following morning the sky presented an uninterrupted blaze of light. It was unusually near; the forked lightning passed between the masts several times, and the zenith occasionally presented a fiery mass of short curved lines, which shot off in different directions like as many arrows; while the heavy peals of thunder which generally accompany these storms were subdued by crackling discharges not unlike the report of musketry from a long line of infantry. About the commencement of this storm the temperature fell four degrees, but gradually rose again to its former height. The sympeisometer was not sensibly affected.

On the following day fine weather was restored, the trade took its proper direction ; and the sea, which had been much agitated by the changeable winds, abated, and we pursued a steady course. About four days afterwards a brilliant meteor was discharged from the zenith towards the north-west, in the direction of some heavy clouds (nimbi), which were pouring down torrents of rain. It presented a long bright liquid flame of a bluish cast, and was followed by a train of sparks, until it had reached within 15° of the horizon, when it exploded, and three distinct fragments, having the appearance of being red hot, were discharged. They gradually lost their brilliancy as they fell, and were quite extinguished before they came in contact with the water. With the exception of the nimbi in the north-west, the sky was perfectly clear, particularly at the zenith, whence the meteor appeared first to be discharged. After these meteorological disturbances we had fine weather almost all the remainder of the passage.

At two o'clock on the 15th we were within a few leagues of the situation of Wake's Island, and the ship was brought to until daylight; but seeing no land at that time we bore away again, and at noon were exactly on the spot where the island is placed in Arrowsmith's chart. A few tern and a gannet were seen about eight o'clock in the morning, but we had no other indications of land: still in the expectation of falling in with it, we continued the course due west, and ran throughout the night, which was clear and fine, but without being more successful. I afterwards learned that the

CHAP. IV. March, 1827.

master of an American trader landed upon a coral island, nearly in the same longitude, in the latitude 19° 18′ N. which is about twenty-three miles to the northward of the island in Arrowsmith's chart, and in all probability is the same place.

With fine weather and a fair wind we pursued our course, without experiencing any inconvenience except that occasioned by a long swell from the northward, which made the ship roll heavily almost all the passage. On the 25th we saw the island of Assumption, and the next day passed close to it, in order to determine its position. The island is about a league in circumference, and rises from the sea in the perfect form of a cone to the height of 2,026 feet. Time must have made an agreeable alteration in the appearance of this island since it was visited by La Perouse. Instead of a cone covered with lava and volcanic glass, and presenting the forbidding aspect he describes, we traced vegetation nearly to the summit, and observed woods of palm trees skirting its base; particularly in the south-west side. We were more fortunate than La Perouse in obtaining a view of the crater formed at the apex of the cone; it appeared to be very small and perfect, and to emit no smoke. La Perouse, in sailing to leeward of this island, experienced a strong sulphurous odour. There was none, however, when we visited the spot; but it is very probable that the volcano may have been in action when he passed, which might also account for the desolation of which he speaks.

There appeared to be no danger near this island, but, on the contrary, judging from the deep blue colour of the sea, there was deep water close to the base of the island. The south-west side is the least abrupt, but even in that direction La Perouse informs us ships are obliged to comevery close to the shore before they can find anchorage, and then only with a very long scope of cable. This bank is formed of lava and scoriæ, and, being on the leeward side of the island, has probably been raised by frequent eruptions of the volcano. There were no projections in any part of the island, that we could perceive, sufficient to afford protection to a boat attempting to land, and the sea in consequence broke heavily against it in every direction.

The day being clear, we looked to the southward for the island of Agrigan, which on Arrowsmith's chart is placed within twelve miles of the Mangs, but no land could be discerned in that direction, and from the state of the weather, I should think there could not have been any within at least twelve leagues distance of us. This would make the channel between Assumption and Agrigan about forty miles wide: the jesuits extend it to sixty; but this cannot be the case, as it would place Agrigan near the latitude of 18° 45′ N. in which parallel Ybargottia, according to Espinosa, has placed the island of Pagon. It seems necessary, therefore to contract the channel between Assumption and Agrigan as marked in the jesuits' plan, and to reduce the size of Agrigan in order to reconcile the position of the islands. Arrowsmith has incorrectly placed the Mangs on the south side of Assumption; by our astronomical bearings they are situated N. 27° 7′ 30″ W. (true) from the south-east end of that island, and are in latitude 19° 57′ 02″ N. They consist of three high rocks, lying in a south-easterly direction.*

* It is somewhat remarkable that in passing to the southward of the island of Assumption, at the distance of four miles and a half, we did not discover the rocks which Captain Freycinet has

CHAP. IV. April, 1827.

From what I saw of the island of Assumption it appears to be a very proper headland for ships coming from the eastward and bound to Canton to steer for. It is high, and may be safely approached in the night if the weather is clear; and there is a wide channel to the southward of it. It is far preferable to adopt this channel than to pass to the northward of the Mariana group, which is sometimes done; as I am credibly informed that there is much broken ground in that direction. We have as yet no good chart of this group of islands. The geographical position of Assumption and of the Mangs will be found in the table at the end of this work.

Under the lee of the island we observed a great many birds, principally of the pelican tribe, of which there was a species supposed by our naturalists to be new. It is described as being smaller than the frigate-bird, and of a dark brown colour, with the exception of the belly and breast, which were white, and the bill, which was either white or of a light lead colour.

From the Ladrones, I directed the course for the Bashee Islands, and on the 7th of April, after experiencing light and variable winds, got sight of the two northern islands of that group. The long northerly swell, which had attended us almost all the way from the Sandwich Islands, ceased immediately we were to the westward of the Ladrones; and indeed

supposed to be the Mangs, situated in latitude 19° 32′ N. Our latitude when in the meridian of Assumption was 19° 36′. N. by which it is evident that we must have passed within four miles of these rocks, provided both latitudes be correct. Had I known of their existence at the time, I should certainly have stood to the southward, in order to connect them by triangulation with the Assumption and the Mangs; but Captain Freycinet's discoveries were not then published.

the sea between them and the Bashee Islands was so smooth that its heave was scarcely perceptible. We found by our observations that the magnetic meridian intersects the channel between these two groups of islands in the meridian of 226° 48′ W. in the latitude of 20° 12′ N.

The Bashee Islands, so called by the Buccaneers, in consequence of a drink of that name, which was extracted by the natives from the sugar-cane, form a long group very similar to the Ladrones, and extend in the same direction nearly from north to south. Until these islands were surveyed by Captain Horsburgh their positions were as incorrectly determined as those of the Ladrones are at present. A contrary wind, which rendered it necessary to beat through the channel between them and Botel Tobago Xima, afforded an opportunity of connecting these islands trigonometrically, and of obtaining transit bearings when in intermediate stations between them. The longitude also was afterwards measured backwards and forwards between them and Macao, and we thus had an opportunity of examining the chart of Captain Horsburgh, which appeared to be constructed with great truth and with his usual accuracy.

I regret not having seen the Cumbrian reef; we stood purposely towards it until sun-set, and were within six miles of its situation when we were obliged to go about by the approach of night.

The next day we stood toward the island of Formosa and tacked within four miles of the Vele rete rocks, the largest of which has the appearance of a vessel under sail. They lie off the south end of the island of Formosa,* and are surrounded by breakers,

* The large rock bears S. 29° 09′ 15″ E. from the west end of Lamay Island.

CHAP. IV.

April, 1827.

which in thick weather could not be approached with safety. We observed strong ripples in the water near them, but the wind did not permit us to enter any for the purpose of sounding; late in the evening, however, when we were several leagues from them, the weather being nearly calm, we were drawn into one of these ripples and continued in it several hours, during which time we tried for soundings with a hundred fathoms of line without success. Upon trial a current was found to set S. E. seven furlongs per hour; this experiment, however, was made from the ship by mooring a buoy, and was probably incorrect, as the water was much agitated; and had a vessel seen it, or even heard it in the night-time (for it made a considerable noise), she would have taken it for breakers and put about. A peculiar smell was detected in the atmosphere while we remained unmanageable in this local disturbance of the water, which some ascribed to sea-weed, and others to dead fish, but it was never ascertained whence it arose. Some seamen have an idea, though it is not very general, that this peculiar odour precedes a change of weather, and sometimes a storm, particularly in the Mediterranean. On the present occasion nothing of the kind occurred immediately, though about twenty-six hours afterwards, when crossing the channel between Formosa and the mainland, the temperature fell sixteen degrees from the average height of the preceding day, and the wind blew strong from the northward.

Before daylight on the 10th, while we were crossing the channel to the westward of Formosa, going at the rate of ten miles an hour, we found ourselves surrounded by Chinese fishing boats, and narrowly escaped running over several of them, as it was very dark, and they were so thick that in trying to escape one we endangered another, and were obliged to lie to until daylight. These boats are large vessels, and would endanger a small merchant ship were she to run foul of any of them. We were informed that they were upon their usual fishing ground, and vessels therefore in approaching the spot should be cautious how they proceed, as these boats carry only a large paper lanthern, which cannot be seen far off, and I believe they only show this when they perceive a strange vessel. They were fishing in pairs, one vessel being attached by cables to each end of an enormous net, which kept them both broadside to the sea; they were constantly covered with the spray, and being light, were washed about in so violent a manner that it scarcely appeared possible for people to stand upon their decks. Still the crews of several which we passed consisted principally of females, who did not appear to be in the least inconvenienced by their situation.

In the forenoon we passed Piedra Branca, and in the evening entered the channel between the Great Lemma and Potoy. As no pilot offered, I stood on, guided by the chart of Lieutenant Ross, which was extremely accurate, and at ten at night brought up in the Lantao passage, and at nine o'clock next morning anchored in the Typa. In entering this harbour we found less depth of water than is marked in the plan of Captain King; and by the survey which we subsequently made, it appeared that at low water a ship cannot depend upon a greater depth than two fathoms, until after she passes the rocky head on her right.

Immediately after we were anchored, I visited the late Sir William Fraser, who was then chief officer of

CHAP. IV. April, 1827.

the company's factory at Canton, and we both waited upon the Portuguese governor. He gave us a very ungracious reception, for which we could account in no other way than by supposing he felt annoyed at our unceremonious entry of the Typa, without either pilot or permission; for the Portuguese at Macao, I understand, claim the Typa as their own, under the emperor's original grant of Macao to them for their services to China. Some Portuguese officers who came on board during my absence intimated that the ship would not be allowed to remain in the harbour. We heard nothing more of the matter, however, for several days, when a mandarin waited upon Sir William Fraser to inquire into the business of the man of war anchored in the Typa. About the same time several war junks, two of which had mandarin's flags, came down the river, beating their gongs, and anchored not far from us.

The mandarin received a satisfactory answer from Sir William Fraser, but some days after, the Hoppo finding the ship did not go away, addressed the following letter to the Hong merchants:—

"Wan, by imperial appointment, commissioner for foreign duties of the port of Canton, an officer of the imperial household, cavalry officer, &c. &c. &c. raised three steps, and recorded seventeen times,

"Hereby issues an order to the Hong merchants.

"The Macao *Wenguin* have reported, that on the 18th of the 13th moon, the pilot *Chinnang-Kwang* announced that on the 17th an English cruiser, Peitche,* arrived, and anchored at *Tausae*.

* The Chinese call their vessels by the names of the persons who command them.

"On the pilot's inquiring, the said captain affirmed that he came from his own country to cruise about other parts, but gales of wind forced him in here, where he would anchor awhile till the wind was fair, and then he would take his departure. I could only in obedience ascertain these circumstances, and also the following particulars:

"There are in the ship 120 seamen, 26 guns, 60 muskets 60 swords, 700 catties of powder, and 700 balls.

"This information is hereby communicated to higher authority.

"Coming before me, the Hoppo, I have inquired into the case, and since the said vessel is not a merchant ship nor convoy to merchantmen, it is inexpedient to allow pretexts to be made for her anchoring, and creating a disturbance. I, therefore, order her to be driven out of the port, and on the receipt of this order, let the merchants, in obedience thereto, enjoin the said nations, foreigners, to force her away. They will not be allowed to make glossing pretexts for her lingering about, and creating a disturbance which will implicate them in crime. Let the day of her departure be reported. Haste! Haste! a special order.

"Taou Kwang,

"7th year, 3d month, 24th day."

The Hong Merchants transmitted this bombastic letter of the Hoppo to the British factory with the following letter: but I must observe that the pilot was incorrect in saying that he derived his information from me, or that such a pretext for putting into the Typa was made.

"We respectfully inform you that on the 23d inst.

CHAP. IV. April, 1826.

we received an edict from the Hoppo concerning Peitche's cruiser anchoring at Tausae, and ordering her away. We send a copy of the document for your perusal, and beg your benevolent brethren of the committee to enforce the order on the said Peitche's cruiser to go away and return home. She is not allowed to linger about.

"We further beg you to inform us of her departure, that we may with evidence before us report the same to government.

"We write on purpose about this matter alone, and send our compliments, wishing you well in every respect.

"To the chiefs:	"We the merchants:
Mr. Fraser,	Wooshowchang, (Howqua's son),
Mr. Toone,	and others."
Mr. Plowden."	

The officers of the factory were aware of the ground upon which the Chinese founded their appeal, it being understood, I believe, that a vessel of war is not to enter the Chinese territory except for the purpose of protecting their own trading ships. At the same time they were sufficiently acquainted with the Chinese style of writing to know that this was only a common remonstrance, however strong the language used might appear, and they amicably arranged the business until near the time of our departure, when another letter arrived, to which they were able to give a satisfactory answer by our moving out of the Typa.

As our object was to procure the stores we required, and to proceed to sea as quickly as possible, our movements were not in any way influenced by this order of the Hoppo; and had it not been necessary to proceed to Canton to ascertain what was in the market there, we should have sailed before this dispatch reached its destination. It appeared that we had arrived at an unfortunate period, as there were very few naval stores in the place, and the Chinese were either so dilatory, or so indifferent about delivering some that had been bargained for in Canton, that we were obliged to sail without them. We, however, procured sufficient supplies to enable us to prosecute the voyage, and on the 30th of April took our departure.

During our stay at Macao we received the greatest attention from the officers of the Company's establishment, who politely gave us apartments in their houses, and in every way forwarded our wishes; and I am happy to join in the thanks expressed in my officers' journals for the hospitality we all experienced.

Soon after our arrival in the Typa, a febrile tendency was experienced throughout the ship, and before we sailed almost every officer and seaman on board was affected with a cold and cough, which in some cases threatened aneumonia; but the officers who resided in the town were free from complaint until they returned to the ship. The probable causes of this were the humid state of the air, the cold heavy dews at night, and the oppressively hot weather during the day, added to the currents of air which made their way between the islands into the Typa, where the atmosphere, penned in on all sides by hills, was otherwise excessively close. On this account I think the Typa very objectionable, and should recommend the anchorage off Cabreta Point in preference.

By a plan of the Typa, which we contrived to make during our visit, it appears that the depth of water is diminishing in the harbour, and that in some parts of

CHAP. IV. May, 1827.

the channel there is not more than ten feet and a half at low water spring tides; the rise of the tide at this time being seven feet one inch. The channel has shifted since the surveys of Captains King and Heywood, and new land-marks for entering, which I have given in my Nautical Remarks, are become necessary.

On leaving Macao we hoped that the S. W. monsoon would set in, and carry us expeditiously to the northward; instead of this, however, we were driven down upon the island of Leuconia in the parallel of 17° 16′ N. where we perceived the coast at a great distance. Here it fell calm, and the weather, which had been increasing in temperature since our departure from Macao, became oppressively hot, the thermometer sometimes standing at 89° in the shade, and the mean height for the day being 85°,7 of Fahrenheit.

About this time we saw several splendid meteors, which left trains of sparks as they descended On the 6th a parhelion was visible at 21° 50′ on the south side of the sun, when about 2° of altitude, and as we passed Orange Island we felt a sudden shock, accompanied by a momentary gust of wind which threatened the masts: the sky at this time was quite clear and cloudless.

On the 7th we saw the south Bashee Islands, celebrated as one of the resorts of the Buccaneers, and the day following made the Island of Botel Tobago Xima. While off the Bashee Islands we noticed a great rippling in the Balingtang Channel, and during the night we experienced so strong a current to the north west that instead of passing the Cumbrian Reef ten miles to the eastward, as we expected, on the following morning we found, greatly to our surprise, that we had been set on the opposite side of it, and much closer than was consistent with security in a dark night. These currents render precaution very necessary; that by which we were affected ran N. 56° W. twenty-six miles during the night, or about two miles and a half per hour. We continued to feel this effect until we were a full day's sail from Botel Tobago Xima, and we were obliged in consequence to beat through the channel between that island and Formosa. In doing this we had an opportunity of examining the shores of Botel Tobago Xima, and of constructing a tolerably good plan of its northern and western sides, besides determining its position more accurately than had been done when we passed it on the former occasion.

The aspect of this part of the island is both agreeable and picturesque. The mountains are covered with wood and verdure to their summit, and are broken by valleys which open out upon plains sloping rather abruptly from the bases of the hills to the sea coast.

Almost every part of this plain is cultivated in the Chinese manner, being walled up in steep places, like the sides of Dane's Island in the Tigris. Groves and tufts of palm trees break the stiffness which this mode of cultivation would otherwise wear, and by their graceful foliage greatly improve its appearance. In a sandy bay on the north side of the island there is a large village consisting of low houses with pointed roofs.

There are several rocky points on the north-west side, and some detached rocks lie off the northern extremity, which are remarkable for their spire-like form. The coast is rocky in almost every part, and probably dangerous to land upon, as these needle rocks are seen in many parts of the island. With the exception, however, of those off the north extreme, they are

CHAP. IV. May, 1827.

attached to the island by very low land, but the shore under water often assumes the character of that which is above, in which case a vigilant look out for rocks would here be necessary in rowing along the coast. At three miles distance from the land we had no bottom with 120 fathoms of line.

After beating two days off Tobago Xima without being able to make much progress against the current, which on the average ran a mile and a quarter per hour, on the 10th a change of wind enabled us to steer our course. We took our departure, from Samsanna, an island to the northward of Tobago Xima, situated, by our observations, nearly in latitude 22° 42′ N., and exactly 8′ west of the eastern extreme of the Little Tobago Xima.

I intended, on leaving Macao, to explore the sea to the eastward of Loo Choo, particularly that part of it where the Yslas Arzobispo, the Malabrigos, and the Bonin Islands, are laid down in various charts. It was, however, no easy matter to reach thus far, and what with light, variable winds, and contrary currents, our progress was extremely slow, so that on the 15th, we found ourselves not far from the Great Loo Choo with a contrary wind.

About this time it was discovered that the water we had taken on board at Macao was extremely bad, owing to the neglect of the *comprador* in filling the casks, and as I had no object in reaching Kamschatka for nearly two months, I determined upon proceeding to Napakiang in Loo Choo. I was further induced to do this, on account of the longitude of the places we might meet between it and Petropaulski. We therefore bore away to the westward, and in the evening saw the island bearing W. by N. ten leagues distant.

The following morning we were close to the reefs by which the Island of Loo Choo is nearly surrounded, and steered along them to the southward, remarking as we passed the excellent harbours which appear to be formed within them; and planning a chart of them as correctly as our distance from the shore, and other circumstances, would permit. The sea rolled furiously over the reefs, which presented a most formidable barrier to encounter in a dark night, but we were glad to find that this danger was lessened by soundings being found outside them, in a depth of water which would enable a vessel to anchor in case of necessity. This depth gradually increased to seventy-five fathoms, at four miles distance from the reefs.

Daylight had scarcely dawned the following morning before several fishermen paddled towards the ship, and fastened their canoes alongside. They had taken several dolphins, which they exchanged for a very small quantity of tobacco, tying the fish to a rope, and without the least mistrust contentedly waiting until the price of it was handed to them. Their canoes were capable of holding five or six persons each, but there were seldom more than two or three in any of them. They were hollowed out of large trees, and rather clumsily made; but it was evident, from the neat manner in which the inside was fitted with bambo gratings, that the constructors of them were capable of much better workmanship. They had no outriggers, and their sail was made of grass.

After remaining alongside some time they ventured upon deck, and saluted us in the Japanese manner, by bowing their heads very low, and clasping their hands to their breasts. They appeared to be a very diminutive race, and were nearly all bow-legged, from the

CHAP. IV. May, 1827.

habitual confinement of their canoes. Many of them were naked, with the exception of a maro; but those who were clothed wore coarse cotton gowns with large sleeves; and almost every person had a pipe, tobacco-pouch, and match fastened to his girdle. As the Loo-Chooans are reputed to be descended from the Japanese, we naturally sought in the countenances of these people features characteristic of that nation, but found that they bore a much nearer resemblance to those of the Malay tribe. Their manners, however, were very different from those of the Malays; and they were marked with a degree of courtesy and good breeding, which we certainly should not have expected to find in persons of their humble occupation, and inferior condition in life.

Having obtained permission to look over the ship, they examined attentively those things which interested them, and when their curiosity was satisfied they made a low bow, and returned to their canoes, leaving us well pleased with their manners. About this time several dolphins swam round the ship, and the fishermen threw over their lines, and met with tolerable success. Our lines had for some time been towing overboard with various devices of flying-fish, pieces of cloth, &c. attached to them, and springing from the water with the rise of the ship, in imitation of the action of the flying-fish, but without any success, and we were happy to take a lesson from our new acquaintances. Their lines were similar to ours, but their snœuds were made of wire, and their hooks, when probably baited, were quite concealed in the body of a flying-fish which had one side of the flesh cut away. Several lines thus prepared were allowed to run out to the length of about ten fathoms, and when the dolphins were near, speed was given to the canoe, that the bait might have the appearance of a fish endeavouring to escape pursuit. In this manner several were taken at no great distance from us. If the fish happened to be large, the line was carefully drawn in, and they were harpooned with an instrument which every canoe carried for the purpose.

We stood towards Loo Choo, accompanied by several of these canoes, until within a few miles of the land, when fearing to be seen from the shore, they quitted us, first making signs for us to go round to the other side of the island.

About sunset the wind left us close off the south extremity of the Great Loo Choo; and all the next day it was so light that the boats were obliged to tow the ship toward the harbour. This slow progress would have been far less tedious had we been able to see distinctly the country we were passing, and the villages situated in the bays at the back of the reefs; but this prospect was unfortunately destroyed by a dense haze which rendered every distant object indistinct, and tantalized our expectations by the variety of fallacious appearances it created. Our course, until four o'clock in the afternoon, was along the western side of Loo Choo, between it and a reef lying about midway between this western shore and the Kirrama islands. About that time we arrived off Abbey Point, and were entering the harbour of Napakiang, guided by our charts, when we were obliged to drop the anchor to avoid striking upon a coral bank, with only seven feet water on its shallowest part. Upon examination we found that this bank, which had hitherto escaped

CHAP. IV. May, 1827.

observation, had a deep channel on both sides of it; we therefore weighed, and steered through the southern passage. It afterwards became necessary to beat up to the anchorage, in doing which we discovered another rock, and had a still narrower escape*. We reached our destination a little before sunset, and then came to an anchor off the town of Napa.

* The position of these rocks are given in the plan of Napa-kiang, which we constructed during our stay here.

CHAPTER V.

Appearance of Loo Choo—Visits of the Natives—Deputation—Permission given to land—Excursions into the Country—Discover Money in Circulation—Mandarin visits the Ship—Departure of a Junk with Tribute—Visit of the Mandarin returned—Further Intercourse—Transactions of the Ship—Departure—Observations upon the Religion, Manners, and Customs of the People; upon their Laws, Money, Weapons, and Punishments; their Manufactures and Trade—Remarks upon the Country, its Productions and Climate—Directions for entering the Port—Historical Sketch of the Kingdom of Loo Choo.

Loo Choo from the anchorage presents a very agreeable landscape to the admirers of quiet scenery. The land rises with a gradual ascent from the sea-coast to something more than five hundred feet in height, and in almost every part exhibits a delightful picture of industry. The appearance of formality is just removed by a due proportion of hill and valley, and the monotonous aspect of continued cultivation is broken by rugged ground, neatly executed cemeteries, or by knots of trees which mingle the foliage of the temperate zone with the more graceful vegetation of the tropics. The most remarkable feature is a hill named Sumar, the summit of which commands a coup-d'œil of all the country round it, including the shores of both sides of the island. Upon this hill there is a town apparently of greater importance than Napa, called Shui or Shoodi,

CHAP. V. May, 1827.

supposed both by Captain Hall and ourselves to be the capital of Loo Choo. With our telescopes it appeared to be surrounded by a wall, and it had several flags *(hattas)* flying upon tall staffs. The houses were numerous, but the view was so obstructed by masses of foliage which grew about these delightful residences that we could form no estimate of their numbers. Upon a rise, a little above the site of the other houses of the town, there was a large building half obscured by evergreen trees, which some of us imagined might be the residence of the king, who had chosen so elevated a situation, in order to enjoy the luxury of breathing a high current of air in a country occasionally exposed to excessive heat. A rich carpet of verdure sloping to the westward connects this part of the landscape with the bustling town of Napa, or Napa-ching,* of which we could see little more than a number of red roofs turned up at the corners in Chinese style, or at most only a few feet down the chunammed walls which support them, in consequence of a high wall surrounding the town. To the right of the town a long stone causeway stretches out into the sea, with arches to allow the water a free access to the harbour at the back of it, and terminates in a large square building with loop-holes. To this causeway sixteen junks of the largest class were secured: some had prows formed in imitation of animals, and

* Napa is decidedly the name of the village, and the words *ching* and *keang*, which are occasionally subjoined, in all probability are intended to specify whether it is the town, or the river near it, that is intended; *ching* being in chinese language a town, and *keang* a river: and though these substantives are differently expressed in Loo Choo, yet when thus combined, the Chinese expression may probably be used.

Drawn by William Smyth. Pubd. by H. Colburn & R. Bentley. 1831

A VIEW OF NAPAKANG. (LOO CHOO.)
taken from the Fort.

gorgeously coloured; others presented their sides and sterns highly painted and gilt; while, from among their clumsy cordage aloft, and from a number of staffs placed erect along the stern, were suspended variously shaped flags, some indicating, by their colour, or the armorial bearing upon them, the mandarin captain of the junk; some the tributary flag of the celestial empire, and others the ensign of Japan. Many of these were curiously arranged and stamped in gilt characters on silken grounds.

To the left of Napa is the public cemetery, where the horse-shoe sepulchres rise in galleries, and on a sunny day dazzle the eye with the brightness of their chunammed surfaces, and beyond them again, to the northward, is the humble village of Potsoong, with its jos-house and bridge.

The bay in every part is circumscribed by a broad coral ledge, which to seaward is generally occupied by fishermen raising and depressing nets extended upon long bamboo poles, similar to those of the Chinese. Beyond these reefs are the coral islands of Tzee, the more distant islands of Kirrama, and far, in a northern direction, the cone of Ee-goo-sacoo, said to be covered with houses rising in a spiral direction up its sides. The whole when viewed on a fine day, and when the harbour is enlivened by boats passing to and fro, with well-dressed people chanting their harmonious boat-song, has a pleasing effect which it is difficult to describe.

Before our sails were furled the ship was surrounded by boats of various descriptions, and the tops of the houses on shore, the walls, and the forts at the entrance of the harbour, were crowded with spectators watching our operations. Several persons came on

CHAP. V. May, 1827.

board, and with a respectful salutation begged permission to be allowed to look over the ship; but they were interrupted by the approach of a boat with an officer, apparently of rank, whom they endeavoured to avoid. His person underwent a severe scrutiny through our telescopes long before he came on board, and we could distinctly see that he had not the *hatchee-matchee*, or low cylindrical cap worn by persons of rank in Loo Choo, in the same manner as the cap and buttons are by the mandarins of China, yet he was evidently a man of consequence, from the respect paid him by the natives in making room for his approach. When he came along side he was invited upon deck, but for some time he stood minutely examining the outside of the ship, counting the number of port-holes, and apparently forming an estimate of her length and height. At last he ascended the side and made a low salutation on the quarter deck, bowing his head in a respectful manner, and clasping his hands to his breast, as before described. Finding we could not understand his language he waved his hand to seaward, in intimation that we should not be allowed to remain in the port. He then looked down upon the gun deck, and pursued his examination of the inside of the ship with the same rigour that he had bestowed upon the exterior, making notes of what he saw. When he was satisfied, he expressed his thanks for our civility and returned to the shore.

Soon after his departure, several well-dressed persons, with boys holding parasols over them, were observed coming off to us: they were seated in Chinese style upon mats spread in the bottom of the boat, over neat ratan platforms, and were propelled by several persons working at a large oar as a scull, keeping time to a song, of which the chorus was *ya ha mashawdy*, or words very similar. CHAP. V. May, 1827.

They were elegantly dressed in gowns made of grass cloth, of which the texture was fine and open, and being a little stiff, formed a most agreeable attire in a country which was naturally warm. To prevent this robe being incommodious while walking, it was bound at the waist with a girdle, linen or silk, according to the rank of the wearer. They had sandals made of straw, and one of them, whose name was An-yah, had linen stockings. None of them had any covering to the head, but wore their hair turned back from all parts, and secured in a knot upon the crown, with two silver pins, *kamesashe* and *oomesashe*, the former of which had an ornamental head resembling a flower with six petals; the other was very similar to a small marrow-spoon. Each person had a square silken tobacco-pouch embroidered with gold and silver, and a short pipe of which the bowl and mouth-piece were also silver, and one who was secretary to An-yah carried a massy silver case of writing materials.

They saluted us very respectfully, first in the manner of their own country and then of ours, and An-yah, by means of a vocabulary which he brought in his pocket, made several inquiries, which occasioned the following dialogue. "What for come Doo Choo*?" "To get some water, refit the ship, and recover the sick." "How many mans?" "A hundred." "Plenty mans! you got hundred ten mans?' "No, a hundred." "Plenty guns?" "Yes." "How many?" "Twenty-six." "Plenty mans, plenty guns!

* This word is pronounced Doo-Choo by the Natives, but as it is known in England as Loo Choo, I shall preserve that orthography.

CHAP. V.

May, 1827.

What things ship got?" "Nothing, ping-chuen*." "No got nothing?" "No, nothing." "Plenty mans, plenty guns, no got nothing!" and turning to his secretary he entered into a conversation with him, in which it appeared almost evident that he did not wholly credit our statement. It was, however, taken down in writing by the secretary.

In order more fully to explain myself I showed them some sentences written in Chinese, which informed them that the ship was an English man-of-war; that the king of England was a friend of the emperor of China; and that ships of our nation had frequent intercourse with the town of Canton. The secretary, who read these sentences aloud, immediately wrote in elegant Chinese characters† "What is your reason for coming to this place? How many men are there on board your ship?" and was both sorry and surprised to find I could not understand what he had written. Indeed he appeared to doubt my sincerity, particularly after I had shown him the next sentence, which happened to be an answer to his question, but which naturally followed the first, stating that we were in want of water and fresh provisions, and that the sick required to be landed to recover their health, and concluding by specifying our desire to be allowed to pay for every thing that was supplied to us. An-yah received this information with satisfaction, and replied, "I speakee mandarin; Doo Chooman no want pay."

These sentences were kindly furnished me by Dr.

* A man-of-war in China is called ping-chuen or soldier-ship.

† This, as well as several other papers written by the Loo Chooans, was afterwards interpreted by Mr. Hultmann of the Asiatic Society, to whom, and also to Sir William Ousely, I beg permission to be allowed to express my thanks.

Morrison, at my own request, in case circumstances should render it necessary to put into Loo Choo, and they were written in Chinese characters, which Dr. Morrison was well aware would be quite intelligible to the literati of Loo Choo, who express themselves in the same character as the Chinese, though their language is totally different. They contained many interesting inquiries, and afforded the means of asking questions without the chance of misinterpretation. To several of them the negative or affirmative was all that was required, and these are expressions understood by most people. It happened, however, that An-yah had learned enough of the English language to say something more than these monosyllables; so that what with his proficiency, and the help of these sentences, besides a dictionary, vocabulary, and dialogues in both languages, which Dr. Morrison had also very generously given me, we had the means of gaining a good deal of information; more, probably, than we could have done through an indifferent interpreter. As, however, opinions vary concerning the written character of China being in general use in Loo Choo, I shall hereafter offer some observations on the subject.

After our visiters had satisfied their curiosity concerning our object in putting into Loo Choo, they sat down to dinner, which was ready, and with much address and good-humour showed us they had learned to chin-chin, or drink healths in the English manner.

I was very anxious to find out who my guest with the vocabulary was, as it at first occurred to me that it might be Mádera, of whom Captain Hall so frequently speaks in his delightful publication on Loo Choo; but then he did not seem to be so well ac-

CHAP V. May, 1827.

quainted with the English language as Mádera appears to have been, and, besides, he must have been much younger. His objection to answering our inquiries on this head, and disclaiming all knowledge of any vessel having ever been at Loo Choo before, put it out of my power at first to inform myself on the point, and had not his own curiosity overcome his prudence, it would perhaps have long remained a secret.

The manner in which the discovery was made is curious: after the sackee* had gone round a few times, An-yah inquired if "ship got womans?" and being answered in the negative, he replied, somewhat surprised, "other ships got womans, handsome womans!" alluding to Mrs. Loy, with whom the Loo Chooans were so much captivated that, it is thought, she had an offer from a person of high authority in the island. I then taxed him with having a knowledge of other ships, and when he found he had betrayed himself, he laughed heartily, and acknowledged that he recollected the visit of the Alceste and Lyra, which he correctly said was 144 moons ago, and that he was the linguist An-yah whom Captain Hall calls An-yah *Toonshoonfa*, but he disclaimed all right to this appendage to his name. Having got thus far, I inquired after almost all the characters which so much interested me in reading the publication alluded to above; but they either prevaricated, or disclaimed all recollection of the persons alluded to, and I found it extremely difficult to get a word in answer.

At last one of them said Ookoma was at the other end of the island, and another immediately added that

* The Loo Choo name for wine or spirits.

he had gone to Pekin. A third stated that Mádera was very ill at the capital, while it was asserted by others that he was dead, or that he was banished to Pātanjān*. They all maintained they had never any knowledge of such persons as Shangfwee, and Shang-Pungfwee, the names given to the king and prince of Loo Choo in Captain Hall's publication. From this conversation it was very evident that they knew perfectly well who Ookoma and Mádera were, but did not intend to give us any correct information about them.

I was a little vexed to find that neither An-yah nor Isaacha-Sandoo, who was also of our party, and is mentioned by Captain Hall, made the slightest inquiry after any of the officers of the Alceste or Lyra, by whom they had been treated in the most friendly manner, and for whom it might have been inferred, from the tears that were shed by the Loo Chooans on the departure of those ships, that the greatest regard had been entertained. The only time they alluded to them was when Mrs. Loy recurred to their imagination.

When they had drank enough sackee they rose to take their leave, and, emptying the contents of the fruit dishes into their pockets, retired in great good humour; but An-yah, not quite satisfied about the number of men on board the ship, probably imagining, from the number he saw aloft, that there were many more, again asked the question, "how many mans?" and on being answered as before, replied "not got hundred one?" which he wrote down a second time;

* An island situated near Ty-ping-chan, upon which Captain Broughton was wrecked.

CHAP. V. May, 1827. and having satisfied himself on this knotty point shook us by the hand and said, "well, I speakee mandarin, to-morrow come water; Doo Chooman no want pay: fife day you go away." "That," I returned, "will depend upon the health of the sick, who must be allowed to land and walk about." I then desired him to tell the mandarin, that to-morrow I should go on shore and wait on him in his own house. An-yah, alarmed lest the threat might be carried into execution, hastily exclaimed, "No, no, I speakee mandarin, mans go shore, walk about, no go house—no go house." Thus by threatening to do more than was intended, we obtained a tacit consent to that which we wanted without much chance of giving offence. Unwilling to give him any further uneasiness, I permitted him to go, requesting he would deliver to the mandarin an invitation to visit the ship, which he promised to do; and seating himself and his companions on the mat in the boat, he sculled on shore to the musical chorus of "ya-ha-me-shawdy."

Our decks were by this time crowded with spectators, who had been coming off in boat loads. The place did not appear to afford many of these conveyances, and they had to go backwards and forwards between the ship and the shore a great many times, always singing their boat songs as they sculled themselves along. Our visiters had paid us the compliment of putting on their best attire, all of which was made of the grass-cloth in the manner before described; the colours were various, but mostly blue.

The utmost good breeding was manifested by every one of them, not only in scrupulously making their bow when they entered and quitted the ship, but in not allowing their curiosity to carry them beyond what they thought perfectly correct. They all seemed determined to be pleased, and were apparently quite happy in being permitted to indulge their curiosity, which was very great, and bespoke them a people extremely desirous of information. It was amusing to observe which objects attracted the particular attention of each individual, which we thought always accorded with the trade or profession of the party; for, as we had at different times all the population of Napa on board the ship, we must have had persons of all occupations. We observed two of these people, after having gratified their curiosity about the deck, seat themselves in their canoes, and commence drawing a picture of the ship—one selected a broadside view, and the other a quarter, each setting at defiance all rules of perspective. The artist on the quarter had of course the most difficult task, and drew the stern as a continuation of the broadside, by which it appeared like an enormous quarter gallery to the ship. That they might make an exact representation, they took their station at the distance of twenty feet from the side of the ship, and commenced their drawing upon a roll of paper about six feet in length, upon which they pourtrayed not only the outline of the ship, but the heads of all the bolts, the but ends of the planks, and before it was finished, no doubt, intended to trace even the grain of the wood. Whatever merit might have been attached to the drawing, the artists were entitled to commendation for their perseverance, which overcame every difficulty; and they had some few to contend with. A little before sunset they rolled up their paper and paddled on shore.

We were scarcely up the following morning before our ears were assailed by the choruses of the boatmen

CHAP. V. May, 1827.

bringing off new visiters to the ship, who continued to pass between her and the shore the whole of the day, carrying a fresh set at every trip, so that the harbour, if possible, presented a more lively scene than it did the day before: on shore the walls and housetops were occupied by groups who sat for hours looking towards the anchorage. Our visiters as before were well dressed and well bred people, and extremely apprehensive of giving offence or even of incommoding us.

The mandarin, however, fearful we might experience some annoyance from having so many people on board without any person to control them, sent off a trusty little man with a disproportionably long bamboo cane to keep order, and who was in consequence named Master at Arms by the seamen. This little man took care that the importance of his office should not escape notice, and occasionally exercised his baton of authority, in a manner which seemed to me much too severe for the occasion; and sometimes even drew forth severe though ineffectual animadversions from his peaceable countrymen: but as I thought it better that he should manage matters in his own way, I did not allow him to be interfered with.

Among the earliest of our visiters were An-yah, Shtafacoo, and Shayoon; three intelligent, good-natured persons, who, I have no doubt, were deputed to watch our movements. They were the bearers of a present of a pig and some vegetables. As An-yah had promised, several boats commenced supplying the ship with water, bringing it off in large tubs.* In

* This water proved to be bad, for though it had no very unpleasant taste, it was found, upon being analysed, to contain in solution a large proportion of magnesia and some salt; a circum-

reply to my request that the officers and invalids might be allowed to walk about on shore, An-yah said he had spoken to the mandarin, who had sent off a Loo Choo physician to administer to the health of our invalids, and in fact who would see whether our statement concerning them was correct or not. A consequential little man, with a huge pair of Chinese spectacles, being introduced as the Esculapius in question, begged to be permitted to visit the sick and to feel their pulse. The surgeon says—" he gravely placed his finger upon the radial artery first of one wrist and then of the other, and returned to the first again, making considerable pressure for upwards of a minute upon each. To one patient affected with a chronic liver complaint, and in whom the pulsations are very different in the two arms, in consequence of an irregular distribution of the arteries, he recommended medicine: of another person affected with dyspepsia whose pulse was natural, he said nothing; no other part of the animal economy attracted his notice. He appeared to be acquainted with quicksilver and moxa, but not with the odour of cinnamon."

After this careful examination he returned to the cabin and wrote in clumsy Chinese characters that one of the patients had an affection of the stomach and required medicine; and inquired of another if he were costive. This report, which we did not understand at the time, was satisfactory to An-yah, who immediately gave us permission to land at Potsoong and Abbey Point, but with an understanding that we were not to go into the town. He then produced a list of

stance which should be borne in mind by vessels obtaining a supply at this place.

CHAP. V.

May, 1827.

inquiries, which he had been ordered to make, such as the dimensions of the ship, the time we had been from England, Canton, &c., and lastly, what weather we had experienced, as he said Loo Choo had been visited by a violent tyfoong in April, which unroofed the houses and did much other mischief.

The permission to land was immediately taken advantage of by several of the officers, who went to Potsoong, and were received in a very polite manner by a great concourse of spectators, who conducted them to the house in which Sir Murray Maxwell and his officers had been entertained; and regaled them with *(tsha)* tea, and *(amasa)* sweet cakes. Some of the party, instead of entering the house, strolled inland to botanize, and to look at the country; but they had not proceeded far before two or three persons ran towards them, and intimated that their company was expected at the house where the other officers were assembled drinking tea, and were waiting for them. This was the Loo Choo polite manner of preventing their proceeding inland, or of making themselves acquainted with the country; and thus, whenever any parties landed afterwards, they were shown to this house, where there was always tea ready prepared, and kept boiling in a kettle, inclosed in a neatly japanned wooden case; there were also trays of charcoal for lighting pipes, and a box to receive the ashes when they were done with: the natives endeavoured, by every possible means, to engage their attention at this place, by putting a thousand inquiries, offering pipes, and pressing them to smoke, and to drink tsha, which was always poured out in small cups, and drank without milk or sugar, which, as it was quite new, and not of the best kind, or much improved by being kept boiling, had a very insipid taste; it, however, served to quench the thirst on a hot day.

On no account would these people receive any present, nor would they sell any of their property in public; but if they thought we desired to possess any thing they could spare, they would offer it for our acceptance. I one day made a present to a person who had been very civil in showing me over his grounds, which he at first refused, and when I insisted on his taking it, and placed it in his pocket, he gave it me back again; but finding I would not receive it, he threw it after me; and it was not until after I had returned it in the same manner, that he was prevailed upon to accept it. Upon doing this, he first exhibited it to the crowd around him, and then thanked me for it. On another occasion one of the officers offered a man, named Komee, two Spanish dollars for his pouch, which he declined, and could not be prevailed upon to accept; but with perfect good breeding he presented to him the object he desired, and insisted upon his keeping it. In private, however, they had less objection to presents, and even asked for several things: small bargains were also effected.

From this time we visited the shore daily, and made many excursions into the country, confining our rambles within reasonable limits, to avoid giving uneasiness to our guides, who were very much distressed whenever we strayed beyond what they considered strictly within the limits prescribed by their instructions. We met many peasants and other persons in these excursions, all of whom seemed eager to show us attention, and with whom there was less reserve, and less disinclination to our proceeding inland, than

CHAP. V. May, 1827.

was manifested by our guides from Napa, who were evidently acting under much constraint.

Lieutenant Wainwright, who, since leaving Sán Francisco, had been an invalid, having suffered severely from a disease of the heart, was provided with a horse by the natives, and permitted to ride every day for his health. He was attended by a guide, and received much kindness and attention from the humane Loo Chooans, who, though they often gave us many reasons to suspect the purity of their intentions, were, by their acts, certainly entitled to our gratitude.

On the 19th we received a bullock weighing 100lbs., five pigs, a bag of sweet potatoes, some firewood, and some more water. Several of the officers landed and walked into the country, attended by the natives, who endeavoured by every species of cunning, and even by falsehood, to prevent their going near the villages, or penetrating far inland. We had again a ship full of visiters, and the two artists were employed the greater part of the day in completing their drawing, which they refused to part with. After the strangers were gone on shore, a thermometer that was kept upon deck for the purpose of registering the temperature was missed, and the natural conclusion was, that it had attracted the attention of some of our visiters, who, it must be remembered, were of all classes.

It was a curious coincidence, but I believe perfectly accidental, that the day after the instrument was missed not a single person came off to the ship, except those employed in bringing water: when An yar came on board the next morning I made our loss known to him; he was much distressed at hearing of it, and said he would make every inquiry about it on shore, and added—"plenty Doo Choo man teef—plenty mans teef," he also advised us at the same time to look well after our watches, handkerchiefs, and particularly any of the instruments that were taken on shore. These precautions I am almost certain were unnecessary, and I am inclined to believe that An-yah painted his countrymen in such odious colours to make us take proper precautions. Though the Loo Chooans are extremely curious, and highly prize such an instrument, yet the theft is not in character with the rest of their conduct, and however appearances may condemn them, I am inclined to believe them guiltless of taking the thermometer, which, probably, was left in the tub used for drawing up sea-water to try its temperature, and was accidentally thrown overboard. And yet in so large a body of people there must naturally be some who are bad; however, we never heard any thing more of the thermometer.

A little before noon I landed to observe the meridional altitude, and met Shtafacoo and several other Loo Choo gentlemen, who were attended by little boys holding parasols over them, and carrying small japanned cases containing smoked and dried meats, small cups of preserves, and boiled rice, sackee, a spirit resembling the samchew of China, and fresh water. They ordered mats to be spread for us, and we made a good luncheon of the many nice things in their boxes. We afterwards crossed over to Potsoong, where we were met by an elderly gentleman, who made a very low obeisance, and pressed us to come into the house in which the officers of Sir Murray Maxwell's squadron and of the Blossom had been entertained, and which appeared to be set apart entirely for our use. It was situated in a square area laid out in lawn and flower beds, and enclosed by a high wall; the

CHAP V. May 1827.

house was built of wood, and roofed with tiles in the Chinese style; the floor was raised about two feet from the ground, and the rooms, though small, were capable of being thrown into one by means of shifting panels. To the right of the house there was a large brass bell, which was struck with a wooden club, and had a very melodious tone; at the further end of the garden was a jos house, a place of worship, which, as it has been described by Captain Hall, I shall notice only by the mention of a screen that was let down before the three small images on the inside. It was made of canvas stretched upon a frame forming two panels, in each of which was a figure; one representing a mandarin with a yellow robe and hatchee-matchee seated upon a bow and quiver of arrows, and a broad sword; the other, a commoner of Loo Choo dressed in blue, and likewise seated upon a bow and arrows. The weapons immediately attracted my attention, and I inquired of my attendant what they were, for the purpose of learning whether he was acquainted with the use of them, and found that he was by putting his arms in the position of drawing the bow, and by pointing to the sword and striking his arm forward; but he implied that that weapon belonged to the mandarins only. A great many pieces of paper were suspended on each side of the picture, some of them marked with Chinese characters, and were, no doubt, invocations to the deities for some temporary benefits, as all the sects are in the habit of writing inscriptions of this kind, and depositing them in the jos houses, or placing them upon stones, of which there are several in Loo Choo under the name of Karoo. Under a veranda which surrounded the temple there were several wooden forms strewed with flowers, and upon the middle one a drum was suspended by thongs in a handsome japanned stand.

A building in front of this jos house, mentioned by Captain Hall, has been rebuilt, but was not quite finished at the time of our visit: though so near to the temple the panels were scrawled over with groups of figures, some of which were very inappropriate to such a situation.

After we had partaken of tea in the dwelling-house, we determined upon a walk in the interior, much to the discomfiture of the old gentleman, who used every means he could think of to induce us to desist, and produced pipes, sweet cakes, tcha, and masa chorassa, preserves with which they tempted us whenever they feared our walk would be directed inland. Finding he could not detain us, he determined to be our companion, and endeavoured to confine us to the beach by praising the freshness of the breeze, saying how hot we should find it inland, and what bad paths there were in that direction, every word of which proved to be false, as we found the roads very good, and by gaining elevated situations we enjoyed more of the breeze.

We passed some tombs excavated in the cliffs, and in one that was broken down we discovered a corpse lying upon its back, half decayed and covered over with a mat; a jar of tea and some cups were placed by it, that the spirit might drink; but there was nothing to eat, and our guide informed us that it was customary to place tea only by the side of the bodies, and that food was never left there. He turned us away from this shocking spectacle as much disgusted as ourselves, and seemed sorry that we had hit upon it. This discovery seems to strengthen some information which

CHAP. V.

May, 1827.

I afterwards received concerning the manner in which the dead were disposed of, namely, that the flesh is allowed to decay before the bones are placed in jars in the cemetery.

From this place we ascended a hill covered with tombs, which were excavated in the rock in a manner very similar to those near Canton; they had almost all of them niches, wherein bowls of tea, lamps, and cups were placed, and appeared to be kept in good order, as they had a cleanly and decent appearance. We wandered among these some time, without finding any open, but at last we came to one of an inferior kind, in which the door was loosely placed before the entrance; it consisted of a large slab of red pottery, pierced with a number of holes about an inch in diameter. Having removed this, we saw about twenty jars of fine red pottery covered with lids shaped like mandarins' caps; the size of the jars was about twenty inches deep by eight in the broadest part, which was one-third of the way from the mouth; they were also perforated in several places with holes an inch in diameter. We did not remove any of the lids, as it seemed to give offence, but were told that the jars contained the bones of the dead after the flesh had been stripped off or had decayed. On putting the question whether they burned the bones or the flesh off them, it was answered by surprise, and an inquiry whether we did so in England? Therefore, unless the custom has altered, the account of Supoa Quang, a learned Chinese, who visited Loo Choo in 1719, is incorrect.

After visiting the grave of one of the crew of the Alceste who was buried in this island, we were satisfied with this tour of the tombs, and turned off inland,

A LOO CHOO SEPULCHRF.

Published by Henry Colburn and Richard Bentley, New Burlington Street. 1831.

very much to the discomfiture of our guide, and in spite of a great many remonstrances. He was a silent companion until we came to a path that went back to the beach, and there, politely stepping forward, said it was the one that would take us where we wished to go, and, touching our elbow, he would have turned us into it had he not thought it rude; but we pursued our original path, followed by a crowd of persons, who seemed to enjoy the discomfiture of our companion, and laughed heartily as we came to every track that crossed ours, each of which our officious and polite conductor would have persuaded us to take, as being far more agreeable than the other, and as leading to our destination. The mirth of the crowd pretty well satisfied us there was no great danger in advancing, and we went on further than we should otherwise have done; but in a little time they began to drop off, and we were at last left alone with the guide, who really became alarmed. We had reached the foot of the hill on which the capital is situated, and were ascending to have a near view of the houses, when he threw himself on his knees in evident alarm, bowed his head to the dust, and embracing our knees implored us to desist, assuring us that the mandarin would take his head off if we did not. Some of the officers who went in another direction were told by their guide that he would get bambooed if they did not turn back, which is more probable than that the heavy penalty apprehended by our companion should be attached to so light a crime.

To quiet the irritation of the poor old man, who trembled violently, we ascended a hill some distance to the left, which commanded an extensive view of the country, and from whence we could survey the

CHAP. V. May, 1827.

capital with our telescopes. The country was highly cultivated, and the grounds irrigated with Chinese ingenuity and perseverance by small streams of water passing through them, keeping such as were planted with rice thoroughly wet. We noticed in our walk sweet potatoes, millet, wheat, Indian corn, potatoes, cabbages, barley, sugar-cane, pease, tea shrubs, rice, taro, tobacco, capsicums, cucumbers, cocoa-nuts, carrots, lettuces, onions, plantains, pomegranates, and oranges; but amidst this display of agricultural industry there were several eminences topped with pine trees, on which the hand of the farmer might have been advantageously employed, but which were allowed to lie waste, and to be overrun with a rank grass. Such places, however, being usually the repositories of the dead, it may have been thought indecorous by the considerate Loo Chooans to disturb the ground near it with a hoe. These eminences, like the basis of the island, being formed of a very porous calcareous rock, are peculiarly adapted to the excavation of tombs, and the natives have taken advantage of them to dispose of their dead in them. The accompanying view from Mr. Smyth's sketch will convey the best idea of what they are like.

The capital, for such I am disposed to call the town on the hill, notwithstanding the denial of several of the natives, was surrounded by a white wall, within which there were a great many houses, and two strong buildings like forts; with, as already mentioned, several small masts with gaffs, bearing flags of different colours. This space was thickly interspersed with trees, whence we conjectured the houses were furnished with gardens. There seemed to be very few people moving about the island, even between the upper and lower towns, with which it would be supposed there must necessarily be much intercourse. We rested awhile on the eminence that afforded this agreeable view of a country but very little known, and were joined by several persons whom fear or indolence had prevented keeping pace with us. Our guide now lighted his pipe and forgot his apprehension in the consoling fumes of tobacco, while some of the party amused themselves with viewing the capital through a telescope, each preventing the other having a quiet view by their anxiety to obtain a peep. Our clothes in the meantime were undergoing an examination from the remainder of the party, who, after looking closely into the texture of the material, exclaimed—choorassa, choorassa! (beautiful).

While we sat here a Japanese junk bore down from the northward, and according to the information of those around us, which afterwards proved to be correct, she came from an island called Ooshimar, to the northward of Loo Choo, and was laden with rice, hemp, and other articles. Her sails and rigging resembled the drawing of the Japanese junks in La Perouse's voyage. She passed close to the Blossom at anchor, and from the report of the officers her crew had their heads shaved in the fashion of the Japanese. Her arrival excited general interest, brought all the inhabitants to the housetops, and a number of canoes crowded round her before she reached the inner harbour, where she was towed and secured alongside several other junks bearing the same flag.

On our return we passed through a village consisting of a number of square inclosures of low stone walls, separated by lanes planted on both sides, and so overgrown with bamboo and ratans that we could neither see the houses nor the sky; several handsome

CHAP. V. May, 1827.

creepers entwined themselves round the stems of these canes, and a variety of flowers, some of which were new to us, exhaled a delicious fragrance from the gardens which bordered these delightful avenues. A more comfortable residence in a hot climate could not well be imagined, but I am sorry to say that the fascination was greatly lessened by the very filthy state of the dwellings and of the people who occupied them. In one of these huts there was a spinning-wheel and a hand loom, with some grass-cloth of the country in a forward state of preparation for use.

Several little children accompanied us through these delightfully cool lanes, running before us catching butterflies, or picking flowers, which they presented with a low Chinese salam, and then ran away laughing at the idea of our valuing such things. We afterwards crossed two high roads, on which there were several horses and jack-asses bearing panniers; but we saw no carriages, nor the marks of any wheels, nor do I believe there are any in Loo Choo. The horses, like the natives, were very diminutive, and showed very little blood. Several peasants, both male and female, were working in the plantations as we passed through them, neither of whom endeavoured to avoid us, and we had an opportunity of beholding, for the first time, several Loo Choo women. They were of the labouring class, and of course not the most attractive specimens of their sex; but they were equally good-looking with the men, and a few of them were pretty, notwithstanding the assertion of An-yah, that "Loo Choo womans ugly womans." There was nothing remarkable about them to need particular description; they were clothed much in the same manner as the men, and generally in the same colours;

Wm. Smith del.

KOMEE.

E. Finden. sc.

NATIVES OF LOO CHOO.

& Richard Bentley 1831.

their hair, however, was differently dressed, being loosely fastened at the side of the head by a pin resembling a salt-spoon with a very long handle. Their feet were of the natural size, and without shoes or sandals. We noticed some who were tattooed on the back of the hand, which we were told was done to distinguish all those who were married; An-yah said the custom prevailed equally in high life.

I subjoin a sketch of a male and female of Loo Choo, drawn by Mr. Smyth, from which the reader may form a tolerably correct judgment of the general appearance of these people, though Komee was by no means the handsomest of his countrymen.

Upon the high road we met a man with a bundle of firewood, on his way to town; and were much pleased at the confirmation of a fact, which we had no doubt existed, though the natives took every precaution to conceal it. None of our visiters to the ship had as yet shown us any money, and An-yah, if I understood him correctly, said there was none in Loo Choo; our meeting with this peasant, however, disclosed the truth, as he had a string of cash * (small Chinese money) suspended to his girdle, in the manner adopted by the Chinese. I examined the string with much interest, and offered to purchase it with Spanish coin, but my guide would not permit the woodman to part with it, and tucking it into his belt that it might not be seen again, he said something to him in an angry tone, and the poor fellow walked on with his load to the town. We afterwards got some of this money, which

* These coins being of small value, they are strung together in hundreds, and have a knot at each end, so that it is not necessary to count them.

CHAP. V. May, 1827.

was exactly the same as that which is current at Canton, and found that it was also in circulation in Loo Choo. Though they afterwards admitted this fact, they denied having any silver or gold coin in the country.

Our subsequent excursions were nearly a repetition of what has been described, and were made nearly to the same places, with the exception of two or three, which I shall describe hereafter. In all these the same artifice was practised to induce us to confine ourselves to the beach, and particularly to prevent a near approach to the villages. Tobacco, tsha, and chorassa masa were the great temptations held out to us; but neither the tea, nor the masa, which, by the by, was seldom produced, had sufficient charms to dissuade some of our young gentlemen from gratifying their curiosity, though it was at the expense of the convenience of the natives, whose dresses were very ill adapted to speed; and thus, by outrunning them, they saw many places which they would not otherwise have been permitted to enter, and got much nearer to the town than I felt it would be right for me to do in consequence of my promise to An-yah. I shall, therefore, give such extracts from their journals as are interesting, but in a few pages in advance, that I may not disturb the order of the narrative.

On the 21st, An-yah came off to say, that the mandarin had accepted my invitation to visit the ship, and would come on board that day: we consequently made preparation to receive him. As it appeared to me that Napa-keang possessed no boat sufficiently good for so great an occasion, I offered to send one of ours to the town for his accommodation, which, in addition to obliging the mandarin, would afford an opportunity of seeing the place; but An-yah would not permit it, and fearful that we might really pursue this piece of politeness further, got out of the ship as fast as he could, saying the mandarin was at Potsoong, and not in the town. About two o'clock he pushed off from that place with his party in two clumsy punts, sculled by several men singing a chorus, which differed, both in words and air, from that used by the boatmen in general. The mandarin was seated in the largest of these boats, under a wide Chinese umbrella, with two or three mandarins of inferior rank by him; the other boat contained An-yah, Shtafacoo, Sandoo, and others, with whom we were well acquainted, and who rowed on before the mandarin, and announced his approach by presenting a crimson scroll of paper, exactly a yard in length, on which was elegantly written in Chinese characters, "Ching-oong-choo, the magistrate of Napa, in the Loo Choo country, bows his head to the ground, and pays a visit." By this time the other boat with the great man was alongside the ship, and four domestics with scarlet hatchee-matchees ascended the side, one of them bearing a large square hatchee-matchee box, in which there was an old comb. They pulled up the side ropes, and carefully inspected them, to see whether they were strong enough to hold their master, and let them down again for the mandarin, who, very little accustomed to such feats, ascended the side with difficulty.

He was received with a guard under arms, and a mandarin's salute was fired as he put his foot upon the deck, with which he was much gratified, and he shook every officer by the hand with unaffected pleasure. The yards had been manned as he was coming off, and when the pipe was given for the seamen to

CHAP. V. May, 1827.

come down, the evolution produced a little surprise, and must have impressed the Loo Chooans with the decided advantage of our dress over theirs, where activity is required. Ojee, one of the party, who also styled himself Jeema, and is mentioned by Captain Hall, followed, and then the rest of the mandarins in yellow hatchee-matchees and gowns.

To persons who had visited a fine English frigate, disciplined by one of the ablest officers in the British Navy, the Blossom could have presented nothing extraordinary; and as the greater part of our visiters were familiar with the Alceste, they were very little interested in what they saw; but Ching-oong-choo had not been long from Pekin, and never, probably, having put his foot on the deck of a ship before, a Chinese junk excepted, examined every thing very attentively, and made many inquiries about the guns, powder, and shot.

None of the natives offered to seat themselves in the cabin in the presence of the mandarin until dinner was brought in, but they then dispensed with formalities, and those who were familiar with European customs chinchinned each other with wine, and reversed their glasses each time, to the great amusement of their superior. During dinner the fate of Mádera was inquired into, but we got no satisfactory answer, and a mystery seemed to hang over his fate, which made us suspect he had in some way or other been disgraced. Jeema took the opportunity of showing he recollected his visits to the Alceste and Lyra, but he did not make any inquiry after his friends in either vessel.

As we had lately been at Canton, we were provided with many things which were happily to the taste of

Wm. Smyth delt. E. Finden sc.

Pubd. by H. Colburn & R. Bentley, 1831.

our guests, who would otherwise have fared badly, as they did not appear to relish our joints of meat; nor did some bottled porter accord better with their taste, for after occasioning many wry faces, it was put aside as being bitter; a flavour which I have observed is seldom relished for the first time. Not so some noyeau, which was well adapted to the sweet palate of the Loo Chooans; nor some effervescing draughts, which were quite new to them, and created considerable surprise. They, however, seemed to enjoy themselves a great deal; were jovial without being noisy, and with the exception of a disagreeable practice of eructation, and even worse, they were polite people; though I cannot say I approved of their refinement upon our pocket handkerchief. An-yah often intimated to me that he thought it was a disagreeable practice to use a handkerchief and carry it about all day, and thought it would be better for us to adopt their custom of having a number of square pieces of paper in our pockets for this purpose, any one of which could be thrown away when it had been used. I did not at first think he was in earnest, and when I observed my guests pocket these pieces of paper, I sent for some handkerchiefs, but they declined using them, saying paper was much better.

While we were at dinner a large junk which we had observed taking in a cargo the day before, was towed out of the harbour by an immense number of boats, making the shores echo with her deep-toned gong. She grounded off the entrance of the harbour, but was soon got off, and placed outside the reefs. A more unwieldy ark scarcely ever put to sea, and when she rolled, her masts bent to that degree that the people on her deck seemed to be in imminent danger of their

CHAP. V. May, 1827.

lives. She was decorated with flags of all sorts and sizes: at the fore there was hoisted the white flag of the emperor; at the main, the Loo Choo colours, a triangular flag, red and yellow, with a white ball in it, denoting, I believe, a tributary state; there were besides several others, and a great many mandarins' flags upon staffs along the stern. Ching-oong-choo said she was the junk with tribute which was sent every second year from Loo Choo to Fochien. Her cargo, before it was stowed, was placed upon the wharf in square piles, with small flags upon sticks, stuck here and there upon the bales of goods, which were apparently done up in straw matting; for it was only with our telescopes that we were allowed to see this.

After dinner was over, the mandarin went on shore, and begged to have the pleasure of our company to dinner at Potsoong the next day; but the rest of the company obtained permission to stay and enjoy a little more sackee, after which they pocketed the remains of the dessert as usual, and as a token of their friendship, they each threw down their pipe and tobacco-pouch, and begged my acceptance of them; but as I knew these articles were valuable in Loo Choo, and was conscious that with some of them it was only a matter of form, I declined accepting them.

The next day it rained heavily, but An-yah came off to keep us to our engagement, saying the mandarin was at Potsoong in readiness to receive us; we accordingly went, and were met at the landing-place by Jeema and a great crowd of Loo Chooans, with umbrellas, who accompanied us to the house, where we were received by the mandarin in a most cordial and friendly manner. For convenience both apartments were thrown into one, by the removal of shifting panels, and the servants were regaled upon the floor in the inner room, while we were seated at a table in the outer apartment. Our table, which had been made in Japan, was nicely lacquered, and had Chinese characters gilt upon its edges and down the sides of the legs, recording the date and place where it was made, as well as the name of the workman, &c. It was covered with dishes containing a variety of eatables, principally sweetmeats, and two sorts of spirits, sackee and mooroofacoo. The former resembles the samscheu of China, and the other is a dark coloured cordial possessing a bitter-sweet taste. We were seated on one side of the table, myself in an old-fashioned chair, and the other officers upon camp-stools with japanned backs, and the host, Jeema, and the other mandarins, on the other side: and each person was provided with a small enamelled cup, and a saucer with a pair of chopsticks laid across it; the crowd all the while surrounding the house, and watching through its open sides every motion we made. Pipes and mooroofacoo were first offered to us, and then each dish in succession; of which we partook, according to our different tastes, without being aware of the Chinese custom of giving the sweets first, and reserving the substantial part of the dinner for the last.

Among the dishes, besides some sweet cakes made very light, were different kinds of pastry, one of a circular form, called *hannaburee*, another tied in a knot, hard and disagreeable, called *matzakai*, and a third called *kooming*, which enclosed some kind of fish. There was also a mamalade, called *tsheeptang*, a dish of hard boiled eggs without the shells, painted red, and a pickle which was used instead of salt, called *dzeeseekedakoonee;* besides a small dish of sliced cold

CHAP. V. May, 1827.

liver, called *watshaingo,* which in this course was the only meat upon the table. We ate more plentifully of these sweet things than we liked, in consequence of our ignorance of what was to follow, and partly from our not being aware that their politeness prevented them from sending away any dish as long as we could be prevailed upon to partake of it—a feeling which induced them continually to press us to eat, and offer us part of every dish on the end of their chopsticks. The next course induced us to regret that we had not made the tasting more a matter of form, for it consisted of several good dishes, such as roast pork, hashed fowls, and vermicelli pudding, &c. After these were removed they brought basins of rice, but seeing we would eat no more, they ordered the whole to be taken away.

During the whole time we were closely plied with sackee in small opaque wine glasses, which held about a thimblefull, and were compelled to follow the example of our host and turn our glasses down; but as this spirit was of a very ardent nature, I begged to be allowed to substitute port and madeira, which was readily granted, and we became more on a footing with our hosts, who seemed to think that hospitality consisted in making every person take more than they liked, and argued that as they had been intoxicated on board, we ought to become so on shore.

After dinner was removed, Jeema favoured us with two songs, which were very passable, and much to the taste of the Loo Chooans, who seemed to enjoy them very much. Nothing could exceed the politeness and hospitality of the mandarin throughout, who begged that dinner might be sent off to one of the officers, whose health would not permit him to risk a wetting.

and that all the boats' crews might be allowed to come to the house and partake of the feast. Though there was a little ceremony in receiving and seating us, yet that almost immediately wore off, and Ching-oong-choo to make every person at his ease took off his hatchee-matchee, and with the rest of the mandarins sat without it. By this piece of politeness we discovered that his hair was secured on the top of the head by a gold hair pin, called *kamesache*, the first and the only one we saw made of that precious metal.

We afterwards took a short walk in the garden, when I was surprised to find An-yah and Shtafacoo in the dress and hatchee-matchee of mandarins of the second class: whether this was intended as a trick, or, following Mádera's example, they preferred making their first acquaintance in disguise, is not very clear; but as they both possessed a great deal of influence, and were much respected by the lower orders of the inhabitants, it was probably their proper dress.

As soon as Ching-oon-choo permitted us, we took our leave, and were accompanied to the boat by a great crowd of persons, who opened a passage as we proceeded, and were officiously anxious to be useful in some way or other; and we then parted with Jeema and the rest amidst the greetings and salutations of hundreds of voices.

On the 21st, one of the officers made an excursion to the southward of Abbey Point, and was attended as usual by a concourse of boys and young men, who were extremely polite and respectful. They used every artifice and persuasion to deter him from proceeding, said they were tired, tempted him with tsha, and declared that they were hungry, but he ingeniously silenced the latter complaint by offering his

CHAP. V. May, 1827.

guide a piece of bread which he had in his basket. It was thankfully accepted, but with a smile at the artifice having failed. At a village called Aseemee he surprised two females standing at a well filling their pitchers; they scrutinized him for some time, and then ran off to their homes.

The village contained about fifty houses; and was almost hid from view by a screen of trees, among which were recognised the acacia, the porou of the South Seas, and the hibiscus rosa sinensis, but the greater part of the others appeared to be new; they formed a lively green wood, and gave the village an agreeable aspect. In one of the cottages a boy of about six years of age was seated at a machine made of bamboo resembling a small Scotch muckle wheel, spinning some very fine cotton into a small thread. Though so young, he appeared to be quite an adept at his business, and was not the least embarrassed at the approach of the strangers. A quantity of thread ready spun lay in the house; there was a loom close by, and some newly manufactured cloth, which appeared to have been recently dyed, was extended to dry outside the house. Near this cottage there were broken parts of a mill, which indicated the use of those machines, and circular marks on the earth, showing that this one had been worked by cattle. About a mile and a half to the southward of Abbey Point, near a steep wooded eminence, which we christened Wood Point, there was another village named Oofoomee, through which Mr. Collie passed, preceded by his guide, who warned the female part of the inhabitants of his approach in order that they might get out of his way. His guide was delighted when he directed his steps toward the ship, as he was very tired, and even had a horse brought to him before he got to the beach. This animal was eleven hands and a half in height, and would hardly have kept a moderately tall person's feet off the ground; but his guide, though there was not much necessity for bracing his feet up very high, obviated the possibility of this inconvenience by riding with his knees up to his breast. The stirrups were massy, and made of iron curiously inlaid with brass, and shaped something like a clumsy Chinese shoe. At Abbey Point he visited some sepulchres hewn out of the rock or formed of natural caverns; one of these happened to be partly open, and he discovered four large red earthen jars, one of which was fortunately broken, and exhibited its contents, consisting of bones of the human skeleton.

In another excursion made by this gentleman to the north-east of Potsoong, he visited a temple of Budh, situated in a romantic copse of trees. The approach to it was along a path paved with coral slabs, partly overgrown with grass, and under an archway in the formation of which art had materially assisted the hand of nature. After resting a short time in this romantic situation he descended the paved way, passed some tall trees, among which was a species of erethrina of large growth, and arrived at the house of a priest, who invited him to smoke and partake of tea and rice. Three young boys were in the house, who, as well as the priest, had their heads shaved according to the custom of the priesthood in China.

By the 25th May, we had completed the survey of the port, replenished our water, received a little fresh stock, and obtained some interesting astronomical and magnetical observations; the day of departure was

CHAP. V. May, 1827.

consequently near at hand. This event, after which many anxious inquiries were made by the natives, was, I believe, generally contemplated with pleasure on both sides; not that we felt careless about parting with our friends, but we could not enjoy their society without so many restrictions, and we were daily exposed to the temptation of a beautiful country without the liberty of exploring it, that our situation very soon became extremely irksome. The day of our departure, therefore, was hailed with pleasure, not only by ourselves but by those to whom the troublesome and fatiguing duty had been assigned of attending upon our motions: and they must moreover have looked with suspicion on the operations of the survey that were daily going forward, even had they not suspected our motives for putting into their port.

I was very anxious before this day arrived to possess a set of the pins that are worn by the natives in their hair. From their conduct it appeared that these ornaments had some other value attached to them than that of their intrinsic worth, or there would not have been so much difficulty in procuring them. Seeing they set so much value upon them that none of the natives could be induced to part with them, I begged An-yah would acquaint the mandarin with my desire, and if possible, that he would procure me a set. An-yah replied he would certainly deliver my message to the mandarin, and the next morning brought a set of the most inferior kind, made of brass. As the mandarin had received some liberal presents from me, I observed to An-yah that this conduct was ungenerous, and that I expected a set made of silver; his opinion he said very much coincided with mine, and added that he would endeavour to have them changed, but the following morning he met me on shore and said—"mandarin very bad man, no give you silver kamesache:" but An-yah, determined that my request should be complied with, had by some means succeeded in procuring a set for me, which he presented in his own name. I rewarded his generous behaviour by making him a present of some cut glass decanters and wine glasses, which are more esteemed in Loo Choo even than a telescope.

On the 27th we made preparations for weighing by hoisting our sails, and An-yah, Shtafacoo, and Shayoon, who had been our constant attendants, came off to take leave. These good people had been put to much trouble and anxiety on our account, and had so ingratiated themselves with us, that as the moment approached I really believe the desire for our departure was proportionably lessened; and when the day arrived they testified their regret in a warm but manly manner, shook us heartily by the hand, and each gave some little token of regard which they begged us to keep in remembrance of them. As we moved from the anchorage, the inhabitants assembled on the house-tops, as before, upon the tombs, in the forts, and upon every place that would afford them a view, of our operations, some waving umbrellas and others fans.

Having brought to a conclusion the sketch of our visit to Loo Choo, I intend in the few pages that follow to embody what other information was collected from time to time, and to offer a few remarks on the state of the country as we found it, as compared with that which has been given by Captain Hall and the late Mr. M'Cleod, surgeon of the Alceste. In the foregoing narrative I have avoided entering minutely into

CHAP. V. May, 1827.

a description of the manners and persons of the inhabitants; and I have omitted several incidents and anecdotes of the people, as being similar to those which have already been given in the delightful publications above mentioned.

Loo Choo has always been said to be very populous, particularly the southern districts, and we saw nothing in that part of the island which could induce us to doubt the assertion. On the contrary, the number of villages scattered over the country, and the crowds of persons whom we met whenever we landed, amply testified the justness of the observation. We were, certainly, in the vicinity of the capital, and at the principal seaport town of the island; but in forming our estimate of the population, it must be borne in mind that we were very likely to underrate its amount, in consequence of the greater number of persons who crowd into Chinese towns than reside in villages of the same size in countries from which we have taken our standard.

The people are of very diminutive stature, and according to our estimation their average height does not exceed five feet five inches. As might be expected, from the Loo Chooans being descendants of the Japanese, and numerous families from China having settled in the island, there is a union of the disposition and of the manners, as well as of the features of both countries. The better classes seemed by their features to be allied to the Chinese, and the lower orders to the Japanese; but, in each, the manners of both countries may be traced. Their mode of salutation, their custom of putting to their foreheads any thing that is given to them, their paper pocket handkerchiefs, and some parts of their dress, are peculiarly

Japanese. In other respects they resemble the Chinese. The hatchee-matchee and the hair-pins are, I believe, confined to their own country, though smaller metal hair-pins are worn by the ladies of Japan*. On the whole they appear to be a more amiable people than either the Chinese or Japanese, though they are not without the vices natural to mankind, nor free from those which characterise the inhabitants of the above mentioned countries. They have all the politeness, affability, and ceremony of the Chinese, with more honesty and ingenuousness than is generally possessed by those people; and they are less warlike, cruel, and obsequious than the Japanese, and perhaps less suspicious of foreigners than those people appear to be. In their intercourse with foreigners their conduct appears to be governed by the same artful policy as that of both China and Japan, and we found they would likewise sometimes condescend to assert an untruth to serve their purpose; and so apparent was this deceitfulness, that some among us were led to impute their extreme civility, and their generosity to strangers, to impure motives. They are exceedingly timorous and effeminate, so much so that I can fancy they would be induced to grant almost any thing they posssess rather than go to war; and, as one of my officers justly observes in his journal, had a party insisted upon entering the town, they would probably have submitted in silence, treated them with the greatest politeness, and by some plausible pretext have got rid of them as soon as they could.

They appear to be peaceable and happy, and the lower orders to be as free from distress as those of any

* See Langsdorff's Travels, vol. ii.

CHAP. V. May, 1827.

country that we know of; though we met several men working in the fields who were in rags, and nearly naked. The most striking peculiarity of the people is the excessive politeness of even the lowest classes of inhabitants: on no account would they willingly do any thing disagreeable to a stranger, and when compelled, by higher authorities than themselves, to pursue a certain line of conduct, they did it in the manner that was the least likely to give offence; and it was quite laughable to notice the fertility of their invention in order to obtain this end, which was seldom gained without a sad sacrifice of integrity. Their reluctance to receive remuneration for their trouble, or for the provisions which they supply to foreigners, is equally remarkable. Captain Broughton and Captain Hall have noticed their conduct in this respect. In the case of a whale ship which put into Napa-keang in 1826, and received nearly two dozen bullocks and other supplies, the only remuneration they would receive was a map of the world. And in our own instance (though we managed by making presents to the mandarins and to the people to prevent their being losers by their generosity), An-yah's reply to my question, whether we should pay for the supplies we received in money or goods? was, "Mandarin *give* you plenty, no want pay." But with all this politeness, as is the case with the Chinese, they cannot be said to be a polished people.

Our means of judging of their education were very limited: a few only of the lower orders could read the Chinese characters, and still fewer were acquainted with the Chinese pronunciation; even among the better classes there were some who were ignorant of both. Schools appear to have been established in

Loo Choo as far back as the reign of Chun-tien, about the year 1817, when characters were introduced into the country, and the inhabitants began to read and write. These characters were said to be the same as those of the Japanese alphabet *yrofa**. In the year 1372, other schools were established, and the Chinese character was substituted for that of the Japanese; and about the middle of the seventeenth century, when the Mantchur dynasty became fixed upon the throne of China, the Emperor Kang-hi built a college in Loo Choo for the instruction of youth, and for making them familiar with the Chinese character. An-yah intimated that schoolmasters had recently been sent there from China; and one day while I was making some observations, several boys who were noticed among the crowd with books, and who seemed proud of being able to read the Chinese characters, were pointed out by An-yah as being the scholars of those people.

I am of opinion that the inhabitants of Loo Choo have no written character in use which can properly be called their own, but that they express themselves in that which is strictly Chinese. We certainly never saw any except that of China during our residence in the country. The manuscripts which I brought away with me were all of the same character precisely, and some were written by persons who did not know that I was more familiar with the Chinese character than with any other.

It is very probable that the Japanese character was in use formerly; but it is now so long since schools have been established in Loo Choo for teaching the

* Recueil de Père Gaubil.

CHAP. V. May, 1827.

Chinese character, viz. since 1372, and the Chinese, whose written character is easier to learn than the other, have always been the favourite nation of the Loo Choo people, that it is very probable the Japanese character may now be obsolete. An-yah would give us no information on this subject, nor would he bring us any of the books which were in use in Loo Choo. One which I saw in the hands of a boy at Abbey Point appeared to be written in Chinese characters, which are so different from those of the Japanese that they may be readily detected.

M. Grosier on this subject, quoting the Chinese authors, says that letters, accounts, and the king's proclamations are written in Japanese characters; and books on morality, history, medicine, astronomy, &c. in those of China. One of the authors whom he quotes adds, that the priests throughout the kingdom have schools for teaching the youth to read according to the precepts of the Japanese alphabet Y-ro-fa. As we may presume they teach morality in these schools, it would follow, as books on those subjects are all written in Chinese characters, that the boys must be taught both languages; but had this been the case, I think we should have seen the Japanese character written by some of them. It is to be observed, that the invocations in the temples and on the kao-roo stones are all in the character of China.

While upon this subject, I must observe, that the idea of Mons. P. S. Du Ponceau,* "that the meaning of the Chinese characters cannot be understood alike in the different languages in which they are used," is not strictly correct, as we found many Loo Choo

* See a letter from this gentleman to Captain Basil Hall, R. N., published in the Annals of Philosophy for January, 1829.

people who understood the meaning of the character, which was the same with them as with the Chinese, but who could not give us the Chinese pronunciation of the word. And this is an answer to another observation which precedes that above mentioned, viz. that "as the Chinese characters are in direct connexion with the Chinese spoken words, they can only be read and understood by those who are familiar with the spoken language." The Loo Choo words for the same things are very different from those of the Chinese, the one being often a monosyllable, and the other a polysyllable; as in the instance of *charcoal*, the Chinese word for it being *tan*, and the Loo Chooan *chá-ehee-jing*, and yet the people use precisely the same character as the Chinese to express this word; and so far from its being necessary to be familiar with the language to understand the characters, many did not know the Chinese words for them. Their language throughout is very different from that of the Chinese, and much more nearly allied to the Japanese. The observation of M. Klaproth, in Archiv fur Asiatische Litteratur, p. 152, that the Loo Choo language is a dialect of the Japanese with a good deal of Chinese introduced into it, appears to be perfectly correct, from the information of some gentlemen who have compared the two, and are familiar with both languages. The vocabulary of Lieutenant Clifford, which we found very correct, will at any time afford the means of making this comparison.

The inhabitants of Loo Choo are very curious on almost all subjects, and seem very desirous of information; but we were wholly unable to judge of their proficiency in any subject, in consequence of the great disadvantages under which we visited their country.

CHAP. V. May, 1827.

Like the Japanese, they have always shown a determination to resist the attempts of Europeans to trade with them, partly, no doubt, in consequence of orders to that effect from China, and partly from their own timidity; and whenever a foreign vessel arrives it is their policy to keep her in ignorance of their weakness, by confining the crew to their vessel, or, if they cannot do that, within a limited walk of the beach, and through such places only as will not enlighten them on this point; and also to supply her with what she requires, in order that she may have no pretext for remaining.

Mr. Collie in his journal has given a phrenological description of the heads of several Loo Chooans which he examined and measured, in which proportions he thinks the lovers of that science will find much that is in accordance with the character of the people. The article, I am sorry to say, is too long for insertion here, and I only mention the circumstance that the information may not be lost.

We had but few opportunities of seeing any of the females of this country, and those only of the working class. An-yah said they were ugly, and told us we might judge of what they were like from the lower orders which we saw. They dressed their hair in the same manner as those people, and were free from the Chinese custom of modelling their feet.

The Loo Choo people dress extremely neat and always appear cleanly in their persons; they observe the Chinese custom of going bareheaded, and when the sun strikes hot upon their skulls, they avert its rays with their fans, which may be considered part of the dress of a Loo Chooan. In wet weather they wear cloaks and broad hats similar to those of the

Japanese, and exchange their straw sandals for wooden clogs. They have besides umbrellas to protect them from the rain. Of their occupations we could not judge; it was evident that there were a great many agriculturists among them, and many artisans, as they have various manufactures, of which I shall speak hereafter.

They appear to be very temperate in their meals, and indulge only in tea, sweatmeats, and tobacco, of which they smoke a great quantity; it is, however, of a very mild quality and pleasant flavour. Their pipes are very short, and scarcely hold half a thimbleful; this is done that they may be the oftener replenished, in order to enjoy the flavour of fresh tobacco, which is considered a luxury.

For further information on the manners, the dress, and minor points of interest belonging to these people, I must refer to the publications of Captain Hall and Mr. Macleod, who have so interestingly described all the little traits of character of the simple Loo Chooans, and who have pourtrayed their conduct with so much spirit, good feeling, and minuteness. These descriptions, though they have been a little overdrawn from the impulse of grateful recollections, from the ignorance in which the authors were kept by the cautious inhabitants, and from their desire to avoid giving offence, by pushing their inquiries as far as was necessary to enable them to form a correct judgment upon many things, are, upon the whole, very complete representations of the people.

The supposition that the inhabitants of Loo Choo possessed no weapons, offensive or otherwise, naturally excited surprise in England, and the circumstance became one of our chief objects of inquiry. I

CHAP V. May, 1827.

cannot say the result of the investigation was as satisfactory as I could have wished, as we never saw any weapon whatever in use, or otherwise, in the island; and the supposition of their existence rests entirely upon the authority of the natives, and upon circumstantial evidence. The mandarin Ching-oong-choo, and several other persons, declared there were both cannon and muskets in the island; and An-yah distinctly stated there were twenty-six of the former distributed among their junks.* We were disposed to believe this statement, from seeing the fishermen, and all classes at Napa, so familiar with the use and exercise of our cannon, and particularly so from their appreciating the improvement of the flint-lock upon that of the match-lock, which I understood from the natives to be in use in Loo Choo; and unless they possessed these locks it is difficult to imagine from whence they could have derived their knowledge. The figures drawn upon the panels of the joshouse, seated upon broadswords and bows and arrows, may be adduced as further evidence of their possessing weapons; and this is materially strengthened by the fact of their harbour being defended by three square stone forts, one on each side of the entrance, and the other upon a small island, so situated within the harbour, that it would present a raking fire to a vessel entering the port; and these forts having a number of loop-holes in them, and a platform and parapet formed above with stone steps leading up to it in several places. This platform would not have been wide enough for our cannon, it is true; but unless it were built for the reception of those weapons, there

* There were none on board the junk which sailed for China.

is apparently no other use for which it could have been designed. I presented the mandarin with a pair of pistols, which he thankfully accepted, and they were taken charge of by his domestics without exciting any unusual degree of curiosity. Upon questioning An-yah where his government procured its powder, he immediately replied from Fochien.

It is further extremely improbable that these people should have no weapons, considering the expeditions which have been successively fitted out by both China and Japan against Loo Choo. and the civil wars which unfortunately prevailed in the island, more or less, during the greater part of the time that the nation was divided into three kingdoms.* Besides, the haughty tone of the king to the commander of an expedition which was sent, in A. D. 605, to demand submission to his master the Emperor of China, viz. "That he would acknowledge no master," is not the language of a people destitute of weapons. Loo Choo has been subdued by almost every expedition against it, yet it is not likely the country could have made even a show of resistance against the invaders had the inhabitants been unarmed; they nevertheless resisted the famous Tay Cosama, and though conquered, threw off the yoke of Japan soon afterwards, and returned under the dominion of China. It was afterwards retaken by Kingtchang with 3,000 Japanese, who imprisoned the king, and killed Tching-hoey, his father, because he refused to acknowledge the sovereignty of Japan.† They are, besides, said to have sent swords

* From its division under Yut-ching in 1300, until it was united under Chang-pat-chi, about a century afterwards.

* Report of Supao-Koang, a learned Chinese physician, sent by the Emperor of China to Loo Choo in 1719, to report upon the country.—Lettres Edifiantes et Curieuses, vol. xxviii.

CHAP V. May, 1827.

as tribute to Japan. In 1454 the king Chang-tai-keiou had to sustain a civil war against his brother, who was at first successful, and beat Chang-tai-keiou in a battle, in which he fought at the head of *his troops*. It is not probable that all this warfare and bloodshed should have transpired without the Loo Chooans being possessed of arms; besides, it is expressly stated by Supao-Koang, that arms were manufactured in the island. I am, therefore, disposed to believe that the Loo Chooans have weapons, and that they are similar to those in use in China. And with regard to the objection which none of them having ever been seen in Loo Choo would offer, I can only say, that while I was in China, with the exception of the cannon in the forts, I did not see a weapon of any kind, though that people is well known to possess them.

It was also thought that the Loo Choo people were ignorant of the use of money. But this point has now been satisfactorily determined by our having seen it in circulation in the island, and having some of it in our own possession. The coin was similar to the *cash* of China. An-yah declared that there were no gold or silver coins in the country, not even ingots, which are in use in China; but this will hereafter, perhaps, prove to be untrue, as he even denied the use of the cash until it was found in circulation. There is very little doubt that money has been long known to, if not in use among, the Loo Chooans. About the year A.D. 1454, in the reign of Chang-tai-keiou, we are told that so large a quantity of silver and brass coin was taken from China to Loo Choo, that the provinces of Tche-Kiang and of Fochien complained to the emperor of the scarcity it had occasioned in those places; * and

* Recueil de Père Gaubil.

Père Gaubil, quoting Supao-Koang,* after enumerating several articles of trade, says " tout cela se vende et s'achète, ou par échange ou en deniers de cuivres de la Chine."

Our countrymen were further led to believe, from what they saw of the mild and gentle conduct of the superior orders in Loo Choo towards their inferiors, that the heaviest penalty attached to the commission of a crime was a gentle tap of a fan. Our friend with his bamboo cane, who was put on board to preserve order among his countrymen, afforded the first and most satisfactory evidence we could have had of this being an error, and had we possessed no other means of information, his conduct would have favoured the presumption of more severe chastisement being occasionally inflicted. It happened, however, fortunately, that I had purchased in China a book of the punishments of that country, in which the refined cruelty of the Chinese is exhibited in a variety of ways. By showing these to the Loo Choo people, and inquiring if the same were practised in their country, we found that many of their punishments were very similar. Those which they acknowledged were death by strangulation upon a cross, and sometimes under the most cruel torture; and minor punishments, such as loading the body with iron chains; or locking the neck into a heavy wooden frame; enclosing a person in a case, with only his head out, shaved, and exposed to a scorching sun; and binding the hands and feet, and throwing quicklime into the eyes. I was further assured that confession was sometimes extorted by the unheard-of cruelty of dividing the joints of the fingers

* Ibid. p. 402, Lettres Edifiantes.

CHAP. V. May, 1827.

alternately, and clipping the muscles of the legs and arms with scissars. Isaacha Sando took pains to explain the manner in which this cruelty was performed, putting his fingers to the muscles in imitation of a pair of sheers, so that I could not be mistaken: besides, other persons at Potsoong told me in answer to my inquiry, for I was rather sceptical myself, that it was quite true, and that they had seen a person expire under this species of torture. However, lest it should be thought I may have erred in attaching such cruelties to a people apparently so mild and humane, I shall insert some questions that were put to the Loo Chooans out of Dr. Morrison's Dictionary, and their answers to them respectively.

"Do the Loo Choo people torture and interrogate with the lash?" "Yes."—"Do they examine by torture?" "Yes."—"Do they give false evidence through fear of torture?" "Yes."—"Are great officers of the third degree of rank and upwards, who are degraded and seized to be tried, subjected to torture?" "No."—"Is torture inflicted in an illegal and extreme degree?" "Not illegal."—"Do you torture to death the real offender?" "Yes, sometimes."—"What punishment do you inflict for murder?" "Kill, *by hanging or strangulation*."*—"For robbery?" "The same."—"For adultery?" "*Banish to* Patanjan" (probably Pat-chong-chan, an island to the south-west of Typing-san.)—"For seduction?" "The same." Minor offences we were told were punished with a bambooing or a flagellation with a rod. Crimes are said to be few in number, and speaking generally there appears to be very little vice in the people.

* The words in italics were implied by signs.

I was assured by An-yah that marriages in Loo Choo were contracted as they are in China, by the parents or by a friend of the parties, without the principals seeing each other. Only one wife, I believe, is allowed in Loo Choo, though to the question, whether a plurality of wives was permitted? both An-yah and Shtafacoo said that the mandarin had five, and that the king had several.* They, however, afterwards declared that in their country it was customary to have only one wife. Perhaps it is the same in Loo Choo as in China, where a man may have only one lawful wife; but with her permission he may marry as many more as he can provide for. These wives are as much respected as the first wife, but they do not inherit their husbands' property.

In Loo Choo, as in China, there is no religion of the state, and every man is allowed freely to enjoy his own opinion, though here, also, a distinction is made between the sects, one being considered superior to the other. The sects in Loo Choo are Joo, Taou, and Foo, or Budh; but the disciples of the latter consist almost entirely of persons of the lowest order, and An-yah appeared to think very lightly of its votaries, saying they were "no good." It is upon record that it is 1011 years since this sect passed from China to Loo Choo. For several centuries its doctrines appear to have been advocated by the court as well as by the common people: but with the latter classes they have since been supplanted by those of Confucius. We are told that in the year 1372 several families from Fochien settled near Napa-kiang, and introduced ceremonies in honour of the great Chinese philosopher, whose me-

* Supao-Koang says a plurality of wives is permitted.

CHAP. V. May, 1827.

mory was further honoured by a temple being erected to him in Loo Choo, in 1663, by the Manshur Tartar, Emperor Kang-hi. Confucius is now honoured and revered by all classes in Loo Choo. The sect Taou, which is equally corrupt with that of Foo, has but few advocates among the better classes of society.

Like the Chinese, the Loo Chooans are extremely superstitious, and invoke their deities upon every occasion, sometimes praying to the good spirit, and at others to the evil. Near the beach to the northward of Potsoong, upon the shore which faces the coast of China, there were several square stones with pieces of paper attached to them. The natives gave us to understand they were the prayers of individuals; but we could not exactly understand the nature of them. A label similarly placed to those upon the beach was carried away by Captain Hall, and found to contain a prayer for the safe voyage of a friend who had gone from Loo Choo to China; it is very probable, therefore, that those which we saw were for similar purposes. At the Jos House at Potsoong I have mentioned pieces of paper being suspended between the panels, and have also suggested the probability of their being supplications of a similar nature. Indeed one of these also was taken to Macao by Lieutenant Clifford, and found to be an invocation of the devil.*

In a natural cave near Abbey Point, I found a rudely carved image, about three feet in height, of the goddess Kwan-yin (pronounced Kwan-yong by the Loo Chooans). In front of the deity there were several square stone vessels for offerings, and upon one of them some short pieces of polished wood were

* Hall's Loo Choo, 4to. p 206.

E.W. Beechey

KWAN-YIN,

GODDESS OF MERCY.

London, Published by Henry Colburn & Richard Bentley 1831.

placed, which I conjectured to be for the purpose of deciding questions, in the manner practised by the Foo sect in China, by being tossed in the air, or rattled in a bamboo case until one falls to the ground with its mark uppermost; when it is referred to a number in the book of the priest, and an answer is given accordingly. The natives were very unwilling to allow me to approach this figure, and pulled me back when I stepped into a small stone area in front of i , for the purpose of examining these pieces of wood. In China there are fasts in honour of this goddess, and no doubt there are the same in Loo Choo.

The following answers to several questions which I put to the natives of Loo Choo will fully explain the religion of the people.

"How many religions are there in Loo Choo?" "Three."—"What are these religions?" "Joo, Shih, Taou. Shih is the same as Foo."—"Are there many persons of the religion of Joo?" "Plenty."—"Foo?" "No good."—"Taou?" "Few."—"Does the sect Joo worship images?" "Sometimes kneel down to heaven, sometimes pray in heart, sometimes go priest house (temple)."—"Do they go to the temple of Kwan-yin?" "Yes."—"Do they go to the temple of Pih-chang?" "Sometimes."—"Do they go to the temple of Ching-hwang?"* "No."—"Do Joo, Shih, and Taou believe that heaven will reward the good aud punish the bad?" "Yes."

To the sentence, "At heart the doctrine of the three religions is the same; and it is firmly believed that heaven will do justice by rewarding the good and punishing the bad," An-yah did not assent. To the

* Ching-hwang is the goddess of Canton.

CHAP. V. May, 1827.

following sentence, "Both in this life and in the life to come there are rewards and punishments; but there is regard to the offences of men, whether heinous or not: speedy punishments are in this life; those that are more remote in the world to come," An-yah replied, "Priest say so."

"God created and constantly governs all things?" "Englishman's God, yes."—"When God created the great progenitor of all men, he was perfectly holy and perfectly happy?" "No."—"The first ancestor of the human race sinned against God, and all his descendants are naturally depraved, inclined to evil, and averse from good." "Good."—"If men's hearts be not renewed, and their sins atoned for, they must after death suffer everlasting misery in hell." "Priest say so: An-yah not think so."—"Do the three sects believe in metempsychosis?" This was not understood. —"Do they believe that all things are appointed by heaven?" "Yes."—"Are there any atheists in Loo Choo?" "Many."

In Loo Choo the priesthood are as much neglected and despised as in China, notwithstanding their being consulted as oracles by all classes. Several of them visited me in the garden at Potsoong, and remained while I made my magnetical observations. As these occupied a long time, I had an opportunity of particularly remarking these unfortunate beings, and certainly I never saw a more unintellectual and careworn class of men. Many persons crowded round the spot to observe what was going forward, and the poor priests were obliged to give way to every new comer, notwithstanding they were in their own garden. Their heads were shaved, similar to those of the Bodzes in China. I am not aware in what this practice originated, but as an observer I could not help noticing that the same operation is performed on the heads of criminals, or of persons who are disgraced in China; and from l'Abbé Grosier it appears to be considered a similar disgrace in Loo Choo.*

I endeavoured to distribute amongst the inhabitants some religious books which Dr. Morrison had given me in China, but there was a very great repugnance among the better part of the community to suffer them even to be looked into, much less to being carried away; and several that were secretly taken on shore by the lower orders were brought back the next day. However, I succeeded in disposing of a few copies, and Mr. Lay, I am glad to find, was equally fortunate with some which he also obtained from the same gentleman.

It has been shown, in the course of the narrative, that the present manner of disposing of the dead differs from that described by Père Gaubil, who says they burn the flesh of the deceased, and preserve the bones. It is not improbable that the custom may have changed, and that there is no mistake in the statement, as there is no reason to doubt the veracity of the Chinese author whom he quotes.

They pay every possible attention and respect to their departed friends by attending strictly to their mourning, frequently visiting the tombs, and, for a certain time after the bodies are interred, in supplying the cups and other vessels placed there with tea, and the lamps with oil, and also by keeping the tombs exceedingly neat and clean. We have frequently seen persons attending these lamps, and Lieutenant Wainwright noticed an old man strewing flowers and shells

* Description de la Chine, vol. II. p. 143.

CHAP. V.

May, 1827.

upon a newly made grave, which he said contained his son, and watching several sticks of incense as they burned slowly down to the earth in which they were fixed.

The trade of this island is almost entirely confined to Japan, China, and Formosa; Manilla is known as a commercial country, and it is recorded that a vessel has made the voyage to Malacca. In China their vessels go to Fochien, which they call Wheit-yen, and sometimes to Pekin. Commerce between Japan and Loo Choo is conducted entirely in Japanese vessels, which bring hemp, iron, copper, pewter, cotton, culinary utensils, lacquered furniture, excellent hones, and occasionally rice; though this article when wanted is generally supplied from an island to the northward belonging to Loo Choo, called Ooshima; but this is only required in dry seasons. The exports of Loo Choo are salt, grain, tobacco, samshew spirit, rice, when sufficiently plentiful, grass hemp, of which their clothes are made, hemp, and cotton. In return for these they bring from China different kinds of porcelain, glass, furniture, medicines, silver, iron, silks, nails, tiles, tools, and tea, as that grown upon Loo Choo is of an inferior quality. Several other articles of both export and import are mentioned by Supao-Koang, such as gold and silver from Formosa, and iron from China; among the former, mother of pearl, tortoise shell, bezoar stone and excellent hones. The last-mentioned articles, however, if found in Loo Choo, are certainly not very plentiful, as they are carried thither from Japan; and An-yah denied there being any mother of pearl there. This trade is conducted in two junks belonging to Loo Choo, which go annually to China: and they have besides these their tribute vessel.

The trade with Japan appears formerly to have been limited at 125 *thails* (tael of Canton), beyond which nothing was allowed to be sold. The goods carried to that country consisted of silks and other stuffs, with Chinese commodities, and the produce of their own country, such as corn, rice, pulse, fruits, spirits, mother of pearl, cowries, and large flat shells, which are so transparent that they are used in Japan for windows instead of glass.*

Their manufactures do not appear to be numerous, and are probably only such as are necessary for their own convenience. I have spoken of the rude hand-looms in use, the spinning-wheel, and the mills worked by cattle; these were the only machines we saw, though it may be inferred they have others. A short distance to the southward of Napa-kiang I was told there was a paper manufactory, and had a quantity of paper given me said to have been made there. It closely resembled that of China, but appeared to be more woolly. Grass-cloth, of a coarse texture, and coarse cottons are also wove upon the island; but I believe all the finer ones come from China, as well as the broad cloth of which their cloaks are made. Red pottery moderately good, a bad porcelain, and tiles, are among their manufactures, and also paper fans, of which the skeleton is bamboo; pipes, hair pins, and wicker baskets, and two sorts of spirits distilled from grain; moroofocoo already described; and another called sackee, resembling the samshew of China; salt, from the natural deposition of the sea, is collected in pans.

Supao-Koang mentions, among the manufactures of this country, silk, arms, brass instruments, gold and

* Kæmpfer's History of Japan, p. 381.

CHAP. V. May, 1827.

silver ornaments, a paper even thicker than that of Corea, made of *les cocons*, and another made of bamboo, besides that manufactured from the bark of the paper tree. He states they have woods fit for dyes, and particularly esteem one made from a tree, the leaves of which resemble those of the citron tree; and mentions brass, pewter, saddles, bridles, and sheaths as being manufactured with considerable taste and neatness upon the island, and as forming part of the tribute to China, from which it might be inferred that they were better executed than those in Pekin.

Previous to our departure I offered An-yah a patent corn-mill and a winnowing machine, and showed him the use of them. He was extremely thankful for them at first, but after a little consideration he declined the present, without assigning any reason. He probably imagined the introduction of foreign machinery might be disapproved by his superiors.

It has been observed that drums and tambourines were the only musical instruments among these people; we saw a flute, and were told that the inhabitants possessed violins and other stringed instruments; yet they do not appear to be a musical nation.

Among our numerous inquiries there was not one to which we got such contradictory answers as that concerning the residence of the king of Loo Choo. It was evident that there was a person of very high authority upon the island, whom they styled *wang*, which in Dr. Morrison's Dictionary is translated king, and that his residence was not far from Napa-kiang; but An-yah provoked me much by always evading this question. Sometimes he said it was four days to the north-east, at others that it was only one, and at last that it was at a place called Sheui, or Shoodi. Some of the natives whom I interrogated on this subject declared it was at Ee-goo-see-coo, about nine leagues to the northward; others, however, told me the name of his residence was Shoodi, or Sheui, as before. Mr. Collie was also informed it was at Shoodi; therefore, Sheui, or Shoodi, is in all probability the correct name of the place. As the natives pointed out to me the town upon the hill at the back of Napa-kiang as Shoodi, and as another party named it to Mr. Collie Shumi, we may presume that this town is the capital of Loo Choo; and this is the conclusion, as already remarked, that Captain Hall came to, after many inquiries on the same subject. Indeed I should think there could not be much doubt about it, as it answers very well both in name and position to the capital described by Supao-Koang, who remarks that the king holds his court in the south-west part of the island. The ground it stands upon is called Cheuli,* and that near this place the palace of the king is situated upon a hill. In another part he says that the space between Napa-kiang and the palace is almost one continued town.† Mr. Klaproth, however, has published extracts from some Chinese documents, which place the capital twenty *lis* (ten miles ?) east of Napa-kiang.

In the journals of my officers, I find that some of them were informed by the inhabitants that tribute was sent to China only once in *seven* years, and others, that it was paid every year. Kæmpfer also says that tribute is sent *every* year to the Tartarian monarch, in token of submission. By the Chinese accounts it is demanded every *second* year, as I have already stated.

* Cheuli by the Loo Chooans would be pronounced Cheudi, in the same way as they call Loo-Choo Doo-Choo.

† Lettres Edifiantes, p. 340.

M. J. Klaproth, quoting one of these authors, says, in 1654 Loo Choo sent Chang-Chy, the king's son, with an ambassador to Pekin, when it was arranged that every *second* year an ambassador should be sent to that court with tribute, which should consist of 3,000 lbs. of copper, 12,600 lbs. of sulphur, and 3,000 lbs of a strong silk; and that the number of his suite should not exceed a hundred and fifty persons.

Lord Macartney, when on his embassy to the court of China, met the mandarins from Loo Choo, who were going with this tribute to Pekin, and who informed him their chief sent delegates every *two years* to offer tribute.* And when we were at Loo Choo, both Ching-oong-choo and An-yah informed me to the same effect, viz. that it was sent every second year. We may therefore conclude, that this is the period agreed upon between the two countries.

M. Klaproth, p. 164, informs us, that notwithstanding tribute is paid to the court of China, Loo Choo is also compelled to acknowledge the sovereignty of Japan, to send ambassadors there from time to time, and to pay tribute in swords, horses, a species of perfume, ambergris, vases for perfumes, and a sort of stuff, a texture manufactured from the bark of trees, lacquered tables inlaid with shells or mother of pearl, and madder, &c. I shall merely observe upon this passage, that some of the articles which are said to be carried as tribute to Japan are actually taken from thence, and from China to Loo Choo, such as the vases and lacquered tables; and that mother of pearl is said by the natives not to be found upon the shores of their island.

* Embassy to China, by Sir George Staunton, vol. ii. p. 459.

The highest point of Loo Choo which we saw was a hill situated at the back of Barrow's Bay, in about the latitude of 26′ 27″ N., answering in position nearly to a mountain which appears on the chart of Mr. Klaproth, under the name of Onnodake. The height of this mountain is 1089 feet. The next highest point to this, which was visible from the anchorage, was the summit of the hill of Sumar, on which the capital is built; the highest point of this is 540½ feet. Abbey Point is 98½, and a bluff to the northward of Potsoong 99¾ feet. The Sugar-Loaf (Ee-goo-see-coo) was too far distant for us to determine its height; but I think Mr. Klaproth is wrong in saying it may be seen twenty-five sea leagues, as our distance from it was only ten leagues, and it was scarcely above the horizon.* It is certainly not so high as Onnodake, which, to a person at the surface of the sea, would be just visible at the distance of thirty-four miles. He is also mistaken in supposing it the only peak on the island.

These heights appear to be gained by ascents of moderate elevation only. In no part did we perceive any hills so abrupt that they could not be turned to account by the agriculturist. The centre of the island, or perhaps a line drawn a little to the westward of it, is the most elevated part of the country. Still the island is not divided by a ridge, but by a number of rounded eminences, for the most part of the same elevation, with valleys between them; so that when viewed at a distance the island appears to have a very level surface. In a Chinese plan of Loo Choo all these eminences are occupied by palaces and by courts of the king. The higher parts of the island are, in general, surmounted by trees, generally of the pinus mas-

* Klaproth's Mémoires relatifs à l'Asie, tom. ii. p. 173.

CHAP. V. May, 1827.

soniana, and the cycas; though they are sometimes bare, or at most clothed with a diminutive and useless vegetation. It not unfrequently happens that small precipices occur near the summits of the hills, and that large blocks of a coral-like substance are seen lying as if they had been left there by the sea. This substance, of which all the rocky parts of the island that we examined were composed, is a cellular or granular limestone, bearing a great resemblance to coral, for which it might easily be mistaken. It has a very rugged surface, not unlike silex maclière. Lieutenant Belcher found sandstone of a loose texture, enclosing balls of blue marl, and in one instance interstratified with it in alternate seams with the coral formation. This formation constituted part of a reef, dry at low water. In the marl he found cylindrical and elongated cones, similar to the belemnite, of a light colour, and occasionally crystallizations of calcareous spar.

The precipices inland, as well as those which form cliffs upon the coast, are hollowed out beneath, as if they had been subjected to the action of the waves. Upon the sea-coast this has no doubt been the case, and the Capstan Rock, spoken of before, presents a curious instance of its effect; but it is not quite so evident that the sea has reached the cliffs near Abbey Point, as they are separated from it by a plain covered with vegetation, and the violence of the waves is broken by reefs which lie far outside them.

The soil in the vicinity of Napa-kiang is generally arenaceous and marly, but to the south-east of Abbey Point there is a stratum of clay, which, in consequence of its retaining moisture better than other parts of the soil, is appropriated to the cultivation of rice.

The greater part of the island is surrounded by reefs

CHAP. V. May, 1827.

of coral. These are of two sorts; one in which the animals have ceased to exist, and the other which is still occupied by them. Both are darker-coloured than the reefs in the middle of the Pacific, owing, probably, to the various depositions which the rains have washed from the land. The shells found upon them are very much incrusted. About eight miles to the northward of Napa-kiang there is a deep bay, the shores of which are very flat, and have been converted into salt-pans by the natives. A river which appears to have its rise near the capital, after passing at the back of some hills, about five miles inland, empties itself into this bay. There is also another stream at Potsoong. The natives would not permit us to ascertain how far inland the water flowed up the harbour; nor would they inform us whether it was a division of the island, as its appearance induced us to suppose. In the Chinese plan already alluded to, the island is divided by such a channel; but it is doubtful whether this division may not be intended for the channel which separates Loo Choo from the Madjico-sima group, as the island to the southward has *Ta-ping-chan* written upon it, and there is a small island close to the eastward of it called *Little Lew-Kew*.* The relative positions of these are correctly given in the plan, but, if intended for those places, there is an egregious violation of all distance and proportion.

It has been already mentioned that the vegetable productions of the torrid and temperate zones are here found combined. The palmæ, boerhaavia, scævola, tournefortia, and other trees and shrubs recall the

* Formosa, notwithstanding its is considerably larger than Loo Choo, was called Little Lieou-Kieou, from there being so few inhabitants upon it.—*Recueil de P. Gaubil.*

CHAP. V. May, 1827. Coral Islands of the tropical regions to our view, while the rosaceæ, onagrariæ, etc. remind us of the temperate shores of our own continent. The remarkable genus of clerodendrum is here peculiarly abundant. Among the trees and shrubs which adorn the heights, the bamboo, hibiscus tiliaceus, thespesia popularia, hibiscus rosa sinensis, pandanus, piscidium, and several other trees and shrubs, some of which were new to us, were found uniting their graceful foliage; while in the gardens we noticed plantain, banana, fig, and orange trees, though the latter were apparently very scarce. We were told that they had pomegranates, but that they had neither pine-apples, plums, nor lēches, though they were perfectly acquainted with them all. The lē-chē is a fruit which is said to be peculiar to China: indeed Père J. B. Duhalde, in his Description de la Chine, vol. i. p. 104, says it grows only in two provinces of that great empire, Quang-tong, and Fokien. Père Gaubil, however, affirms that it is at Loo Choo, and that there are also there citrons, lemons, raisins, plums, apples, and pears, none of which we saw.

We were informed that the tea plant was tolerably abundant, and that the mild and excellent tobacco which was brought on board was the growth of the island. Gaubil affirms they have ginger, and a wood which they burn as incense, as well as camphor trees, cedars, laurels, and pines. Among the vegetable productions the sweet potatoe appears to be the most plentiful; the climate seemed so favourable to its growth, that we observed the tops rising from a soil composed almost entirely of sand. Both the root and the leaf are eaten by the natives.

The soil appears to be cultivated entirely with the hoe, and there are very few places on which this kind of labour has not been bestowed. Streams of water are not very abundant, and it is highly interesting to notice the manner in which the inhabitants have turned those which they possess to the greatest advantage, by conducting them in troughs from place to place, and at last allowing them to overflow flat places near the beach, for the purpose of raising rice and taro, which require a soil constantly wet.

The principal animals which we saw at Loo Choo were bullocks, horses, asses, goats, pigs, and cats; all of very diminutive size: a bullock which was brought to us weighed only 100 lbs. without the offal, and the horses were so low that a tall person had difficulty in keeping his feet off the ground; yet these animals must be esteemed in Japan, as they are said to have formed part of the tribute to that place. The poultry are also small: we heard dogs, but never saw any. Klaproth, p. 187, asserts there are bears, wolves, and jackals. A venomous snake is also said to exist in the interior. But the only other animals we saw were mice, lizards, and frogs; the latter somewhat different to those of our own country.

The insects are grasshoppers, dragon-flies, butterflies, honey-bees, wasps, moskitos of a large size, spiders, and a mantis, probably peculiar to the island.

There appeared to be very few birds, and of these we could procure no specimens, in consequence of the great objection on the part of the natives to our firing at them, arising probably from their belief in transubstantiation. Those which we observed at a distance resembled larks, martins, wood-pigeons, beach-plovers, tringas, herons, and tern. An-yah said there were no partridges in the island.

Fish are more abundant though not large, except-

CHAP. V. May, 1827.

ing sharks and dolphins, which are taken at sea, and guard-fish, which are often seen in the harbour. Those frequenting the reefs belong principally to the genera chætodon and labrus. A chromis, a beautiful small fish, was noticed in the waters which inundated the rice fields.

Upon the reefs are several *asteriæ*. These animals are furnished with long spiny tentaculæ, and are in the habit of concealing their bodies in the hollow parts of the coral, and leaving their tentaculæ to be washed about and partake of the waving motion of the sea; and to a person unacquainted with the zoophytes which form the coral, they might be supposed to be the animals connected with its structure. Lieutenant Belcher remarks of these reefs that a great change must have taken place in them since they were visited by the Alceste and Lyra, as he never observed any coral reefs apparently so destitute of animation as those which surround Loo Choo. The sea anemone and other zoophytes were very scarce.

We saw no shells of any value. A few cardium, trochus, and strombus were brought me by An-yah, and the haliotis was seen on the beach; but the history of this island states that the mother of pearl, large flat shells nearly transparent, and cowries, formed part of the tribute to Japan. An-yah, however, assured me there were no pearl shells upon the coast.

The Climate of Loo Choo must be very mild, from the nature of the dwelling-houses and the dress of the people; the mean temperature of the air, for the fortnight which we passed in the harbour, was 70° Unlike the Typa, we here experienced no great transitions, but an almost uniform temperature, which dissipated all the sickness the Typa had occasioned. We had, however. a good deal of rain in this time, which was about the change of the monsoon. By An-yah's account this island is occasionally visited by violent ta-foongs (mighty-winds), which unroof the houses and destroy the crops, and do other damage. They had experienced one, only the month previous to our arrival, which we were told had destroyed a great deal of rice, and was the cause of so many Japanese vessels being in the port. In 1708 it appears that one of these hurricanes did incalculable mischief, and occasioned much misery. The inhabitants seem to entertain a great dread of famine, and it is not improbable that these ta-foongs may occasion the evil. April, May, June, July, August, and September are the months in which these winds are liable to occur.

The harbour of Napa-kiang, though open to winds from the north, by the west to south-west is very secure, provided ships anchor in the Barnpool; a bay formed by the coral, to the northward of the Capstan Rock. In the outer anchorage, at high water, there is sometimes a considerable swell; and were it to blow hard from the westward at the time of the spring tides I have no doubt it would be sensibly felt. The reefs which afford protection to the harbour are scarcely above the sea at low water neap tides, and some remain wholly covered. In general they are much broken, and have many knolls in their vicinity, which ought to make ships cautious how they stand towards them. There are two entrances to the outer harbour, one from the northward, and the other from the westward. The former is narrow, and has several dangerous rocks in the channel, which, as they are not in general visible, are very likely to prove injurious to vessels; and as it can seldom happen that there is a

CHAP. V. May, 1827.

necessity for entering the harbour in that direction, the passage ought to be avoided. The western entrance is divided into two channels by a coral bank, with only seven feet water upon it, which, as it was discovered by the Blossom, I named after that ship. The passage on either side this rock may be made use of as convenient; but that to the southward is preferable with southerly winds and flood tides, and the other with the reverse. A small hillock to the left of a cluster of trees on the distant land, in the direction of Mount Onnodake, open about 4° to the eastward of a remarkable headland to the northward of Potsoong will lead through the south channel; and the Capstan Rock, with the highest part of the hill over Napakiang, which has the appearance of a small cluster of trees, will lead close over the north end of Blossom Rock. This notice of the dangers of entering the harbour will be sufficient in this place, and if vessels are not provided with a chart, or require further directions, it will be prudent to anchor a boat upon the rock.

Though the inhabitants of Loo Choo show so much anxiety for charts, they do not appear to have profited much by those which have been given to them, nor by those published in China and Japan. Their knowledge of geography is indeed extremely limited, and, with the exception of the islands and places with which they trade, they may be said to be almost ignorant of the geography of every other part of the globe. I did not omit to inquire about Ginsima, Kinsima, and Boninsima, islands which were supposed to exist at no great distance to the eastward of Loo Choo. The two first have never been seen since their discovery, but the other group has long been known to Japan;

and if we can credit the charts of the Japanese, it has been inhabited some time, as several villages and temples are marked therein. The Loo Chooans, however, could give me no information of it, or of any other islands lying to the eastward of their own, and were quite surprised at hearing a Japanese vessel * had been cast away upon an island in that direction.

The groups of islands seen in the distance to the westward of Loo Choo are called by the natives Kirrama and Agoo-gnee. Kirrama consists of four islands, Zammamee, Accar, Ghirooma, and Toocastchee, of which all but the last are very small. Agoognee consists of two small islands, Aghee and Homar. Both groups are peopled from and are subject to Loo Choo. Kirrama has four mandarins, one of the higher order, and three inferior; and Agoo-gnee two of the latter. The islands are very scantily peopled: in Toocastchee, which is the largest, there are but five hundred houses. The small coral islands off Napakiang are called Tzee.

To the northward of Loo Choo there are two islands, from which supplies are occasionally received; Ooshima,† of which I have spoken before as being subject to Loo Choo, and Yacoo-chima, a colony of Japan. Ooshima produces an abundance of rice, and as in dry seasons in Loo Choo this valuable grain sometimes fails, Yacoo-chima junks, which appear to be the great carriers to Loo Choo, go there and load. Yacoo-chima is said to be an island of great extent, but the chart which An-yah drew to show its situation

* See Kæmpfer's History of Japan.

† Probably O-foushima of Supao-Koang, situated in latitude 30° N.

CHAP. V. May, 1827.

was too rude for me even to conjecture which of the islands belonging to Japan it might be.

In my narrative of Loo Choo I have made allusion to the works of several Chinese and Japanese authors,* who have written upon that island. As their accounts generally wear the appearance of truth, and as they are the only records we have of the early history of a country so little visited by Europeans, I shall give a sketch of them, that my reader may become acquainted with what is known of the history of that remote country, without having to search different books, only one of which has as yet been published in England.

The inhabitants of Loo Choo are extremely jealous of their antiquity as a nation. They trace their descent from a male and a female, who were named Omo-mey-keiou, who had three sons and two daugters. The eldest of these boys was named Tien sun (or the grandson of heaven). He was afterwards the first king of Loo Choo, and from the first year of his reign to the first of that of Chun-tien, who ascended the throne A. D. 1187, they reckon a period of no less than 17,802 years. The kings were supposed to be descended from the eldest son, the nobility from the second, and the commoners from the youngest. The eldest daughter was named Kun-kun, and had the title of Spirit of Heaven; the other, named Tcho-tcho, was called the Spirit of the Ocean.

* The works of these authors will be found in Lettres Edifiantes et Curieuses, tom. xxiii. 1811; Grosier sur la Chine, tom. ii.; M. J Klaproth, Memoires sur la Chine; Kæmpfer's History of Japan, vol. i.; P. J. B. Duhalde. For other information on Loo Choo, the reader is referred to the Voyages of Benyowsky, Broughton, and of H. M. ships Alceste and Lyra.

We are told that five-and-twenty dynasties successively occupied the throne of Loo Choo, from the death of Tien-sun to the reign of Chun-tien; but nothing further was known of the history of the country until the year A. D. 605, when the Emperor of China, of the dynasty of "Soui," being informed there were some islands to the eastward of his dominions named Loo Choo, became desirous of reconnoitring their situation, and of becoming acquainted with the resources of the islands. He accordingly fitted out an expedition, but it did not effect what the emperor desired. It, however, brought back a few natives; and an ambassador from Japan happening to be at the court of China at that time, informed the emperor that these people belonged to Loo Choo, and described their island as being poor and miserable, and the inhabitants as barbarians. Being informed that in five days a vessel could go from his dominions to the residence of the king of these islands, the emperor, Yang-tee, sent some learned men with interpreters to Loo Choo to obtain information, and to signify to the king that he must acknowledge the sovereignty of the Emperor of China, and do him homage. This embassy succeeded in reaching its destination, but, as might have been expected from the ruler of an independent people, it was badly received, and was obliged to return with the haughty answer to their sovereign, that the prince of Loo Choo would acknowledge no chief superior to himself. Indignant at being thus treated by a people who had been described as barbarians, he put ten thousand experienced troops on board his junks, and made a successful descent upon the Great Loo Choo. The king, who appears to have been a man of great courage, placed himself at the head of his troops, and disputed the ground with the Chinese; but unfortunately he was killed; his troops gave way; and the victorious invaders, after pillaging and setting fire to the royal abode, and making five thousand slaves, returned to China.

It is said that at this time the inhabitants of Loo Choo had neither letters nor characters, and that all classes of society, even the king himself, lived in the most simple manner. It does not, however, appear that the people were entitled to the appellation of barbarians, which was given to them by the ambassador of Japan in China, nor that they merited the title of *poor devils,* which the word lieu-kieu implies in Japanese; as they had fixed laws for marriages and interments, and paid great respect to their ancestors and other departed friends; and they had other well regulated institutions which fully relieved them from the charge of barbarism. Their country was not so poor nor so destitute of valuable productions, or even of manufactures, but that Chinese merchants were glad to open a trade with it, and to continue it through five dynasties which successively ruled in China after the conquest of Loo Choo, notwithstanding the indifference of the emperors who, during that period, ceased to exact the tribute that had been made to their predecessors. It is not improbable, therefore, that this stigma, which ought properly to belong to Formosa—which, though a much larger island, was then called Little Loo Choo—may have been attached to the island we visited from the similarity of names.

Chun-tien was said to be descended from the kings of Japan, but it is not known at what period his family settled in Loo Choo. Before he came to the throne, he was governor of the town of Potien. On his accession his title was disputed by a nobleman named Li-yong; but he being defeated and killed, Chun-tien was acknowledged King of Loo Choo by the people. Having reigned fifty-one years, and bestowed many benefits upon his subjects, whose happiness was his principal care, he died at the age of seventy-two. In this reign reading and writing are said to have been first introduced from Japan, the character being that of Y-rofa.

Very little mention is made of the son and successor of Chun-tien; but the reign of his grandson Y-pen is marked by the occurrence of a famine and a plague, which nearly desolated the island; and by his abdication in favour of any person whom the people might appoint to succeed him. The choice fell upon Ynt-sou, the governor of a small town; but the king, desirous of ascertaining whether he was a competent person to succeed him, first made him prime minister; and being at length satisfied that the choice of the people was judicious, he abdicated in his favour, reserving a very moderate provision for himself and family. Ynt-sou ascended the throne A. D. 1260, and reigned forty years. He is said to have been the first to levy taxes, and to have introduced useful regulations for the cultivation of the soil. In his reign Ta-tao, Ki-ki-ai, and other islands to the north-east and north-west came under the dominion of Loo Choo. This reign was also marked by an attempt of the Emperor of China to renew his demand of tribute, which had not been made for so many generations that the Loo Chooans began to consider themselves absolved from the obligation. The Emperor of China, however, determining not to relinquish the advantages which had been gained by his predecessor Yang-ti, equipped a fleet for the purpose of compelling payment; but about this time China having suffered a serious defeat from the Japanese, and from the kingdoms of Tonquin and Cochin China, and lost 100,000 men in her expeditions against those places, disaffection spread throughout the troops, and the expedition returned without even having reached its destination.

Ynt-sou was succeeded by his son Ta-tchin, who was followed by his son Ynt-see. two princes much esteemed for their wisdom and benevolence. Not so Yut-ching, a prince of avaricious and voluptuous disposition, who ascended the throne of his father in 1314; during whose reign the state fell into considerable disorder. The governor of Keng-koaey-gin revolted and declared himself King of Chanpe, the northern province of the island. The governor of Tali also revolted, and became king of the southern province Chan-nan, leaving Yut-ching to govern only the centre of the island, which was called Tchong-chan. Thus was this island, not sixty miles in length, divided into three independent kingdoms. The greatest animosity prevailed between these three principalities; and long and bloody wars ensued. About sixty years after the country had been thus divided, Tsay-tou, a prince beloved by his people and esteemed for his valour, came to the throne of the middle province. It was in his reign that Hong-vou, the Emperor of China, renewed overtures of protection; and the embassy which he sent to the court of Tsay-tou acquitted itself so creditably, that the offer was accepted. The kings of the other districts of Loo Choo were no sooner apprised of the conduct of Tsay-tou, than they also put themselves under the protection of China; and thus Loo Choo once more became tributary to the Celestial Empire.

The Emperor Hong-vou was so much pleased with this conduct of the kings of Loo Choo, that he sent them large presents of iron, porcelain, and other articles which he knew to be scarce in their dominions; and also settled in the middle province thirty-six families from Fochien, who established themselves at a place called Kūmi, a little to the northward of Napa-kiang. These people introduced into Loo Choo the Chinese written character, and ceremonies in honour of Confucius. On the other hand, the kings of Loo Choo sent several youths to Pekin, among whom were the sons and brothers of Tsay-tou, who were educated and brought up at the expense of the emperor.

The best understanding now existed between the kings of Loo Choo and the court of China; and while the emperor was receiving ambassadors from Loo Choo, that country had the satisfaction of seeing several islands to the northward and southward of its own position added to its dominions. On the death of Tsay-tou, which happened in 1396, his son Au-ning was installed king by the emperor in the place of his father. He reigned ten years, and was succeeded by his son Is-tchao. The reigns of these two princes were not distinguished by any remarkable events; but that of her successor, Chang-patché, will ever be remembered by the Loo Chooans from the advantageous union of the free provinces, which for nearly a century had been agitated by a continued state of warfare; and from the estimation in which the king of the island was held by Suent-song, then Emperor of China, who made him large presents of silver, and bestowed upon him the title of *Chang*, which has ever since been the patronymic of the royal family of Loo Choo.

The three following reigns present no occurrences

CHAP. V. May, 1827.

worthy of notice. In 1454, the Chang-tai-kieou ascended the throne amidst difficulties and disaffection. His ambitious brother disputed the elevated rank he had obtained, and enlisted in his cause so powerful a body of the islanders, that the king was defeated, his palace burned, and his magazines reduced to ashes. In this state of affairs he solicited the protection of the Emperor of China, who readily assisted him; and not only restored tranquillity to the island by his interference, but caused the king to be remunerated for all his losses.

The commerce of Loo Choo with China afterwards daily increased; and under the reign of this prince so great a trade was carried on between the two countries, that the provinces of Tche-kiang and Fochien were distressed by the quantity of silver and copper coin that was carried away to Loo Choo. The people even complained to the Emperor of the scarcity, who ordered that in future the trade between these two places should be confined within certain limits.

After a short reign of seven years, Chang-tai-kieou was succeeded by his son Chang-te, a prince whose name was rendered odious by the acts of cruelty he committed, and who was so much detested, that after his death the people refused to acknowledge as king the person whom he had appointed to succeed him; and elected in his stead Chan-y-ven, a nobleman of the island of Yo-pi-chan. Though the reign of this prince is distinguished in history only by the regulation of the number of persons who should accompany the ambassadors to Pekin, yet he is said to have been a great prince. His son, Chang-tching, was a minor at the death of his father, and his paternal uncle was chosen to be his protector. In this reign Loo

Choo became a comparatively great commercial nation. Many vessels were sent to Formosa, to the coasts of Bungo, Fionga, Satzuma, Corea, and other places. Her vessels became the carriers of Japanese produce to China, and vice versâ; and one of them even made the voyage to Malacca.

By this extensive trade, and by being the entrepôt between the two empires of China and Japan, Loo Choo increased in wealth and rose into notice; especially as it was found convenient by both these two great nations to have a mediator on any differences arising between them. The advantage thus derived by Loo Choo was particularly manifested on the occasion of a remonstrance on the part of China against robberies and piracies committed upon the shores of that country by a prodigious number of vessels manned by resolute and determined seamen, principally Japanese, who landed upon all parts of the coast, and spread consternation along the whole of the western shore of the Yellow Sea, even down to Canton. The Emperor of China on this occasion sent ambassadors to Loo Choo; and a representation was made to the Court of Japan of the numerous piracies committed in the dominions of the Emperor of China by the subjects of that country; and succeeded so far that the sovereign of Japan gave up to the King of Loo Choo a number of vessels and slaves which had been captured; but as none of these marauding vessels had been fitted out by his command, and as they were the property of individuals over whom he had no control, it was out of his power to put a stop to the depredations. The Emperor of China rewarded the King of Loo Choo for this important service by sending him large presents of silk, porcelain, and silver, and brass money; and

CHAP. V. May, 1827.

granted to his subjects very great privileges in their commercial transactions with China.

The Japanese pirates, among which there were a great many vessels manned by Chinese, continued their depredations in spite of the efforts and remonstrances of the Emperor of China; and latterly occasioned such alarm in that country, that the famous Tay Cosama, who was then secular ruler of Japan, determined to avail himself of the panic, and premeditated an attack upon the coast of that mighty empire. It was necessary to the success of this bold enterprise that the assault should be conducted with the utmost secrecy; and Tay-Cosama, fearing that the frequent intercourse between China and Loo Choo, which country could not remain in ignorance of the preparations, might be the means of divulging his intentions to China, sent ambassadors to Chang-ning, who was then King of Loo Choo, haughtily forbidding him to pay tribute to China, and desiring him to acknowledge no other sovereign than that of Japan. It is said that he also sent similar notices to the governor of the Philippines, to the King of Siam, and to the Europeans in India.

Chang-ning, however, was not easily intimidated, and remained deaf to the menaces of the Emperor of Japan. He saw through the designs of Tay-Cosama; and by means of a rich Chinese merchant, who happened to be at Napa-kiang at that time, he apprised Ouan-li, then Emperor of China, of his designs. Ouan-li immediately increased his army, fortified his coasts, and made every preparation for a vigorous defence against the invading army of Japan, whenever it might arrive. He also apprised Corea of the danger with which that state was threatened; but the king, misled probably by the designing Emperor of Japan, and imagining the immense preparations making by that prince were intended for the invasion of China, neglected to strengthen his defences, and was at length surprised by the Japanese, who invaded his dominions.

Chang-ning, notwithstanding the invasion with which he was also threatened, continued his tribute to China; and Ouan-li received his ambassadors with the greatest possible respect, and rewarded their sovereign for his fidelity. Some years after, in 1610, the Japanese renewed their menaces against Chang-ning, who, as on the former occasion, acquainted the Emperor of China with his situation, and implored assistance; but China at that time was fully occupied with her own troubles, and unable to render him any service. In this state of things, a nobleman of Loo Choo, named King-tchang, taking advantage of the situation of Chang-ning, revolted, and retired to Satzuma, where he fitted out an expedition consisting of 3000 Japanese, and took Chang-ning prisoner, killed his father, Tching-hoey, because he would not acknowledge his dependency to Japan, pillaged the royal palace, and carried away the king prisoner to Satzuma.

The conduct of the King of Loo Choo throughout all these disturbances is said to have been so magnanimous and spirited, that it even appeased King-tchang, and prepossessed the Japanese so much in his favour, that after two years' captivity they restored him to his throne with honour. He was scarcely reinstated, when, always faithful to China, notwithstanding the danger he had escaped, and the helpless condition of the emperor, he sent ambassadors to that country to

CHAP. V. May, 1827.

declare his submission as heretofore; and to apprise the emperor of an attack which was intended to be made on Formosa by the Japanese, who had conceived the project of reinstating themselves in that country, and fortifying their settlements there.

Chang-ning left no son to succeed him; and Chang-yong, a descendant of the brother of his predecessor, was installed by the Emperor of China in his stead. This prince, notwithstanding the unsettled state of affairs, and the danger he had to apprehend from Japan, paid the usual tribute to China, and introduced into his country from thence the manufacture of delft-ware, and an inferior kind of porcelain.

About eighty years afterwards, A.D. 1643, the famous revolution occurred in China, which fixed the Tartar dynasty on the throne of that empire; and Chang-tché, who at that time was King of Loo Choo, sent ambassadors to pay homage to the new sovereign; when King Chang-tché received a sign manual from the Tartar monarch, directing that Loo Choo should not pay tribute oftener than once in two years, and that the number of the embassy should not exceed a hundred and fifty persons.

In 1663 the great Emperor Kang-hi succeeded to the throne of China, and received the tribute of Chang-tché on the occasion. This magnanimous prince sent large presents of his own to the King of Loo Choo, in addition to some of an equally superb quality which were intended for that country by his father. His ambassadors passed over to Loo Choo, and according to custom confirmed the king in his sovereignty, the ceremony on this occasion being distinguished by additional grandeur and solemnity.

Kang-hi, probably foreseeing the advantages to be

derived from an alliance with Loo Choo, which had so long continued faithful to the empire of China, turned his attention to the improvement of the country with great earnestness and perseverance. He built a palace there in honour of Confucius, and a college for the instruction of youth in the use of the Chinese character, and established examinations for different branches of literature. Several natives of Loo Choo were sent to Pekin, and educated at the expense of the emperor, among whom was the king's son. The tribute was better adapted to the means of the people; and those articles only, which were either the produce of the soil, or the manufactures of the country, were in future to be sent to Pekin for this purpose. In short, Kang-hi lost no opportunity of gaining the friendship and esteem of his subjects. On the occasion of great distress in Loo Choo, which occurred in 1708, when the palace of the king was burned, and hurricanes did incalculable mischief, and when the people were dying daily with contagious diseases, Kang-hi used every endeavour to mitigate their distress, and, by his humanity and generosity, secured to himself the lasting gratitude of the inhabitants of Loo Choo.

In 1719 he sent Supao-koang, a learned physician, to make himself acquainted with the nature and productions of the island, and to inform himself of every particular concerning the government and the people. Since that period nothing is mentioned of Loo Choo in Chinese history, beyond the periodical payment of the tribute, and the arrival of ambassadors from that country at the court of Pekin.

In 1771 the well-known Count Benyowsky touched

CHAP. V. May, 1827.

at an island belonging to Loo Choo, named Usmay-Ligon, where he found that almost all the inhabitants had been converted to Christianity by a jesuit missionary. If we can credit his statement, he was treated by the natives with the greatest hospitality and unreserve. Contrary to the custom of the eastern Asiatic nations, these people brought their daughters to the count and his associates, and pressed them to select wives from among them. In short, the conduct of the inhabitants is described as being so engaging, that some of Benyowsky's crew determined to remain with them, and were actually left behind when the count put to sea. And the natives, on the other hand, are asserted to have been so attached to their visitors, that they made them promise to return and form a settlement among them, and signed a treaty of friendship with the count. This veracious traveller found muskets with matchlocks in use with these people; and to add to their means of defence, on his departure he presented them with 80 muskets of his own, 600 swords, and 600 pikes, besides 20 barrels of powder and 10 barrels of musket-balls.

Loo Choo in 1796 was visited by Captain Broughton, and in 1803 by the ship Frederick of Calcutta, which made an unsuccessful effort to dispose of her cargo. The inhabitants on both these occasions were, as usual, extremely civil and polite, but resisted every attempt at opening a commerce. The next mention of this interesting island is in the well-known publications of Captain Basil Hall, and of Mr. M'Cleod, the surgeon of the Alceste.

Thus Loo Choo, like almost every other nation, has been disturbed by civil wars, and the state has been endangered by foreign invasion: her towns have been plundered, her palaces consumed, and her citizens carried into captivity. Situated between the empires of China and Japan, she has been mixed up with their quarrels, and made subservient to the interests of both; at one time suffering all the miseries of invasion, and at another acting as a mediator. Allied by preference to China, and by fear and necessity, from her proximity, to Japan, she is obliged, to avoid jealousy, to pay tribute to both, though that to the latter country is said to be furnished by the merchants who are most interested in the trade to that empire. Their conduct to strangers who have touched at their ports has ever been uniformly polite and hospitable; but they would rather be exempt from such friendly visits: and though extremely desirous of obtaining European manufactures, particularly cloth, hosiery, and cutlery, they would oppose any open attempt to introduce them. The most likely means of establishing a communication with them would be through Chinese merchants at Canton, who might be persuaded to send goods there in their own names, and under the charge of their own countrymen.

Whale-ships have occasionally touched at Loo Choo when distressed for provisions. It is satisfactory to find that these interviews have been conducted without giving offence to the natives. It is to be hoped that any vessel which may hereafter be under the necessity of putting in there will preserve the same conduct, and give the inhabitants no cause to regret having extended their hospitality to foreigners.

I have perhaps entered more minutely upon several questions connected with Loo Choo than may be con-

CHAP. V. May, 1827. sidered necessary, after what has already been given to the public; but it appeared desirable to remove doubts upon several points of interest, which could not perhaps be effectually accomplished without combining my remarks with a short notice of the history of the country.

CHAPTER VI.

Passage from Loo Choo eastward—Arrive at Port Lloyd in the Yslas del Azobispo—Description of those Islands—Passage to Kamtchatka—Arrival at Petropaulski—Notice of that Place—Departure—Pass Beering's Strait—Enter Kotzebue Sound—Prosecute the Voyage to the Northward—Stopped by the Ice—Return to the Southward—Discover Port Clarence and Grantley Harbour—Description of these Harbours—Return to Kotzebue Sound—Ship strikes upon a Shoal.

CHAP. VI.

May, 1827.

On the 25th of May we took our departure from Loo Choo, and steered to the eastward in search of some islands which were doubtfully placed in the charts. On the third day we arrived within a few miles of the situation of Amsterdam Island without seeing any land, and passed it to the northward, as near as the wind would permit. The weather was very unfavourable for discovery, being thick and rainy, or misty, with very variable winds. On the 3d of June we regretted exceedingly not having clear weather, as the appearance of plover, sandlings, flocks of shearwaters, and several petrel and albatrosses, created a belief that we were near some island.

June.

Three days afterwards we were upon the spot where the Island of Disappointment is placed in the latest charts. The weather was tolerably clear, but no land could be seen; and as we were so near the situation of a group of islands which, if in existence, would oc-

CHAP. VI. June, 1827.

cupy several days in examining, I did not wait to search for Disappointment Island, which is said to be very small. I have since been informed that this island, which in all probability is the same as the island of Rosario, was seen by a whaler, who, not being able to find it a second time, bestowed upon it the name of Invisible Island. It is said to lie ninety miles N.W. from port Lloyd, a place which I shall presently notice.

The next evening we reached the situation of the Bonin Islands in Arrowsmith's chart, and the following morning made sail as usual, without seeing any land. We were almost on the point of declaring them invisible also, when, after having stood to the eastward a few hours, we had the satisfaction to descry several islands, extending in a north and south direction as far as the eye could discern. They all appeared to be small, yet they were high and very remarkable; particularly one near the centre, which I named after Captain Kater, V. P. R. S., &c.

As the islands to the southward appeared to be the largest, I proposed to examine them first; and finding they were fertile, and likely to afford good anchorage, Lieutenant Belcher was sent in shore with a boat to search for a harbour. In the evening he returned with a favourable report, and with a supply of fourteen large green turtle.

We stood off and on for the night with very thick weather; and at daylight, when by our reckoning the ship should have been seven miles from the land, we unexpectedly saw the rocks beneath the fog, about a fifth of a mile distant, and had but just room to clear them by going about. The depth of water at the time was sixty fathoms; so that had it been blowing strong and necessary to anchor, there would have been but an indifferent prospect of holding on any length of time. The great depth of water, and the strong currents which set between the islands must make the navigation near them hazardous during thick weather. On the evening preceding this unexpected event, we found so strong a current setting to the south-west, to windward, that though the ship was lying to, it was necessary frequently to bear away, to prevent being drifted upon the land.

When the fog cleared away on the 9th, we discovered a distant cluster of islands bearing S. 5° E. true: I therefore deferred anchoring in the bay which Lieutenant Belcher had examined the preceding evening, in the hope of being able to examine the newly discovered islands; but finding both current and wind against us, and that the ship could scarcely gain ground in that direction—as there was no time to be lost, I returned to those first discovered. In running alongshore we observed an opening, which, appearing to afford better security than the before-mentioned bay, the master was sent to explore; and returned with the welcome intelligence of having found a secure harbour, in which the ship might remain with all winds.

We were a little surprised, when he came back, to find two strangers in the boat, for he had no idea that these islands had been recently visited, much less that there were any residents upon them; and we concluded that some unfortunate vessel had been cast away upon the island. They proved to be part of the crew of a whale-ship belonging to London, named the William. This ship, which had once belonged to his majesty's service, had been anchored in the harbour in deep water, and in rather an exposed situation (the

CHAP. VI. June, 1827.

port then not being well known, and had part of her cargo upon deck, when a violent gust of wind from the land drove her from her anchors, and she struck upon a rock in a small bay close to the entrance, where in a short time she went to pieces. All the crew escaped, and established themselves on shore as well as they could, and immediately commenced building a vessel from the wreck of the ship, in which they intended to proceed to Manilla; but before she was completed, another whaler, the Timor, arrived, and carried them all way except our two visiters, who remained behind at their own request. They had been several months upon the island, during which time they had not shaved or paid any attention to their dress, and were very odd-looking beings. The master, Thomas Younger, had unfortunately been killed by the fall of a tree fifteen days previous to the loss of the ship, and was buried in a sandy bay on the eastern side of the harbour.

We entered the port and came to an anchor in the upper part of it in eighteen fathoms, almost land-locked. This harbour is situated in the largest island of the cluster, and has its entrance conspicuously marked by a bold high promontory on the southern side, and a tall quoin shaped rock on the other. It is nearly surrounded by hills, and the plan of it upon paper suggests the idea of its being an extinguished crater. Almost every valley has a stream of water, and the mountains are clothed with trees, among which the areca oleracea and fan-palms are conspicuous. There are several sandy bays, in which green turtle are sometimes so numerous that they quite hide the colour of the shore. The sea yields an abundance of fish; the rocks and caverns are the resort of crayfish and other shellfish; and the shores are the refuge of snipes, plovers, and wild pigeons. At the upper part of the port there is a small basin, formed by coral reefs, conveniently adapted for heaving a ship down; and on the whole it is a most desirable place of resort for a whale-ship. By a board nailed against a tree, it appeared that the port had been entered in September, 1825, by an English ship named the Supply, which I believe to be the first authenticated visit made to the place.

Taking possession of uninhabited islands is now a mere matter of form; still I could not allow so fair an opportunity to escape, and declared them to be the property of the British government by nailing a sheet of copper to a tree, with the necessary particulars engraved upon it. As the harbour had no name, I called it Port Lloyd, out of regard to the late Bishop of Oxford. The island in which it is situated I named after Sir Robert Peel, His Majesty's Secretary of State for the Home Department.

As we rowed on shore towards the basin, which, in consequence of there being ten fathoms water all over it, was named Ten Fathom Hole, we were surrounded by sharks so daring and voracious that they bit at the oars and the boat's rudder, and though wounded with the boat-hook returned several times to the attack. At the upper end of Ten Fathom Hole there were a great many green turtle; and the boat's crew were sent to turn some of them for our sea-stock. The sharks, to the number of forty at least, as soon as they observed these animals in confusion, rushed in amongst them, and, to the great danger of our people, endeavoured to seize them by the fins, several of which we noticed to have been bitten off. The turtle

CHAP. VI. June, 1827.

weighed from three to four hundred-weight each, and were so inactive that, had there been a sufficient number of men, the whole school might have been turned.

Wittrein and his companion, the men whom we found upon the island, were living on the south side of the harbour, in a house built from the plank of the William, upon a substantial foundation of copper bolts, procured from the wreck of the ship by burning the timbers. They had a number of fine fat hogs, a well-stocked pigeon-house, and several gardens, in which there were growing pumpkins, water-melons, potatoes, sweet potatoes, and fricoli beans; and they had planted forty cocoa-nuts in other parts of the bay. In such an establishment Wittrein found himself very comfortable, and contemplated getting a wife from the Sandwich Islands; but I am sorry to find that he soon relinquished the idea, and that there is now no person to take care of the garden, which by due management might have become extremely useful to whale ships, the crews of which are often afflicted with scurvy by their arrival at this part of their voyage. The pigs, I have since learned, have become wild and numerous, and will in a short time destroy all the roots, if not the cabbage-trees, which at the time of our visit were in abundance, and, besides being a delicate vegetable, were no doubt an excellent antiscorbutic.

We learned from Wittrein, who had resided eight months upon the island, that in January of 1826 it had been visited by a tremendous storm, and an earthquake which shook the island so violently, and the water at the same time rose so high, that he and his companion, thinking the island about to be swallowed up by the sea, fled to the hills for safety. This gale, which resembled the typhoons in the China sea, began at north and went round the compass by the westward, blowing all the while with great violence, and tearing up trees by the roots: it destroyed the schooner which the crew of the William had began to build, and washed the cargo of the ship, which since her wreck had been floating about the bay, up into the country. By the appearance of some of the casks, the water must have risen twelve feet above the usual level.*

We were informed that during winter there is much bad weather from the north and north-west; but as summer approaches these winds abate, and are succeeded by others from the southward and south-eastward, which prevail throughout that season, and are generally attended with fine weather, with the exception of fogs, which are very prevalent. Shocks of earthquakes are frequently felt during the winter; and Wittrein and his companion repeatedly observed smoke issuing from the summits of the hills on the island to the northward. Peel island, in which we anchored, is entirely volcanic, and there is every appearance of the others to the northward being of the same formation. They have deep water all round them; and ships must not allow their safety to depend upon the lead, for although bottom may be gained at great depths between some of the islands, yet that is not the case in other directions.

We noticed basaltic columns in several parts of Port Lloyd, and in one place Mr. Collie observed them divided into short lengths as at the Giant's Causeway: he also remarked at the head of the bay in the bed of a small river, from which we filled our water-casks, a

* The seamen affirmed that it rose twenty.

CHAP. VI. June, 1827.

a sort of tessellated pavement, composed of upright angular columns, placed side by side, each about an inch in diameter, and separated by horizontal fissures. It was the lower part of the Giant's Causeway in miniature. Many of the rocks consisted of tuffaceous basalt of a grayish or greenish hue, frequently traversed by veins of petrosilex; and contained numerous nodules of chalcedony or of cornelian, and *plasma?* The zeolites are not wanting; and the stilbite, in the lamellar foliated form, is abundant. Olivine and hornblende are also common. The drusses were often found containing a watery substance, which had an astringent taste not unlike alum, but I did not succeed in collecting any of it.

The coral animals have raised ledges and reefs of coral round almost all the bays, and have filled up the northern part of the harbour, with the exception of Ten Fathom Hole, which appears to be kept open by streams of water running into it; for it was observed here, that the only accessible part of the beach was at the mouths of these streams.

I have before observed, that the hills about our anchorage were wooded from the water's edge nearly to their summit. There were found among these trees, besides the cabbage and fan-palms, the tamanu of Otaheite, the pandanus odoratissimus, and a species of purau; also some species of laurus, of urtica, the terminalia, dodonæa viscosa, eleocarpus serratis, &c. We collected some of the wood for building boats, and found it answer very well for knees, timbers, &c.

We saw no wild animals of the mammalia class except the vampire bat, which was very tame. Some measured three feet across the wings when fully extended, and were eight or nine in length in the body. We frequently saw them flying; but they were more fond of climbing about the trees, and hanging by their hind claw, which appears to be their natural position when feeding. Some were observed with their young at their breast, concealed by the wide membrane of their wing. The tongue of this animal is unusually large, and furnished with fleshy papillæ on the upper surface. Here we also found another species of vespertilio.

Of birds we saw some handsome brown herons with white crests; plovers, rails, snipes, wood-pigeons, and the common black crow; a small bird resembling a canary, and a grossbeak. They were very tame, and until alarmed at the noise of a gun suffered themselves to be approached.

The sea abounded in fish, some of which were very beautiful in colour. We noticed the green fish mentioned at Gambier Island, and a gold-coloured fish of the same genus, both extremely splendid in their appearance. A dentex resembling our carp, a small rayfish, and some large eels, one of which weighed twenty pounds, were caught in the fresh water. We took forty-four turtles on board for sea stock, besides consuming two a day while we remained in port, weighing each about three hundred-weight.

The weather during our stay was fine, but oppressively warm; and though we had no rain, the atmosphere was generally saturated with moisture. There was a thick fog to windward of the islands almost the whole of the time; but it dispersed on its passage over the land, and the lee side was generally clear.

While our operations at the port were in progress, Lieutenant Belcher circumnavigated Peel's Island in the cutter, and discovered a large bay at the south-

CHAP. VI. June, 1827.

east angle of the island, which afforded very secure anchorage from all winds except the south-east; as this is the prevalent wind during the summer, it is not advisable to anchor there in that season. I named it Fitton Bay, in compliment to Dr. Fitton, late president of the Geological Society. Mr. Elson also was employed outside the harbour, and discovered some sunken rocks to the southward of the entrance to the port, on which account ships should not close the land in that direction, so as to shut in two paps at the north-east angle of Port Lloyd with the south bluff of the harbour. With these objects open there is no danger.

On the 15th of June, we put to sea from Port Lloyd; and finding the wind still from the southward, and that we could not reach the islands in that direction without much loss of time, I bore away to ascertain the northern limit of the group. We ran along the western shore, and at noon on the 16th observed the meridian altitude off the northernmost islet. The group consists of three clusters of islands lying nearly N. by E., and extending from the lat. of 27° 44′ 35″ N. to 26° 30′ N. and beyond, but that was the utmost limit of our view to the southward. The northern cluster consists of small islands and pointed rocks, and has much broken ground about it, which renders caution necessary in approaching it. I distinguished it by the name of Parry's Group, in compliment to the late hydrographer, under whose command I had the pleasure to serve on the northern expedition. The middle cluster consists of three islands, of which Peel's Island, four miles and a fifth in length, is the largest. This group is nine miles and a quarter in length, and is divided by two channels so narrow that they can only be seen when abreast of them. Neither of them are navigable by shipping; the northern, on account of rocks which render it impassable even by boats, and the other on account of rapid tides and eddies, which, as there is no anchoring ground, would, most likely, drift a ship upon the rocks. The northern island I named Stapleton, and the centre Buckland, in compliment to the Professor of Geology at Oxford. At the south-west angle of Buckland Island there is a sandy bay, in which ships will find good anchorage; but they must be careful in bringing up to avoid being carried out of soundings by the current. I named it Walker's Bay, after Mr. Walker of the Hydrographical Office. The southern cluster is evidently that in which a whale ship commanded by Mr. Coffin anchored in 1823, who was the first to communicate its position to this country, and who bestowed his own name upon the port. As the cluster was, however, left without any distinguishing appellation, I named it after Francis Baily, Esq. late President of the Astronomical Society.

These clusters of islands correspond so well with a group named Yslas del Arzobispo in a work published many years ago in Manilla, entitled *Navigaçion Especulativa y Pratica*, that I have retained the name, in addition to that of Bonin Islands; as it is extremely doubtful, from the Japanese accounts of Bonin-sima, whether there are not other islands in the vicinity, to which the latter name is not more applicable. In these accounts, published by M. Klaporth in his Mémoire sur la Chine, and by M. Abel Remusat in the Journal des Savans for September, 1817, it is said, that the islands of Bonin-sima, or Mou-nin-sima, consist of eighty-nine islands; of which two are large,

CHAP. VI. June, 1827.

four are of a middling size, four small, and the remainder of the group consists of rocks. The two large islands are there said to be inhabited, and in the Japanese chart, published in the Journal des Savans, contain several villages and temples. They are stated to be extremely fertile, to produce leguminous vegetables and all kinds of grain, besides a great abundance of pasturage and sugar-canes, and the plains to afford an agreeable retreat to man; that there are lofty palm-trees, cocoa nuts, and other fruits; sandal wood, camphor, and other precious trees.

Setting aside the geographical inaccuracy of the chart, which the Japanese might not know how to avoid, and the disagreement of distances and proportions, their description is so very unlike any thing that we found in these islands, that if the Japanese are at all to be credited they cannot be the same; and if they are not to be believed, it may be doubted whether Bonin-sima is not an imaginary island.

The group which we visited had neither villages, temples, nor any remains whatever; and it was quite evident that they had never been resided upon. There were no cocoa-nut trees, no sugar-canes, no leguminous vegetables, nor any plains for the cultivation of grain, the land being very steep in every part, and overgrown with tall trees. Neither in number, size, or direction will the islands at all coincide; and under such dissimilarities it may reasonably be inquired whether it is possible for these places to be the same. If we compare the number, size, and shape of the islands, or direction of the group, there is a yet wider discrepancy; ports are placed in the Japanese map where none exist in these; rocks are marked to the full number, which seem only to create useless alarm to the navigator; and throughout there is a neglect of the cardinal points. I have therefore, on this ground, presumed to doubt the propriety of the name of Bonin-sima being attached to these islands.

Were the situation of Bonin-sima dependent solely upon the account furnished by Kæmpfer, it might safely be identified with the group of Yslas del Arzobispo; but the recent notice of that island by the Japanese authors is so very explicit, that great doubt upon the subject is thereby created. Kæmpfer's account stands thus:—In 1675 a Japanese junk was driven out of her course by strong winds, and wrecked upon an island three hundred miles to the eastward of Fatsissio. The island abounded in arrack-trees (areca ?) and in enormous crabs (turtle ?), which were from four to six feet in length; and was named Bune-sima, in consequence of its being uninhabited. In this statement the distance, the areca-trees, the turtle, and the island being unoccupied agree very well with the description of the island I have given above; and it is curious that Wittrein, whom we found upon the island, declared he had seen the wreck of a vessel in which the planks were put together in a manner similar to that which was noticed by Lieutenant Wainwright in the junk at Loo Choo.

It is remarkable that this group should have escaped the observation of Gore, Perouse, Krusenstern, and several others, whose vessels passed to the northward and southward of its position. In the journals of the above-mentioned navigators we find that when in the vicinity of these islands they were visited by land-birds; but that they never saw land, the three small islands of Los Volcanos excepted, which may be considered the last of the group. The consequence of its

CHAP. VI. June, 1827.

having thus escaped notice was, that all the islands, except the three last-mentioned, were expunged from the charts; and it was not until 1823 that they reappeared on Arrowsmith's map, on the authority of M. Abel Remusat.

Near these islands we found strong currents, running principally to the northward; but none of them equalled in strength that which is said by the Japanese to exist between Bonin-sima and Fatsisio, which indeed was so rapid that it obtained the name of Kourosi-gawa, or Current of the Black Gulf;* nor did their directions accord, as the kourosi-gawa is said to set from east to west. At particular periods, perhaps, these currents may be greater than we found them, and may also run to the westward, but they are certainly not constant. To the southward of Jesso, Captain Broughton experienced a set in the opposite direction—that is, from west to east, and so did Admiral Krusenstern. With us, as has been mentioned before, the set was to the northward.

June 16th. I had spent as much time in low latitudes, fixing the positions of all these islands, as was consistent with my orders, and it became necessary to make the best of our way to the northward; which we did, in the hope of being more successful in our search for the the land expedition than we were the preceding year. At first we stood well to the eastward, in order to get nearly into the meridian of Petropaulski, that we might not be inconvenienced by easterly winds, which appear to be prevalent in these seas in the summer time; and

* Description d'un Grouppe d'Iles peu connu, par M. Remusat.

having attained our object, directed the course for that port. CHAP. VI. June, 1827.

Our passage between corresponding latitudes was very similar to that of the preceding year. Between the parallels of 30° and 35° we experienced light and variable winds, and in 39° of latitude took a southerly wind, which continued with us nearly all the way. We entered the region of fog nearly in the same latitude as before, and did not lose it until the day before we made the land, when, as before, it was dispersed by strong winds off the coast. The currents were similar to those of the preceding year; but when near the Kurile Islands we were impeded by a strong southerly current from the Sea of Okotsk. About this time we noticed so material a change in the colour of the sea that we were induced to try for soundings, but without gaining the bottom. Captain Clerke off the same place observed a similar change, and also tried for soundings without success. It is probable that the outset from the Sea of Okotsk, the shores of which are flat and muddy, may bring down a quantity of that substance, and occasion the alteration.

As we had very little to interest us in this passage, beyond that which always attends a material change of climate, we watched the birds which flew around us, and found that the tropic birds deserted us in 35° N. The brown albatross and shearwaters fell off in 40° N. In 41° we saw the wandering albatross and black divers; some petrel in 45°; puffins, fulmar petrels, and gannets in 49°, and as we approached Kamschatka, lummes, dovekies, and small tern. About the latitude of 42° we saw many whales, but they did not accompany us far. We observed

CHAP. VI. July, 1827.

driftwood occasionally, but it was not so plentiful as in the preceding year.

On the 2d of July we made the snowy mountains of Kamschatka, but did not reach the Bay of Awatska before the evening of the next day, when, after experiencing the difficulties which almost always attend the entry and egress of the port, we came to an anchor off the town of Petropaulski nearly in the same situation as before.

We found lying in the inner harbour the Okotsk Packet, a brig of 200 tons, commanded by a Russian sub-lieutenant, on the point of sailing with the mail for St. Petersburgh, and availed ourselves of the favourable opportunity of transmitting despatches and private letters by her. I received some official letters which had been too late for the ship the preceding year; but neither in them nor in the Petersburgh Gazette, which finds its way occasionally to Kamschatka, was there intelligence to influence our proceedings, and we consequently began to refit the ship for her northern cruize. While this duty was in progress, we were also employed sounding and surveying the capacious bay and the harbours of Tareinski, Rakovya, and Petropaulski, the plans of those places which had been constructed by Captain King being by no means complete.

Before the ship was at an anchor we received from the governor, Captain Stanitski, a very acceptable present of some new potatoes, fresh butter, curds, and spring water—a mark of attention and politeness for which we were very thankful. On landing I had the pleasure to find all the colony in good health, but a little chagrined to learn the ship was not one of the periodical vessels from St. Petersburgh. As these vessels bring out every kind of supply for the inhabitauts, they are most anxiously looked for; and if they are detained they occasion great inconvenience.

We endeavoured to supply some of the deficiencies of the place by presents of flour, rice, tea, and bottled porter, and three large turtle, with some water-melons. Both the last-mentioned were great curiosities, as they had never been brought to the place before, or indeed seen by any of the inhabitants, except those in the government service. Much curiosity was consequently excited when the turtle were landed; and very few would at first believe such forbidding animals were intended to be eaten. As no person knew how to dress them, I sent my cook on shore, and they were soon converted into an excellent soup, some of which was sent round to each of the respectable inhabitants of the place; but, as may be imagined, after having brought the animals so far, we were mortified at hearing several persons declare their preference for their own dishes made of seals' flesh. These turtle were the last of the supply we had taken on board at Port Lloyd, three having died upon the passage, and the ship's company having continued to consume two every day, which on an average was about five pounds a man. This lasted for about three weeks, during which time we saved half the usual allowance of provisions.

The season at Petropaulski was more backward than the preceding year; and though it was the beginning of July, the snow lay deep upon some parts of the shore, and the inhabitants were glad to keep on their fur dresses.

The little town, which has been repeatedly de-

CHAP. VI. July, 1827.

scribed since King's visit, has been removed from the spit of land which forms the harbour, to a valley at the back of it, where there are several rows of substantial log-houses, comfortably fitted up inside, and warmed with large ovens in the centre, furnished with pipes for the conveyance of hot air. Glass for windows has partly superseded the laminæ of talc, before used for that purpose. Neat wooden bridges have been thrown over the ravines which intersect the town, and a new church has been built. A guard-house and several field-pieces command the landing; and a little to the northward there are magazines for powder and stores. Among other buildings in the town there is a hospital and a school. The yourts and balagans of which Captain King speaks are now only used as store-houses for fish.

The greater part of the houses are furnished with gardens; but being badly attended to, they produce very little. That attached to the government-house was in better order, and was planted with pease, beans, cabbages, lettuces, potatoes, radishes, cucumbers, and a few currant-trees which were blighted; barley and a small quantity of wheat were also growing in its vicinity. Some new houses were erecting in the town in expectation of the arrival of some exiles from St. Petersburgh, as it was understood that several persons concerned in the conspiracy against the emperor were to be banished to this place. The town, upon the whole, was much neater than I expected to find it; and I by no means agree with Captain Cochrane, that it is a contemptible place, and a picture of misery and wretchedness. Considering the number of years it has been colonized, and that it is part of the Russian Empire, it ought certainly to have become of much more importance; but it does not differ so materially from the accounts of it that have been published, as to create disappointment on visiting the place, and it appeared to me that nothing is promised in those accounts which the place itself does not afford.

It was with much pleasure we noticed in the governor's garden the monument of our departed countryman Captain Clerke, which for better preservation had been removed from its former position by the late governor. It was on one side of a broad gravel walk, at the end of an avenue of trees. On the other side of the walk, there was a monument to the memory of the celebrated Beering. The former, it may be recollected, was erected by the officers of Captain Krusenstern's ship; and the latter had been purposely sent from St. Petersburgh. This mark of respect from the Russians toward our departed countryman calls forth our warmest gratitude, and must strengthen the good understanding which exists and is daily increasing between the officers of their service and our own. The monument will ever be regarded as one of the greatest interest, as it marks the places of interment of the companions of the celebrated Cook and Beering, and records the generosity of the much-lamented Perouse, who placed a copper plate over the grave of our departed countryman Captain Clerke; and of the celebrated Admiral Krusenstern, who erected the monument, and affixed a tablet upon it to the memory of the Abbe de la Croyère. Such eminent names, thus combined, create a regret that the materials on which they are engraved are not as imperishable as the memory of the men themselves.

Since Admiral Krusenstern visited Kamschatka,

CHAP. VI. July, 1827.

several alterations have been made, probably in consequence of the suggestions in his publication. The seat of government is now fixed at Petropaulski, the town is considerably improved, and the inhabitants are better supplied than formerly. Still much remains to be accomplished before Petropaulski can be of consequence in any way, except in affording an excellent asylum for vessels. In this respect it is almost unequalled, being very secure, and admirably adapted to the purpose of any vessel requiring repair; but for this she will have to depend entirely upon her own resources, as there is nothing to be had in the country but fish, wood, water, and fresh beef.

The population of the town at the beginning of the winter of 1826 was not more than three hundred and eighty-five persons, exclusive of the government establishment: the occupation of the people consists principally in curing fish and providing for a long winter, during which, with the exception of those persons who go into the interior for furs, there is very little to occupy the inhabitants.

There are no manufactures in the country, nor any establishments which require notice. The inhabitants have an idea that the climate is too cold to produce crops of wheat and other grain, and neglect almost entirely the cultivation of the soil. The consequence of this is, that they occasionally suffer very much from scurvy, and are dependant upon the supplies which are sent from St. Petersburgh every second year for all their farinaceous food; and if these vessels are lost the greatest distress ensues. Many attempts have been made to persuade them to attend to agriculture; rewards have been offered by the government for the finest productions; and seeds are distributed to the people every spring. In the autumn there is a fair, at which those persons who have received seeds are required to attend, and to bring with them specimens of the fruit of their labour. The persons who are most deserving then receive rewards, and the day finishes with a feast and a dance. In spite of these encouragements, the gardens are very little attended to. Hay, though it is got in at the proper season, is in such inadequate proportion to the wants of the cattle, that were it not for wild garlic they would famish before the spring vegetation commences. The flavour that is communicated to the milk and butter by the use of this herbage, appears to be so familiar to the inhabitants that they find nothing unpleasant in it; but it is very much the reverse with strangers. Every family has one or two cows, of which great care is taken during the winter, and, strictly speaking, some of the inhabitants live under the same roof with their animals, with no other partition than a screen of single boards. There are very few oxen in the town, and when required they are driven from Bolcheresk, about ninety miles off, where pasturage is more abundant. Beef is consequently a luxury seldom enjoyed; and sheep and goats cannot exist in the country, in consequence of the savage nature of the dogs, which are very large, and occasionally break away from their fastenings: fish therefore constitutes the principal food of the inhabitants.

Necessarily frugal, and blessed with a salubrious climate, the residents in general enjoy good health, and appear to lead a contented life. They are extremely fond of the amusement of dancing, and frequently meet for this purpose. There are several musicians, and musical instruments are manufactured by

CHAP. VI. July, 1827.

an ingenious exile. As spirituous liquors of any kind in the country are scarce, these meetings are not attended with any inebriety, and serve only to pass away the dull hours of a long winter's evening. The only refreshment we saw produced at them consisted of whortle and cran-berries; these were piled up in two or three plates with a dessert-spoon to each, and passed round the company, almost every body using the same spoon. Society is necessarily very mixed, or there could be none in so small a population, and when strangers are not present it is not unusual to see exiles at the governor's parties.

In the winter sledging is a favourite occupation. The dogs are here very large and swift, and are so much esteemed that they are carried to Okotsk for sale. For a description of this amusement, and other recreations of the Kamschatdales, I must refer the reader to Cook's Voyage, to Captain Cochrane's Pedestrian Journey, and to the entertaining Travels of Mr. Dobell, who quitted Kamschatka a short time before we arrived.

At present the only trade carried on at Petropaulski is in furs, which are exchanged for goods brought annually from Okotsk. Every thing is excessively dear, even the necessary article salt is in great demand, and produces a very high price.

The Bay of Awatska and the harbours which open into it leave nothing to be desired in the way of a port. Awatska has many square miles of ground which may be appropriated to secure anchorage, and Tareinski is the beau ideal of a harbour. Petropaulski, though small, has a sufficient depth of water for a first-rate in every part of it. The ground is good, and the smoothness of the water is never affected by any weather upon the coast. As Awatska is nearly surrounded by high land, gusts of wind are of frequent occurrence, particularly opposite Rakovya harbour: on this account it is advisable to moor or ride with a long scope of cable. The entrance to the port is narrow and about four miles in length, and as the wind almost always blows up or down the channel, ships frequently have to beat in and out, and experience great difficulty in so doing, from the confined space to which they are limited, and the eddy currents, which in the springtime in particular must be carefully guarded against. There are but two shoals in the harbour which it is necessary to notice; one off Rakovya, upon which there is a buoy; and the other off the signal station on the west side of the entrance of Awàtska Bay.

Much has been said of the neglected condition of the settlement, and volumes have been written on the government, inhabitants, productions, and on the actual and prospective state of the country;* still there have been no exertions on the part of the government materially to improve or provide for either one or the other. Its neglected state is probably of very little consequence at present; but should the North Pacific ever be the scene of active naval operations, Petropaulski must doubtless become of immense importance. At present it may be said to be unfortified, but a very few guns judiciously placed would effectually protect the entrance.

On the 18th of July, having completed the survey of the bay of Awatska and its harbours, we took our leave of the hospitable inhabitants, and weighed anchor; but, as on the former occasion, we were obliged

* Cook's Third Voyage, vol. iii.; Perouse's Voyage; Krusenstern's Embassy to Japan; Langsdorff's Travels; Cochrane's Journey; Dobell's Travels, &c.

CHAP. VI. July, 1827.

to make several unsuccessful attempts to get out, and did not accomplish our object until the 20th, when we shaped our course towards Chepoonski Noss. A long swell rolled in upon the shore as we crossed this spacious bay, in the depth of which the port of Awatska is situated, and convinced us of the difficulty that would be experienced in getting clear of the land with a strong wind upon the coast, and of the danger a ship would incur were she, in addition to this, to be caught in a fog, which would prevent her finding the port. Our winds were light from seaward, and we made slow progress, striking soundings occasionally from sixty to seventy fathoms, until the following morning, when we took our departure from the Noss, and entered a thick fog, which enveloped us until we made Beering's Island on the 22d; when it cleared away for the moment, and we distinguished Seal Rock. We had no observation at noon, but by comparing the reckoning with the observations of the preceding and following days, it gave the position of the island the same as before.

We quitted the island with the prospect of a quick passage to the Straits, and, attended by a thick fog, advanced to the northward until the 26th, at which time contrary winds brought us in with the Asiatic coast in the parallel of 61° 58′ N. When we were within a few leagues of the coast the fog cleared away, as it generally does near the land, and discovered to us a hilly country, and a coast apparently broken into deep bays and inlets; but as we did not approach very closely, these might have been only valleys. In this parallel the nearest point of land bearing N. 74° W. true, thirteen miles, the depth of water was 26 fathoms; and it increased gradually as we receded from the coast. The bottom near the shore was a coarse gravel, which, as that in the offing is mud or sand, is a useful distinctive feature. With a northerly wind and a thick fog we stood towards St. Lawrence Island, and on the 1st August were apprised of our approach to it, by the soundings changing from mud to sand, and several visits from the little crested auks, which are peculiar to this island. We made the land about the same place we had done the preceding year, stood along it to the northward, and passed its N. W. extreme, at two miles and a half distance, in fifteen fathoms water, over a bottom of stones and shells, which soon changed again to sand and mud. About midnight the temperature of water fell to 31°, and soon after that of the air was reduced from 42° to 34°. The wind shifted to north-west, and cleared away the fog. On the afternoon of the 2d we passed King's Island, and the wind continuing to the northward, anchored off Point Rodney, for the purpose of hoisting out the barge. We came to anchor in seven fathoms, three miles from the land, King's Island bearing N. 70° 29′ W. true, and Sledge Island S. 65° E. true.

Point Rodney is low, and the water being shallow, it is difficult to land. From the beach to the foot of the mountains there is a plain about two miles wide, covered with lichens and grass, upon which several herds of reindeer were feeding; but the communication is in places interrupted by narrow lakes, which extend several miles along the coast. Upon the beach there was a greater abundance of driftwood than we had noticed on any other part of the coast; some of it was perforated by the terredo, and was covered with small barnacles; but there were several trunks which appeared to have been recently torn up by the roots.

CHAP. VI. Aug. 1827.

Near the spot where we landed were several yourts, and a number of posts driven into the ground, and in the lake we found several artificial ducks, which had been left as decoys: but we saw no natives. About two miles from the coast the country becomes mountainous, and far inland rises to peaked hills of great height, covered with perennial snow.

It was calm throughout the greater part of the day, with very fine weather. The temperature, which increased gradually as we left the snowy coast of Asia, at noon reached to 55°, which was twenty-one degrees higher than it had been on the opposite shore; and the mean for the last twenty-four hours was seven degrees higher than that of the preceding day. Part of this difference was evidently owing to the cessation of the northerly wind and our proximity to the land; but part must also have been occasioned by one coast being naturally colder than the other.

During the time we were at anchor there was a regular ebb and flow of the tide; and there appeared by the shore to be about three or four feet rise of the water. The flood came from the S.E., and ran with greater strength than the ebb, which showed there was a current setting towards Beering's Strait. Captain Cook noticed the same circumstance off this part of the coast.

The equipment of our little tender was always a subject of interest, and preparations for hoisting her out seemed to give the greatest pleasure to all on board. She was again placed under the command of Mr. Elson, who received orders to examine the coast narrowly between our station and Kotzebue Sound, and to search for an opening to the eastward of Cape Prince of Wales, of which the Esquimaux had apprised us the preceding year by their chart upon the sand. Mr. Elson was likewise ordered to look into Schismareff Inlet, and afterwards to meet the ship at Chamisso Island. This little excursion was nearly being frustrated by an accident. In hoisting out the boat the bolt in her keel gave way, in consequence of the copper having corroded the iron of the clench; a circumstance which should be guarded against in coppered boats. Fortunately she was not far off the deck, or the accident might have been of a very serious nature, as her weight was as much as our yards would bear when shored up.

As soon as she was equipped, Mr. Elson proceeded in shore; and a breeze springing up shortly afterwards, the ship weighed, and entered the channel between King's Island and the main. The depth of water from the anchorage off Point Rodney decreased gradually as she proceeded, until nearly mid-channel, when the soundings became very irregular; the alternate casts occasionally varying from nine to six fathoms, and vice versa. As it was blowing fresh at the time, the sudden change of soundings occasioned overfalls; and the channel having been very indifferently explored, it was unpleasant sailing. But although I do not think there is any danger, it would still be advisable in passing through the channel, which is full of ridges, to pay strict attention to the lead, particularly as when Captain Cook passed over the same ground, there was, according to his chart, nothing less than twelve fathoms. The wind increasing, and a thick fog approaching, the course was continued with some anxiety; but finding the same irregularity in the soundings, I hauled out due west to the northward of King's Island, which speedily brought us into

CHAP. VI. Aug. 1827.

twenty-eight fathoms, and showed that there was a bank, tolerably steep at its edge, extending from King's Island to the main. We now resumed our course for the strait; but the fog being very thick we had some difficulty in finding the passage, and were obliged to haul off twice before we succeeded in passing it. In doing this we crossed a narrow channel, with thirty-seven fathoms water, which is deeper soundings than have been hitherto found within a great many miles of the strait. As the depth on each side of the channel is only twenty-four fathoms, it may serve as a guide in future to vessels circumstanced as we were at the mouth of the strait in a thick fog. A little before noon we discerned the Fairway Rock, and passed the straits in confidence before a fresh gale of wind, which had just increased so much as to render our situation very unpleasant.

On the morning of the 5th we passed Cape Espenburg, and in the evening came to an anchor off Chamisso Island, nearly in the same situation we had occupied so long the preceding year. On revisiting this island, curiosity and interest in the fate of our countrymen, of whom we were in search, were our predominant feelings; and a boat was immediately sent to ascertain whether they had been at the island. On her return we learned that no new marks had been discerned upon the rocks; no staff was erected, as had been agreed upon in the event of their arrival; and the billet of wood containing despatches was lying unopened upon the same stone on which it had been placed the preceding year; either of which facts was a conclusive answer to our inquiry.

By some chips of wood which had been recently cut, it appeared that the Esquimaux had not long quitted the island; and on examining the grave of our unfortunate shipmate we found it had been disturbed by the natives, who, disappointed in their search, had again filled in the earth. It would be unfair to impute to these people any malicious intentions from this circumstance, as they must have had every reason to suppose, from their custom of concealing provisions underground, and from having found a cask of our flour buried the preceding year, that they would find a similar treasure, especially as they do not inter their dead. The cask of flour and the box of beads, which had been deposited in the sand, had been unmolested; but a copper coin which we nailed upon a post on the summit of the island was taken away.

The swarms of mosquitos that infested the shore at this time greatly lessened our desire to land. However, some of our sportsmen traversed the island, and succeeded in killing a white hare, weighing nearly twelve pounds, and a few ptarmigan; the hare was getting its summer coat, and the young birds were strong upon the wing.

For several days after our arrival the weather was very thick, with rain and squalls from the south-west, which occasioned some anxiety for the barge; but on the 11th she joined us, and I learned from Mr. Elson that he had succeeded in finding the inlet, and that as far as he could judge, the weather being very foggy and boisterous, it was a spacious and excellent port. He was visited by several of the natives while there, one of whom drew him a chart, which corresponded with that constructed upon the sand in Kotzebue Sound the preceding year. On his putting to sea from the inlet, the weather continued very thick, so much so that he passed through Beering's Strait without seeing land; and was unable to explore Schismareff Inlet.

CHAP. VI. Aug. 1827.

The discovery of a port so near to Beering's Strait, and one in which it was probable the ship might remain after circumstances should oblige her to quit Kotzebue Sound, was of great importance; and I determined to take an early opportunity of examining it, should the situation of the ice to the northward afford no prospect of our proceeding further than we had done the preceding year. In order that Captain Franklin's party might not be inconvenienced by such an arrangement, the barge was fitted, and placed under the command of Lieutenant Belcher, who was ordered to proceed along the coast as in the preceding year, and to use his best endeavours to communicate with the party under Captain Franklin's command, by penetrating to the eastward as far as he could go with safety to the boat; but he was on no account to risk being beset in the ice; and in the event of separation from the ship, he was not to protract his absence from Kotzebue Sound beyond the 1st of September. He was also to examine the shoals off Icy Cape and Cape Krusenstern, and to explore the bay to the northward of Point Hope.

Having made these arrangements, we endeavoured to put to sea, but calms and fogs detained us at Chamisso until the 14th, and it was the 16th before we reached the entrance of the sound. The barge, however, got out, and the weather afterwards being very foggy, we did not rejoin for some time. Before we left the island we were visited by several natives whom we remembered to have seen the preceding year. They brought some skins for sale, as usual, but did not find so ready a market for them as on the former occasion, in consequence of the greater part of the furs which had been purchased by the seamen at that time, having rotted and become offensive on their return to warm latitudes. Our visiters were, as before, dirty, noisy, and impudent. One of them, finding he was not permitted to carry off some deep-sea leads that were lying about, scraped off the greasy arming and devoured it: another, after bargaining some skins for the armourer's anvil, unconcernedly seized it for the purpose of carrying it away; but, much to his surprise, and to the great diversion of the sailors who had played him the trick, he found its weight much too great for him, and after a good laugh received back his goods. A third amused the young gentlemen very much by his humorous behaviour. He was a shrewd, observing, merry fellow. For some time he stood eying the officers walking the deck, and at length appeared determined to turn them into ridicule; seizing therefore a young midshipman by the hand, he strutted with him up and down the deck in a most ludicrous manner, to the great entertainment of all present. They quitted us late at night, but renewed their visit at three in the morning, and seemed surprised to find us washing the decks. They probably expected that we should be fast asleep, and that they would have an opportunity of appropriating to themselves some of the moveable articles upon deck. There was otherwise no reason for returning so soon; and from what we afterwards saw of these people, there is every reason to believe that was their real motive.

Off the entrance of Kotzebue Sound we were met by a westerly wind, which prevented our making much progress; but on the 18th the breeze veered to the south-westward, with a thick fog, and as I had not seen any thing of the barge, I steered to the north-

CHAP. VI. Aug. 1827.

ward to ascertain the position of the ice. At noon Cape Thomson was seen N. 46° E. (true) three leagues distant, but was immediately obscured again by fog. At midnight the temperature of both air and sea fell from 43° to 39°, and rose again soon afterwards to 44°, occasioned probably by some patches of ice; but the weather was so thick that we could see only a very short distance around us. We continued to stand to the north-west, with very thick and rainy weather, until half past one o'clock in the afternoon, when I hauled to the wind, in consequence of the temperature of the water having cooled down to 35°, and the weather being still very thick. In half an hour afterwards we heard the ice to leeward, and had but just room to get about to clear a small berg at its edge. Our latitude at this time was 70° 01′ N., and longitude 168° 50′ W., or about 160 miles to the westward of Icy Cape. The soundings in the last twelve hours had been very variable, increasing at one time to thirty fathoms, then shoaling to twenty-four, and deepening again to thirty-two fathoms, *muddy* bottom; an hour after this we shoaled to twenty one fathoms, *stones*, and at the edge of the ice to nineteen fathoms, *stones*. The body of ice lying to the northward prevented our pursuing this shallow water, to ascertain whether it decreased so as to become dangerous to navigation.

Shortly after we tacked, the wind fell very light, and changed to west. We could hear the ice plainly; but the fog was so thick that we could not see thirty yards distance; and as we appeared to be in a bay, to avoid being beset, we stood out by the way which we had entered. At nine o'clock the fog cleared off, and we returned toward the ice. At midnight, being close to

CHAP. VI. Aug. 1827.

its edge, we found it in a compact body, extending from W. to N.E. and trending N. 68° E. true. As the weather was unsettled, I stood off until four o'clock, and then tacked, and at eight again saw the ice a few miles to the south-eastward of our position the day before. We ran along its edge, and at noon observed the latitude in 70° 06. N.

Occasional thick weather and snow showers obliged us to keep at a greater distance from the pack, and we lost sight of it for several hours; but finding by the increase of the temperature of the water that our course led us too much from it, at nine o'clock I steered N.N.E. true, and at midnight was again close upon it. The ice was compact as before, except near the edge, and extended from W.S.W. to N.N.E. mag. trending N. 56° E. true. We now followed its course closely to the eastward, and found it gradually turning to the southward. At three o'clock the wind veered to south-west, with snow-showers and thick weather; and as this brought us upon a lee shore, I immediately hauled off the ice, and carried a press of sail to endeavour to weather Icy Cape. The edge of the packed ice at this time was in latitude 70° 47′ N. trending south-eastward, and gradually approaching the land to the eastward of Icy Cape. By the information of Lieutenant Belcher, who was off the Cape at this time, though not within sight of the ship, it closed the land about twenty-seven miles east of Icy Cape. The passage that was left between it and the beach was extremely narrow; and judging from the effect of the westerly winds off Refuge Inlet the preceding year, it must soon have been closed up, as those winds blew with great strength about the time we hauled off.

From this it appears that the line of packed ice, in

CHAP. VI. Aug. 1827.

the meridian of Icy Cape, was twenty-four miles to the southward of its position the preceding year, and that it was on the whole much nearer the continent of America. With the ice thus pressing upon the American coast, and with the prevalence of westerly winds, by which this season was distinguished, there would have been very little prospect of a vessel bent upon effecting the passage succeeding even in reaching Point Barrow.

The wind continuing to blow from the S.W., with thick weather and showers of snow, we endeavoured to get an offing, and at ten o'clock tacked a mile off the land near Icy Cape. In the afternoon we stood again to the southward, and the next day fetched into the bay near Cape Beaufort, and at night hove to off Cape Lisburn with thick and cold weather. The next morning, being moderate, afforded us the only opportunity we had hitherto had of depositing some information for Captain Franklin's party. The boat landed near the Cape, and buried one bottle for him and another for Lieutenant Belcher, whom we had not seen since we parted at Chamisso Island. In the evening we stretched toward Point Hope, for the purpose of depositing a bottle there also, as it was a point which could not escape Captain Franklin's observation in his route along shore; but the wind increasing from the westward occasioned a heavy surf upon the beach, and obliged the ship to keep in the offing.

Seeing that we could not remain sufficiently close in shore to be of use to our friends during the westerly winds and thick weather, I determined upon the examination of the inlet discovered by Mr. Elson to the eastward of Cape Prince of Wales, and made sail for Kotzebue Sound, for the purpose of leaving there the necessary information for Captain Franklin and Lieutenant Belcher, in the event of either arriving during our absence.

We passed Cape Krusenstern about sunset on the 25th; and in running along shore after dark our attention was directed to a large fire, kindled as if for the purpose of attracting our notice. As this was the signal agreed upon between Captain Franklin and myself, and as we had not before seen a fire in the night on any part of the coast, we immediately brought to, and, to our great satisfaction at the moment, observed a boat pulling towards the ship. Our anxiety at her approach may be imagined, when we thought we could discover with our telescopes, by the light of the aurora borealis, that she was propelled by oars instead of by paddles. But just as our expectation was at the highest, we were accosted by the Esquimaux in their usual manner, and all our hopes vanished. I fired a gun, however, in case there might be any persons on shore who could not come off to us; but the signal not being answered, we pursued our course for Chamisso.

For the first time since we entered Beering's Strait the night was clear, and the aurora borealis sweeping across the heavens reminded us that it was exactly on that night twelvemonth that we saw this beautiful phenomenon for the first time in these seas. A short time before it began, a brilliant meteor fell in the western quarter. The aurora is at all times an object of interest, and seldom appears without some display worthy of admiration, though the expectation is seldom completely gratified. The uncertainty of its movements, and of the moment when it may break out into splendour, has, however, the effect of keeping the attention continually on the alert; many of us in consequence staid up to a late hour, but nothing was exhibited on this occasion more than we had already repeatedly witnessed.

We were more fortunate the following night, when the aurora approached nearer the southern horizon than it had done on any former occasion that we had observed in this part of the globe. It commenced much in the usual manner, by forming an arch from W.N.W. to E.N.E., and then soared rapidly to the zenith, where the streams of light rolled into each other, and exhibited brilliant colours of purple, pink, and green. It then became diffused over the sky generally, leaving about 8° of clear space between it and the northern and southern horizons. From this tranquil state it again suddenly poured out corruscations from all parts, which shot up to the zenith, and formed a splendid cone of rays, blending pink, purple, and green colours in all their varieties. This singular and beautiful exhibition lasted only a few minutes, when the light as before became diffused over the sky in a bright haze.

We anchored at Chamisso on the 26th, and, after depositing the necessary information on shore, weighed the next morning to proceed to examine the inlet. We were scarcely a league from the land when our attention was again arrested by a fire kindled upon the Peninsula, and eight or ten persons standing upon the heights waving to the ship. The disappointment of the preceding night ought certainly to have put us upon our guard; but the desire of meeting our countrymen induced us to transform every object capable of misconstruction into something favourable to our wishes, and our expectations on this occasion carried us so far that some imagined they could perceive the party to be dressed in European clothes. A boat was immediately despatched to the shore; but, as the reader has already begun to suspect, it was a party of Esquimaux, who wished to dispose of some skins for tobacco.

This disappointment lost us a favourable tide, and we did not clear the sound before the night of the 29th. After passing Cape Espenburg, a strong north-west wind made it necessary to stand off shore, in doing which the water shoaled from thirteen to nine fathoms upon a bank lying off Schismareff Inlet, and again deepened to thirteen: we then bore away for the strait, and at eleven o'clock saw the Diomede Islands, thirteen leagues distant; and about four o'clock rounded Cape Prince of Wales very close, in twenty-seven fathoms water.

This celebrated promontory is the western termination of a peaked mountain, which, being connected with the main by low ground, at a distance has the appearance of being isolated. The promontory is bold, and remarkable by a number of ragged points and large fragments of rock lying upon the ridge which connects the cape with the peak. About a mile to the northward of the cape, some low land begins to project from the foot of the mountain, taking first a northerly and then a north-easterly direction to Schismareff Inlet. Off this point we afterwards found a dangerous shoal, upon which the sea broke heavily. The natives have a village upon the low land near the cape called Eidannoo, and another inland, named King-a-ghe; and as they generally select the mouths of rivers for their residences, it is not improbable that a stream may here empty itself into the sea, which,

CHAP. VI. Aug. 1827.

meeting the current through the strait, may occasion the shoal. About fourteen miles inland from Eidannoo, there is a remarkable conical hill, often visible when the mountain-tops are covered, which, being well fixed, will be found useful at such times by ships passing through the strait. Twelve miles further inland, the country becomes mountainous, and is remarkable for its sharp ridges. The altitude of one of the peaks, which is nearly the highest on the range, is 2596 feet. These mountains, being thickly covered with snow, gave the country a very wintry aspect.

To the southward of Cape Prince of Wales the coast trends nearly due east, and assumes a totally different character to that which leads to Schismareff Inlet, being bounded by steep rocky cliffs, and broken by deep valleys, while the other is low and swampy ground. The river called by the natives Youp-nut must lie in one of these valleys; and in all probability it is in that which opens out near a bold promontory, to which I have given the name of York, in honour of his late Royal Highness. On nearing that part of the coast we found the water more shallow than usual.

Having passed the night off Cape York on the 31st, we steered to the eastward, and shortly discovered a low spit of land projecting about ten miles from the coast, which here forms a right angle, and having a channel about two miles wide between its extremity and the northern shore. We sailed through this opening, and entered a spacious harbour, capable of holding a great many ships of the line. We landed first on the low spit at the entrance, and then stood across, nine miles to the eastward, and came to an anchor off a bold cape, having carried nothing less than five and a half fathoms water the whole of the way.

The following morning, Sept. 1st, we stood toward an opening at the north-east angle of the harbour; but finding the water get gradually shallow, came again to anchor. On examination with the boats, we found, as we expected, an inner harbour, ten miles in length by two and a quarter in width, with almost an uniform depth of two and a half and three fathoms water. The channel into it from the outer harbour is extremely narrow, the entrance being contracted by two sandy spits; but the water is deep, and in one part there is not less than twelve fathoms. At the upper end of the harbour a second strait, about three hundred yards in width, was formed between steep cliffs; but this channel was also contracted by sandy points. The current ran strong through the channel, and brought down a great body of water, nearly fresh (1.0096 sp. gr.). The boats had not time to pursue this strait; but in all probability it communicates with a large inland lake, as described by the natives in Kotzebue Sound. At the entrance of the strait, called Tokshook by the natives, there is an Esquimaux village, and upon the northern and eastern shores of the harbour there are two others: the population of the whole amounted to abour four hundred persons. They closely resembled the natives we had seen before, except that they were better provided with clothing, and their implements were neater and more ingeniously made. Among their peltry we noticed several gray fox and land-otter skins, but they would not part with them for less than a hatchet apiece. In addition to the usual weapons of bows and arrows, these people had short iron spears neatly inlaid with brass, upon all which implements they set great value, and kept them wrapped in skins. Among the inhabitants of the village on the northern

CHAP. VI. Sept. 1827.

shore, named Choonowuck, there were several girls with massive iron bracelets. One had a curb chain for a necklace, and another a bell suspended in front, in the manner described the preceding year at Choris Peninsula.

There are very few natives in the outer harbour. On the northern side there is a village of yourts, to which the inhabitants apparently resort only in the winter. At the time of our visit it was in charge of an old man, his wife, and daughter, who received us civilly, and gave us some fish. The yourts were in a very ruinous condition: some were half filled with water, and all were filthy. By several articles and cooking utensils left upon the shelves, and by some sledges which were secreted in the bushes, the inhabitants evidently intended to return as soon as the frost should consolidate all the stagnant water within and about their dwellings. One of these yourts was so capacious that it could only have been intended as an assembly or banquetting room, and corresponded with the description of similar rooms among the eastern Esquimaux.

There was a burying-ground near the village in which we noticed several bodies wrapped in skins, and deposited upon drift-wood, with frames of canoes, and sledges, &c. placed near them, as already described at the entrance of Hotham Inlet. The old man whom we found at this place gave the same names to the villages at the head of the inner harbour, and to the points of land at its entrance, as we had received from the natives of King-a-ghe whom we met in Kotzebue Sound.

His daughter had the hammer of a musket suspended about her neck, and held it so sacred that she would scarcely submit it to examination, and afterwards carefully concealed it within her dress. She was apparently very modest and bashful, and behaved with so much propriety that it was a pleasure to find such sentiments existing beneath so uncouth an exterior.

Upon the low point at the entrance of the inner harbour, called Nooke by the natives, there were some Esquimaux fishermen, who reminded us of a former acquaintance at Chamisso Island, and saluted us so warmly that we felt sorry their recollection had not entirely failed them. They appeared to have established themselves upon the point for the purpose of catching and drying fish; and from the number of salmon that were leaping in the channel, we should have thought they would have been more successful. They had, however, been fortunate in taking plenty of cod, and some species of salmon trout: they had also caught some herrings.

We were also recognised by a party from the southern shores of the harbour, who, the preceding year, had extended their fishing excursions from this place to Kotzebue Sound. These were some of the most cleanly and well-dressed people we had seen any where on the coast. Their residence was at King-a-gne—a place which, judging from the respectability of its inhabitants, whom we had seen elsewhere, must be of importance among the Esquimaux establishments upon this coast.

These two ports, situated so near Beering's Strait, may at some future time be of great importance to navigation, as they will be found particularly useful by vessels which may not wish to pass the strait in bad weather. To the outer harbour, which for convenience and security surpasses any other near Beering's

CHAP. VI. Sept. 1827.

Strait with which we are acquainted, I attached the name of Port Clarence, in honour of his most gracious Majesty, then Duke of Clarence. To the inner, which is well adapted to the purposes of repair. and is sufficiently deep to receive a frigate. provided she lands her guns, which can be done conveniently upon the sandy point at the entrance, I gave the name of Grantley Harbour, in compliment to Lord Grantley. To the points at the entrance of Port Clarence I attached the names of Spencer and Jackson, in compliment to the Honourable Captain Robert Spencer and Captain Samuel Jackson, C. B., two distinguished officers in the naval service: to the latter of whom I am indebted for my earliest connexion with the voyages of Northern Discovery.

The northern and eastern shores of Port Clarence slope from the mountains to the sea, and are occasionally terminated by cliffs composed of fine and talcy mica slate, intersected by veins of calcareous spar of a pearly lustre, mixed with grey quartz. The soil is covered with a thick coating of moss, among which there is a very limited flora: the valleys and hollows are filled with dwarf willow and birch. The country is swampy and full of ruts; and vegetation on the whole, even on the north side of the harbour, which had a southern aspect, was more backward than in Kotzebue Sound; still we found here three species of plants we had not seen before. Plants that were going to seed when we left that island were here only just in full flower, and berrics that were there over ripe were here scarcely fit to be eaten. On the northern side of Grantley Harbour, Mr. Collie found a bed of purple *primulas*, *anemones*, and of *dodecatheons*, in full and fresh blossom, amidst a covering of snow that had fallen the preceding night.

The southern side of Port Clarence is a low diluvial formation, covered with grass, and intersected by narrow channels and lakes; it projects from a range of cliffs which appear to have been once upon the coast, and sweeping round, terminates in a low shingly point (Point Spencer). In one place this point is so narrow and low, that in a heavy gale of wind, the sea must almost inundate it; to the northward, however, it becomes wider and higher, and, by the remains of some yourts upon it, has at one time been the residence of Esquimaux. Like the land just described, it is intersected with lakes, some of which rise and fall with the tide, and is covered, though scantily, with a coarse grass, *elymus*, among which we found a species of artemesia, probably new. Near Point Spencer the beach has been forced up by some extraordinary pressure into ridges, of which the outer one, ten or twelve feet above the sea, is the highest. Upon and about these ridges there is a great quanty of drift timber, but more on the inner side of the point than the outer. Some has been deposited upon the point before the ridges of sand were formed, and is now mouldering away with the effect of time, while other logs are less decayed, and that which is lodged on the outer part is in good preservation, and serves the natives for bows and fishing staves.

We saw several reindeer upon the hilly ground; in the lakes, wild ducks: and upon the low point of the inner harbour, golden plover, and sanderlings, and a gull very much resembling the larus sabini.

The survey of these capacious harbours occupied us until the 5th, when we had completed nearly all that was necessary, and the weather set in with such seventy that I was anxious to get back to Kotzebue

CHAP. VI. Sept. 1827.

Sound. For the three preceding days the weather had been cold, with heavy falls of snow; and the seamen, the boat's crews in particular, suffered from their exposure to it, and from the harassing duty which was indispensable from the expeditious execution of the survey. On this day, the 5th, the thermometer stood at $25\frac{1}{2}^{\circ}$, and the lakes on shore were frozen. We accordingly weighed, but not being able to get out, passed a sharp frosty night in the entrance; and next morning, favoured with an easterly wind, weighed and steered for the strait. As we receded from Point Spencer, the difficulty of distinguishing it even at a short distance accounted for this excellent port having been overlooked by Cook, who anchored within a very few miles of its entrance.

As we neared Beering's Strait the wind increased, and on rounding Cape Prince of Wales, obliged us to reduce our sails to the close reef. On leaving the port the wind had been from the eastward, but it now drew to the northward, and compelled us to carry sail, in order to weather the Diomede Islands. Whilst we were thus pressed, John Dray, one of the seamen, unfortunately fell overboard from the look-out at the masthead, and sunk alongside a boat which was sent to him, after having had his arms round two of the oars. This was the only accident of the kind that had occurred since the sh p had been in commission, and it was particularly unfortunate that it should have fallen to the lot of so good a man as Dray. Previous to his entry in the ship he resided some time at the Marquesas Islands, and was so well satisfied with the behaviour of the natives of that place that he purposed living amongst them; but being on board a boat belonging to Baron Wrangel's ship, at a time when the islanders made a most unjustifiable attack upon her, he was afraid to return to the shore, and accompanied the Baron to Petropaulski, where I received him and another seaman, similarly circumstanced, into the ship.

Toward night the wind increased to a gale, and split almost every sail that was spread; the weather was dark and thick, with heavy falls of snow; and suspecting there might be a current setting through the strait, we anxiously looked out for the Diomede Islands, which were to leeward, and we were not a little surprised to find, on the weather clearing up shortly after daylight the following morning, that there had been a current running nearly against the wind, at the rate of upwards of a mile an hour, in a N. 41° W. direction.

From the time we quitted Port Clarence the temperature began to rise, and this morning stood four degrees above the freezing point. Change of locality was the only apparent cause for this increase, and it is very probable that the vicinity of the mountains to Port Clarence is the cause of the temperature of that place being lower than it is at sea.

In the morning we saw a great many walrusses and whales, and observed large flocks of ducks migrating to the southward. The coast on both sides was covered with snow, and every thing looked wintry. The wind about this time changed to N.W., and by the evening carried us off the entrance of Kotzebue Sound, when we encountered, as usual, an easterly wind, and beat up all night with thick misty weather.

In our run to this place we again passed over a shoal, with eight and a half and nine fathoms water upon it off Schismareff inlet. After beating all night in very thick weather, on the 9th of September we

CHAP. VI. Sept. 1827.

stood in for the northern shore of the sound, expecting to make the land well to windward of Cape Blossom, where the soundings decrease so gradually that a due attention to the lead is the only precaution necessary to prevent running on shore; but there had unfortunately been a strong current during the night, which had drifted the ship towards Hotham Inlet, where the water shoaling suddenly from five fathoms to two and a half, the ship struck upon the sand while in the act of going about; and soon became fixed by the current running over the shoal. In consequence of this current our small boats experienced the utmost difficulty in carrying out an anchor, but they at length succeeded, though to no purpose, as the ship was immoveable. Looking to the possible result of this catastrophe, we congratulated ourselves on having the barge at hand to convey the crew to Kamschatka, little suspecting, from an accident which had already befallen her, in what a helpless condition each party was at that moment placed. Fortunately we were not reduced to the necessity of abandoning the ship, which appearances at one time led us to apprehend, as the wind moderated shortly after she struck, and on the rising of the next tide she went off without having received any apparent injury.

CHAPTER VII.

Arrive at Chamisso Island—Find the Barge wrecked—Lieutenant Belcher's Proceedings—Conduct of the natives—Approach of Winter—Final Departure from the Polar Sea — Observations upon the probability of the North-West Passage from the Pacific—Remarks upon the Tribe inhabiting the North-West Coast of America—Return to California—Touch at San Blas, Valparaiso, Coquimbo, Rio Janeiro—Conclusion.

Sept. 1827.

AFTER having so narrowly escaped shipwreck, we beat up all night with thick weather, and the next morning steered for Chamisso Island. As we approached the anchorage we were greatly disappointed at not seeing the barge at anchor, as her time had expired several days, and her provisions were too nearly expended for her to remain at sea with safety to her crew; but on scrutinizing the shore with our telescopes, we discovered a flag flying upon the south-west point of Choris Peninsula, and two men waving a piece of white cloth to attract attention. Amidst the sensations of hope and fear, a doubt immediately arose, whether the people we saw were the long looked for land expedition, or the crew of our boat, who had been unfortunate amongst the ice, or upon the coast, in the late boisterous weather. The possibility of its being the party under Captain Franklin arrived in safety, after having accomplished its glorious undertaking, was the first, because the most ardent, wish of our sanguine minds; but this was soon contradicted

CHAP. VII. Sept. 1827.

by a nearer view of the flag, which was clearly distinguished to be the ensign of our own boat, hoisted with the union downwards, indicative of distress. The boats were immediately sent to the relief of the sufferers, with provisions and blankets, concluding, as we saw only part of the crew stirring about, and others lying down within a small fence erected round the flag-staff, that they were ill, or had received hurts.

On the return of the first boat our conjectures as to the fate of the barge were confirmed; but with this difference, that instead of having been lost upon the coast to the northward, she had met her fate in Kotzebue Sound, and we had the mortification to find that three of the crew had perished with her. Thus, at the very time that we were consoling ourselves, in the event of our misfortunes of the preceding day terminating disastrously, that we should receive relief from our boat, her crew were anticipating assistance from us.

From the report of Lieutenant Belcher, who commanded the barge, it appears that after quitting Chamisso Island on the 12th ultimo, he proceeded along the northern shore of the Sound, and landed upon Cape Krusenstern, where he waited a short time, and not seeing the ship, the weather being very thick, he stood on for Cape Thomson, where he came to an anchor, and replenished his stock of water. He met some natives on shore who informed him that the ship had passed to the northward (which was not true), and he therefore pursued his course ; but finding the weather thick, and the wind blowing strong from the S.E., he brought to under the lee of Point Hope, and examined the bay formed between it and Cape Lisburn, where he discovered a small cove, which afforded him a convenient anchorage in two fathoms, muddy bottom. This cove, which I have named after his relation, Captain Marryat, R.N. is the estuary of a river, which has no doubt contributed to throw up the point.

After Lieutenant Belcher had constructed a plan of the cove, he proceeded to Cape Lisburn ; the weather still thick, and the wind blowing at S.W. He nevertheless effected a landing upon the north side of the Cape, and observed its latitude to be 68° 52′ 3″ N., and the variation to be 32° 23′ E. From thence he kept close along the shore, for the purpose of falling in with the land expedition, and arrived off Icy Cape on the 19th, when he landed and examined every place in the hope of discovering some traces of Captain Franklin. He found about twenty natives on the point living in tents, who received him very civilly, and assisted him to fill his water casks from a small well they had dug in the sand for their own use. The yourts, which render this point remarkable at a distance, were partly filled with water, and partly with winter store of blubber and oil.

From Icy Cape he stood E.N.E. ten miles, and then N.E. twenty-seven, at which time, in consequence of the weather continuing thick and the wind beginning to blow hard from the south-west, he hauled off shore, and shortly fell in with the main body of ice, which arrested his course and obliged him to put about. It blew so strong during the night that the boat could only show her close-reefed mainsail and storm-jib, under which she plied, in order to avoid the ice on one side and a lee-shore on the other: the boat thus pressed leaked considerably, and kept the crew at the pumps.

On the 21st August, the weather being more moderate, he again made the ice, and after keeping along

CHAP. VII. Sept. 1827.

it some time, returned to Icy Cape, and found that the edge of the packed ice was in latitude 70° 41 N. in a N.N.W. direction from the cape, extending east and west (true).

On the 23d August another landing was made upon Icy Cape, and its latitude, by artificial horizon, ascertained to be 70° 19′ 28″ N., and variation by Kater's compass 32° 49′ E. Lieutenant Belcher's curiosity was here greatly awakened by one of the natives leading him to a large room used by the Esquimaux for dancing, and by searching for a billet of wood, which his gestures implied had been left by some Europeans, but not finding it, he scrutinized several chips which were in the apartment, and intimated that some person had cut it up. This was very provoking, as Lieutenant Belcher naturally recurred to the possibility of Captain Franklin having been there, and after leaving this billet as a memorial, having returned by the same route. Nothing, however, was found, and Lieutenant Belcher, after depositing a notice of his having been there, embarked and passed the night off the Cape in heavy falls of snow, hail, and sleet. The next day he again fell in with ice in latitude 70° 40′ N. which determined him to stand back to the cape and examine the shoals upon which the ship lost her anchor the preceding year.

On the 26th, the ice was again found in 70° 41′ N., and the next day was traced to the E.S.E. to within five or six miles of the land, and at the distance of about twenty miles to the eastward of Icy Cape. The ice appeared to be on its passage to the southward, and the bergs were large and scattered. Under these circumstances, Lieutenant Belcher, to avoid being beset, stood back to the cape, and had some difficulty in maintaining his station off there, in consequence of the severity of the weather, which cased his sails, and the clothes of the seamen exposed to the spray, with ice.

Three of his crew at this time became invalids with chilblains and ulcers occasioned by the cold: and the necessity of carrying a press of sail strained the boat to such a degree that she again leaked so fast as to require the pumps to be kept constantly at work. It became necessary, therefore, to seek shelter, and he bore up for Point Hope; but before he reached that place the sea broke twice over the stern of the boat, and nearly swamped her. Upon landing at the point he was met by the natives, who were beginning to prepare their yourts for the winter. His crew here dried their clothes for the first time for several days, and Lieutenant Belcher having obtained the latitude, again put to sea; but finding the weather still so bad that he could not keep the coast with safety, and the period of his rendezvous at Chamisso Island having arrived, he pursued his course for that place, where he found the instructions I had left for him before I proceeded to examine Port Clarence.

Among other things he was desired to collect a quantity of drift-timber, and to erect an observatory upon Choris Peninsula; in which he was engaged, when the wind coming suddenly in upon the shore where the barge was anchored, the crew were immediately ordered on board. It unfortunately happened that the weather was so fine in the morning that only two persons were left in the vessel, and the boat belonging to the barge being small could take only four at a time. One boat-load had joined the vessel, but the surf rose so suddenly, that in the attempt to reach her a second time, the oars were broken, and the boat

CHAP. VII. Sept. 1827.

was thrown back by the sea, and rendered nearly useless. Several persevering and unsuccessful efforts were afterwards made to communicate with the vessel, which being anchored in shallow water struck hard upon the ground, and soon filled. Some Esquimaux happened to have a baidar near the spot, and Mr. Belcher compelled them to assist him in reaching the barge; but the sea ran too high, and the natives not being willing to exert themselves, the attempt again failed. The sea was now making a breach over the vessel, and Mr. Belcher desired the cockswain to cut the cable, and allow her to come broadside upon the shore; but whether through fear, or that the cockswain did not understand his orders, it was not done. There were four men and a boy on board at this time, two of whom, finding no hope of relief from the shore, jumped overboard, with spars in their hands, and attempted to gain the beach, but were unfortunately drowned. The others retreated to the rigging; among them was a boy, whose cries were for some time heard on shore, but at length, exhausted with cold and fatigue, he fell from the rigging, and was never seen again.

The party of Esquimaux, who had so reluctantly rendered their personal assistance, beheld this loss of lives with the greatest composure, giving no other aid than that of their prayers and superstitious ceremonies; and seeing the helpless condition of those thrown upon the shore, began to pilfer every thing they could, bringing the party some fish occasionally, not from charitable motives, but for the purpose of engaging their attention, and of affording themselves a better opportunity of purloining the many articles belonging to the boat which were washed ashore. About eleven o clock at night the sea began to subside, and at midnight, after very great exertions, a communication with the vessel was effected, and the two remaining seamen were carried on shore, and laid before the fire, where they recovered sufficiently to be taken to a hut near the fatal scene.

The morning after this unfortunate occurrence, part of the crew were employed collecting what was washed on shore, and preventing the natives committing further depredations. Seeing there was no chance of obtaining any thing more of consequence from the wreck, the party took up its quarters on Point Garnet, where we found them on our return from Port Clarence. Previous to this, several Esquimaux had pitched a tent in the bay close to the party, and lost no opportunity of appropriating to themselves whatever they could surreptitiously obtain. Among these were four persons whom Mr. Belcher had a short time before assisted, when their baidar was thrown on shore, and one of the party drowned. These people did not forget his kindness, and brought him fish occasionally, but they could not resist the temptation of joining their companions in plunder when it was to be had. Mr. Belcher seeing several articles amongst them which must have accompanied others in their possession, searched their bags, and recovered the boat's ensign, and many other things. No opposition was offered to this examination, but, on the contrary, some of the party which had been saved from the wreck of the baidar, intimated to Mr. Belcher that a man who was making off with a bag had part of his property; and on searching him, a quantity of the boat's iron and the lock of a fowling-piece were discovered upon him.

Upon the whole, however, the natives behaved

CHAP. VII. Sept. 1827.

better than was expected, until the day on which the ship arrived. This appears to have been a timely occurrence; for early that morning two baidars landed near the wreck, and the Esquimaux party was increased to twenty-four. The man who had been searched the preceding evening, finding his friends so numerous, and being joined by another troublesome character, came towards our people, flourishing his knife, apparently with the determination of being revenged. It fortunately happened that there was a person of authority amongst the number, with whom Mr. Belcher effected a friendship. He expostulated with the two refractory men, and one of them went quietly away, but the other remained brandishing his weapon; and there is but too much reason to believe that had he commenced an attack, he would have been seconded by his countrymen, notwithstanding the interference of the chief.

When the ship's boat came to the relief of our party, Mr. Belcher ordered the man who had been so refractory to be bound and taken on board the ship, intimating to the others that he should be kept until more of the stolen property was returned. This they appeared perfectly to understand, as the prisoner pointed to his boat, where, upon search being made, the other lock of the fowling-piece, and a haversack belonging to Lieutenant Belcher, were found. The strength of this man was so great, that it required as many of our people as could stand round to pinion his arms and take him down to the boat. As soon as this was effected, all the other Esquimaux fled to their baidars, and did not approach the place again; the chief excepted, who returned almost immediately, and pitched one tent for himself, and another for the prisoner. Lieutenant Belcher, in concluding his account of this disastrous affair, speaks in high terms of commendatian of the conduct of Mr. (now Lieutenant) Rendall, William Aldridge, and George Shields, seamen, and of Thomas Hazlehurst, marine; and it is with much pleasure I embrace the opportunity of giving publicity to their meritorious behaviour.

I must exonerate Lieutenant Belcher from any blame that may attach to him as commander of the vessel; for though her loss was evidently occasioned by her being too close in shore, and by too few a number of persons being left on board, yet it is to be observed, that she was only a boat; that the crew were upon the beach in readiness to assist her; and that had it been a case of ordinary nature, they would no doubt have succeeded in their object. In place of this, however, the wind changed suddenly, and the sea rose so fast that there was no possibility of effecting what, under general circumstances, would have been perfectly practicable: the water, besides, was two feet lower than usual. The strenuous exertions of Lieutenant Belcher to save the crew, and his resolute conduct toward the natives, after he was thrown amongst them unprovided with arms, a brace of pistols excepted, show him to be an officer both of humanity and courage.

After the loss of our favourite boat, parties were repeatedly sent to the wreck, in the hope of being able to raise her, or to procure what they could from her cabin and holds; but she was completely wrecked and filled with sand, and a few days afterwards went to pieces. Mr. Belcher was a great loser by this unfortunate accident, as he was well provided with instruments, books, papers, &c., and had some expensive fowling-pieces and pistols, all of which were lost or

CHAP. VII. Sept. 1827.

spoiled; and this was the more provoking, as some of them had been purchased to supply the place of those he had the misfortune to lose when upset in the cutter at Oeno Island. I am happy to say the government, on the representation of his peculiar case, made him a compensation.

On the 12th the body of one of the seamen, Thomas Uren, was found near the place where the boat was wrecked; and on the Sunday following it was attended to the grave by all the officers and ship's company. The place of interment was on the low point of Chamisso Island, by the side of our shipmate who had been buried there the preceding year.

On the 13th we were visited by two baidars, and among their crews discovered the party who had visited the ship so early in the morning, when she was at the anchorage in August, one of whom drew his knife upon the first lieutenant; they were also of the party which made an attack upon our cutter in Escholtz Bay the preceding year. They had with them a few skins and some fish for sale, but they were very scrupulous about what they took for them; and on being ordered away late in the evening, they twanged their bows in an insolent manner, and pushed off about a couple of yards only. The officer of the watch desired them to go away, and at length presented a musket at the baidar, on whichthey fired an arrow into the sea in the direction of the ship, and paddled to the island, where we observed them take up their quarters.

When the boats landed the next day to fill the casks, Mr. Smyth, who had charge of the party, was desired to arm his people, and to order the Esquimaux off the island if they were offensive to him, or interfered with the duty. On landing, the natives met him on the beach, and were very anxious to learn whether the muskets were loaded, and to be allowed to feel the edges of the cutlasses, and were not at all pleased at having their request refused. The arms were rolled up in the sail for the purpose of being kept dry, but one of the natives insisted on having the canvas unrolled, to see what it contained, and on being refused he drew his knife, and threatened the seaman who had charge of it. Coupling this act with the conduct of the party on the before-mentioned occasions, Mr. Smyth ordered the arms to be loaded; on which the natives fled to their baidar, and placed every thing in her in readiness to depart on a minute's warning, and then, armed with their bows and arrows and knives, they drew up on a small eminence, and twanged their bow-strings, as before, in defiance. A few minutes before this occurred, five of the party, who had separated from their companions, attacked two of our seamen, who were at some distance from Mr. Smyth, digging a grave for their unfortunate shipmate, and coming suddenly upon them, while in the pit, three of the party stood over the workmen with their drawn knives, while the others rifled the pockets of their jackets, which were lying at a little distance from the grave, and carried away the contents, together with an axe. The hostile disposition of the natives on the hill, who were drawn up in a line in a menacing attitude, with their bows ready strung, and their knives in their left hands, obliged Mr. Smyth to arm his people, and, in compliance with his instructions, to proceed to drive them off the island. He accordingly advanced upon them, and each individual probably had singled out his victim, when an aged man of the Esquimaux party made offers of peace, and the arms of both parties were laid aside. The media-

CHAP. VII. Sept. 1827.

tor signified that he wanted a tub, that had been left at the well, which was restored to him, and the axe that had been taken from the grave was returned to our party. The Esquimaux then embarked, and paddled towards Escholtz Bay. I have been thus particular in describing the conduct of these people, in consequence of a more tragical affair which occurred a few days afterwards.

Strong winds prevented the completion of our water for several days; but on the 29th it was in progress, when the same party landed upon the island near our boat. The day being very fine, several of the officers had gone in pursuit of ptarmigan, which were about this time collecting in large flocks previous to their migration; and I was completing a series of magnetical observation in another part of the island. The first lieutenant observing a baidar full of men approach the island, despatched Lieutenant Belcher to the place with orders to send them away, provided there were any of the party among them who had behaved in so disorderly a manner on the recent occasion. On landing, he immediately recognised one of the men, and ordered the whole of the party into the baidar. They complied very reluctantly; and while our seamen were engaged pushing them off, they were occupied in preparations for hostility, by putting on their eider-duck frocks over their usual dresses, and uncovering their bows and arrows. They paddled a few yards from the beach, and then rested in doubt as to what they should do; some menacing our party, and others displaying their weapons. Thus threatened, and the party making no attempt to depart, but rather propelling their baidar sidewise toward the land, Mr. Belcher fired a ball between them and the shore, and waved them to begone. Instead of obeying his summons, they paddled on shore instantly, and quitted their baidar for a small eminence near the beach, from whence they discharged a flight of arrows, which wounded two of our seamen. Their attack was of course returned, and one of the party was wounded in the leg by a musket ball.

Until this time they were ignorant of the effect of fire-arms, and no doubt placed much confidence in the thickness of their clothing, as, in addition to their eider-duck dress over their usual frock, they each bound a deer-skin round them as they quitted their baidar; but seeing the furs availed nothing against a ball, they fled with precipitation to the hills; and the commanding officer of the Blossom observing them running towards the place where I was engaged with the dipping needle, fired a gun from the ship, which first apprised me of anything being amiss. On the arrival of the cutter, I joined Mr. Belcher, and, with a view of getting the natives into our possession, I sent a boat along the beach, and went with a party over land. We had not proceeded far, when suddenly four of the marines were wounded with arrows from a small ravine, in which we found a party so screened by long grass that it was not visible until we were close upon it. The natives were lying upon the ground, peeping between the blades of grass, and discharging their arrows as opportunity offered. In return, one of them suffered by a ball from Mr. Elson; on which I stopped the firing, and endeavoured ineffectually to bring them to terms. After a considerable time, an elderly man came forward with his arms and breast covered with mud, motioned us to begone, and decidedly rejected all offers of reconciliation. Unwilling to chastise them

CHAP. VII. Sept. 1827.

further, I withdrew the party, and towed their baidar on board, which kept them prisoners upon the island. I did this in order to have an opportunity of bringing about a reconciliation, for I was unwilling to allow them to depart with sentiments which might prove injurious to any Europeans who might succeed us; and I thought that by detaining them we should be able to convince them our resentment was unjustifiably provoked, and that when they conducted themselves properly, they should command our friendship. This baidar had a large incision in her bottom, made by the person who last quitted her when the party landed, and must have been done either with a view of preventing her being carried away, or by depriving themselves of the means of escape, showing their resolution to conquer to die. We repaired her as well as we could, and kept her in readiness to be restored to her owners on the first favourable opportunity that offered.

The next morning a boat was sent to bring them to friendly terms, and to return everything that was in the baidar, except some fish which they had brought for sale, in lieu of which some blue beads and tobacco were left, but the natives were averse to reconciliation, and kept themselves concealed. The night was severely cold, with snow showers; and next day, seeing nothing of the party, the baidar was returned. The natives removed her during the night to the opposite side of the island, where she appeared to be undergoing an additional repair; but we saw none of the people, who must have secreted themselves on the approach of the boat. We took every opportunity of showing them we wished to obtain their friendship, but to no purpose; they would not make their appearance, and the next night decamped, leaving a few old skins in return for the articles we had left for them. —On examining the ravine in which they had concealed themselves, we found one man lying dead, with his bow and quiver, containing five arrows, placed under his body, and clothed in the same manner as when he quitted the baidar. The ravine was conveniently adapted to the defence of a party, being narrow, with small banks on each side of it, behind which a party might discharge their arrows without much danger to themselves until they became closely beset; to obviate which as much as possible, and to sell their lives as dearly as they could, we found they had constructed pits in the earth by scooping out holes sufficiently large to contain a man, and by banking up the mud above them. There were five of these excavations close under the edges of the banks, which were undermined; one at the head of the ravine, and two on each side, about three yards lower down; the latter had a small communication at the bottom, through which an arrow might be transferred from one person to another, without incurring the risk of being seen by passing it over the top. The construction of these pits must have occupied the man who presented himself to us with his arms covered with mud: as a defence they were as perfect as circumstances would allow, and while they show the resources of the people, they mark a determination of obstinate resistance. The effect of the arrows was fully as great as might have been expected, and, had they been properly directed, would have inflicted mortal wounds. At the distance of a hundred yards a flesh-wound was produced in the thigh, which disabled the man for a time; and at eight or ten yards another fixed the right arm of a marine to his side; a third buried itself two

CHAP. VII. Sept. 1827.

inches and a half under the scalp. The wounds which they occasioned were obliged to be either enlarged, to extract the arrows, which were barbed, or to have an additional incision made, that the arrow might be pushed through without further laceration. Most of these wounds were inflicted by an arrow with a bone head, tipped with a pointed piece of jaspar.

We were sorry to find our musketry had inflicted so severe a chastisement upon these people, but it was unavoidable, and richly deserved. It was some consolation to reflect that it had fallen upon a party from whom we had received repeated insult, and it was not until after they had threatened our boat in Escholtz Bay, insulted us alongside the ship, defied our party on shore, had twice drawn their knives upon our people, and had wounded several of them, that they were made acquainted with the naturé of our fire-arms; and I am convinced the example will have a good effect, by teaching them that it was forbearance alone that induced us to tolerate their conduct so long.

For the purpose of keeping together the particulars of our transactions with the Esquimaux, I have omitted to mention several occurrences in the order in which they transpired. Many circumstances indicated the earlier approach of winter than we had experienced the preceding year. About the middle of September, therefore, we began to prepare the ship for her departure, by completing the water, taking on board stone ballast, in lieu of the provisions that had been expended, and refitting the rigging. These operations were for several days interrupted by strong westerly winds, which occasioned much sea at the anchorage, and very unaccountably had the effect of producing remarkably low tides, and of checking the rise which on several occasions was scarcely perceptible.

Sept. 1827.

On the 18th a party of the officers landed in Escholtz Bay to search for fossils, but they were unsuccessful, in consequence of an irregularity in the tide, which was on that occasion unaccountably high, and scarcely fell during the day. The cliffs had broken away considerably since the preceding year; and the frozen surface of the cliff appeared in smaller quantities than before, but the earth was found congealed at a less depth from the top. This examination tended to confirm more steadfastly the opinion that the ice forms only a coating to the cliff, and is occasioned by small streams of water oozing out, which either become congealed themselves in their descent, or convert into ice the snow which rests in the hollows.

On the 24th and 28th the nights were clear and frosty, and the aurora borealis was seen forming several arches. On the 28th the display was very brilliant and interesting, as it had every appearance of being between the clouds and the earth; and after one of these displays, several meteors were observed issuing from parts of the arch, and falling obliquely toward the earth. This was also one of the rare instances of the aurora being seen to the southward of our zenith.

Oct.

In the beginning of October we had sharp frosts and heavy falls of snow. On the 4th the earth was deeply covered, and the lakes were frozen; the thermometer during the night fell to 25°, and at noon on the 5th to 24°, and there was every appearance of the winter having commenced. It therefore became my duty seriously to consider on the propriety of continuing longer in these seas. We had received no intelligence of Captain Franklin's party, nor was it

CHAP. VII. Oct. 1827.

very probable that it could now appear; and we could only hope, as the time had arrived when it would be imperative on us to withdraw from him the only relief he could experience in these seas, that he had met with insurmountable obstacles to his proceeding, and had retraced his route up the M'Kenzie River.

Anxious, however, to remain to the last, on the chance of being useful to him, I again solicited the opinions of the officers as to the state of the season, and finding them unanimous in believing the winter to have commenced, and that the ship could not remain longer in Kotzebue Sound with safety, I determined to quit the anchorage the moment the wind would permit. Weighing the probability of Captain Franklin's arrival at this late period in the season, no one on board, I believe, thought there was the smallest chance of it; for, had his prospects the preceding year been such as to justify his wintering upon the coast, the distance remaining to be accomplished in the present season would have been so short that he could scarcely fail to have performed it early in the summer, in which case we must have seen him long before this date, unless, indeed, he had reached Icy Cape, and found it advisable to return by his own route, a contingency authorised by his instructions. Upon the chance of his arrival after the departure of the ship, the provision that had been buried for his use was allowed to remain, and the billet of wood was again deposited on the island, containing a statement of the behaviour of the natives, and of other particulars, with which it was important that he should be made acquainted.

On the 6th, sharp frosty weather continuing, we weighed from Chamisso, and beat out of the sound.

In passing Cape Krusenstern we perceived a blink in the N.W. direction, similar to that over ice, and it is not unlikely that the westerly winds which were so prevalent all the summer had drifted it from the Asiatic shore, where it rests against the land in a much lower parallel than upon the American coast.

As we receded from the sound the wind freshened from the N.W. with every appearance of a gale; we kept at a reasonable distance from the land until day light, and then steered towards Cape Prince of Wales, with a view of passing Beering's Strait. Our depth of water thus far had been about fifteen fathoms, but at eleven o'clock in the forenoon it began to diminish, and the sea being high, the course was altered, to increase our distance from the coast: we had scarcely done this when the water shoaled still more, and a long line of breakers was observed stretching from the land, crossing our course, and extending several miles to windward. The weather was so hazy that we could scarcely see the land; but it was evident that we had run down between the coast and a shoal, and as there was no prospect of being able to weather the land on the opposite tack, the only alternative was to force the ship through the breakers; we accordingly steered for those parts where the sea broke the least, and kept the ship going at the rate of seven knots, in order, as the shoal appeared to be very narrow, that she might not hang, in the event of touching the ground.

The sea ran very high, and we entered the broken water in breathless suspense, as there was very little prospect of saving the ship, in the event of her becoming fixed upon the shoal. Four fathoms and a half was communicated from the channels, a depth in

CHAP. VII. Oct. 1827.

which it may be recollected we disturbed the bottom in crossing the bar of San Francisco; the same depth was again reported, and we pursued our course momentarily expecting to strike. Fortunately this was the least depth of water, and before long our soundings increased to twenty fathoms, when having escaped the danger, we resumed our course for the strait.

This shoal, which appears to extend from Cape Prince of Wales, taking the direction of the current through the strait, is extremely dangerous, in consequence of the water shoaling so suddenly, and having deep water within it, by which a ship coming from the northward may be led down between the shoal and the land, without any suspicion of her danger. Though we had nothing less than twenty-seven feet water, as near as the soundings could be ascertained in so high a sea, yet, from the appearance of the breakers outside the place where the ship crossed, the depth is probably less. It is remarkable that this spit of sand, extending so far as it does from the land, should have hitherto escaped the observation of the Russians, as well as of our countrymen. Cook, in his chart, marks five fathoms close off the cape, and Kotzebue three, but this spit appeared to extend six or seven miles from it. It is true that the weather was very hazy, and we might have been deceived in our distance from the shore: but it is also probable that the spit may be extending itself rapidly.

We passed Beering's Strait about one o'clock, as usual with a close reefed topsail breeze, and afterwards ran with a fresh gale until midnight, when, as I wished to see the eastern end of St. Lawrence Island, we rounded to for daylight. It was, however, of little consequence, as the weather was so foggy the next day that we could not see far around us. As we approached the island, flocks of alca crestatella and of eider and king ducks, and several species of phalaropes, flew about us, but no land was distinguished. About noon the water shoaling gradually to eleven fathoms, created a doubt whether we were not running upon the island; but on altering the course to the eastward, it deepened again, and by the observations of the next day it appeared that the ship had passed over a shoal lying between St. Lawrence Island and the main. It is a curious fact, that this shoal is precisely in the situation assigned to a small island which Captain Cook named after his surgeon, Mr. Anderson; and as that island has never been seen since, many persons, relying upon the general accuracy of that great navigator, might suppose the island to have been sunk by some such convulsion as raised the island of Amnuk in the same sea; while others might take occasion from this fact to impeach the judgment of Cook. I am happy to have an opportunity of reconciling opinions on this subject, having discovered a note by Captain Bligh, who was the master with Captain Cook, written in pencil on the margin of the Admiralty copy of Cook's third voyage, by which it is evident that the compilers of the chart have overlooked certain data collected off the eastern end of St. Lawrence Island, on the return of the expedition from Norton Sound, and that the land, named Anderson's Island, was the eastern end of the island of St. Lawrence; and had Cook's life been spared he would no doubt have made the necessary correction in his chart.

CHAP. VII. Oct. 1827.

Thick weather continued until the 10th, when, after some hard showers of snow, it dispersed, and afforded us an opportunity of determining the position of the ship, by observation, which agreed very nearly with the reckoning, and showed there had been no current of consequence. Two days afterwards we saw the island of St. Paul, and endeavoured to close it, in order to examine its outline, and compare our observations with those of the preceding year; but the wind obliged us to pass at the distance of eight miles to the eastward, and we could only accomplish the latter. The next morning we passed to the eastward of St. George's Island, and fixed its position also. This was the island we were anxious to see the preceding year, as its situation upon our chart was very uncertain, and in some of the most approved charts it is omitted altogether.

Off here we observed a number of shags, a few albatrosses, flocks of ortolans, and a sea otter.

At daylight on the 14th, we saw the Aleutian Islands, and steered for an opening which by our reckoning should have been the same strait through which we passed on a former occasion; but the islands being covered more than half way down with a dense fog, we were unable to ascertain our position correctly; and it was not until the latitude was determined by observation that we discovered we were steering for the wrong passage. This mistake was occasioned by a current S. 34° W. true, at the rate of nearly three miles an hour, which in the last twelve hours had drifted the ship thirty-five miles to the westward of her expected position. Fortunately the wind was fair, and enabled us to correct our error by carrying a press of sail. Before sunset we got sight of the Needle Rock in the channel of Oonemak, and passed through the strait. The strength and uncertainty of the currents about these islands should make navigators very cautious how they approach them in thick weather: whenever there is any doubt, the most certain course is to steer due east, and make the Island of Oonemak, which may be known by its latitude, being thirty miles more northerly than any other part of the chain; and then to keep along its shores at the distance of four or five miles, until the Needle Rock, which lies nearly opposite the Island of Coogalga, is passed; after which the coast on both sides trends nearly east and west, and a ship has an open sea before her.

The Aleutian Islands, when we passed, were covered about two-thirds of the way down with snow, and indicated an earlier winter than they had done the preceding year.

Having taken our final leave of Beering's Strait, all hope of the attainment of the principal object of the expedition in the Polar Sea was at an end; and the fate of the expedition under Captain Franklin, which was then unknown to us, was a subject of intense interest. Amidst the disappointment this failure in meeting with him had occasioned us, we had the consolation of knowing that, whatever vicissitudes might have befallen his party, our efforts to maintain our station in both years had, by the blessing of Providence, been successful, so that at no period of the appointed time of rendezvous could he have missed both the boat and the ship, or have arrived at the

CHAP VII.

Oct. 1827.

appointed place in Kotzebue Sound without finding the anticipated relief.

The enterprising voyage of Captain Franklin down the Mackenzie, and along the northern shores of the continent of America, is now familiar to us all, and, considering that the distance between the extremities of our discoveries was less than fifty leagues, and that giving him ten days to perform it in, he would have arrived at Point Barrow at the precise period with our boat, we must ever regret that he could not have been made acquainted with our advanced situation, as in that case he would have been justified in incurring a risk which would have been unwarrantable under any other circumstances. Let me not for a moment be supposed by this to detract one leaf from the laurels that have been gained by Captain Franklin and his enterprising associates, who, through obstacles which would have been insurmountable by persons of less daring and persevering minds, have brought us acquainted with an extent of country which, added to the discovery it was our good fortune to push so far along the shore to the westward of them, has left a very small portion of the coast unknown.

The extent of land thus left unexplored between Point Turnagain and Icy Cape is comparatively so insignificant that, as regards the question of the north-west passage, it may be considered to be known; and in this point of view both expeditions, though they did not meet, may be said to have been fully successful. From the nature and similarity of the coast at Return Reef and Point Barrow, it is very probable that the land from Franklin Extreme trends gradually to the eastward to Return Reef, leaving Point Barrow in latitude 71° 23′ 30″ N. the northern limit of the continent of America.

The determination of this great geographical question is undoubtedly important; but though it sets a boundary to the new continent, and so far diminishes the difficulties attending an attempt to effect a passage from the Pacific to the Atlantic, yet it leaves the practicability of the north-west passage nearly as doubtful as ever; and it is evident that it cannot be otherwise, until the obstructions set forth in Captain Parry's voyage are removed, as it would avail little to be able to reach Hecla and Fury Strait, provided that channel were always impassable.

From what has been set forth, in the foregoing narrative of our proceedings, it is nearly certain that, by watching the opportunity, a vessel may reach Point Barrow, and in all probability proceed beyond it. Had we been permitted to make this attempt, we should no doubt be able to speak more positively upon this subject; and, as I have always been of opinion that a navigation may be performed along any coast of the Polar Sea that is *continuous*, I can see no insurmountable obstacle to the exploit. In this attempt, however, it is evident that a vessel must be prepared to encounter very heavy pressure from the ice, and must expect, on the ice closing the coast to the westward of Point Barrow, which it unquestionably would with every strong westerly wind, to be driven on shore in the manner in which our boat was in 1826.

As regards the question, whether it be advisable to attempt the passage from the Atlantic to the Pacific, the advantage of being able to pursue the main land with certainty from Icy Cape is unquestionably great; and the recollection that in that route every foot gained to the eastward is an advance towards the point whence supplies and succour may be obtained, is a

CHAP. VII. Oct. 1827.

cheering prospect to those who are engaged in such an expedition. But while I so far advocate an attempt from this quarter, it must not be overlooked that the length of the voyage round Cape Horn, and the vicissitudes of climate to be endured, present material objections to prosecuting the enterprise by that course.

It does not appear that any preference can be given to the western route from prevailing winds or currents, as both are so variable and uncertain, that no dependence can be placed upon them. In 1826, easterly winds prevailed almost throughout the summer, both on the northern coast of America, and in the open sea to the westward of Icy Cape: while in 1827, in the latter situation at least, the reverse took place. And as the coincidence of winds experienced by Captain Franklin and ourselves in 1826 is very remarkable, there is every probability that the same winds prevailed to the eastward of Point Barrow.

The current, though it unquestionably sets to the northward through Beering's Strait, in the summer at least, does not appear to influence the sea on the northern coast of America which is navigable; as Captain Franklin, after the experience of a whole season, was unable to detect any current in either direction. In the sea to the westward of Icy Cape, the current setting through Beering's Strait is turned off by Point Hope, and does not appear to have any perceptible influence on the water to the north-eastward of Icy Cape; for the current there, though it ran strong at times, seemed to be influenced entirely by the prevailing wind. The body of water which finds its way into the Polar Sea must undoubtedly have an outlet, and one of these appears to be the Strait of Hecla and Fury; but as this current is not felt between the ice and the continent of America, the only part of the sea that is navigable, it must rather impede than favour the enterprise, by blocking the ice against both the strait, and the western coast of Melville Peninsula. Upon the whole, however, I am disposed to favour the western route, and am of opinion that could steam vessels properly fitted, and adapted to the service, arrive in good condition in Kotzebue Sound, by the beginning of one summer, they might with care and patience succeed in reaching the western shore of Melville Peninsula in the next. There, however, they would undoubtedly be stopped, and have to encounter difficulties which had repulsed three of the most persevering attempts ever made toward the accomplishment of a similar object.

I shall now offer a few remarks upon the inhabitants whom we met upon this coast.

The western Esquimaux appear to be intimately connected with the tribes inhabiting the northern and north-eastern shores of America, in language, features, manners, and customs. They at the same time, in many respects, resemble the Tschutschi, from whom they are probably descended. These affinities I shall notice as I proceed with my remarks upon the people inhabiting the north-west coast of America, whom, for the convenience of the reader, I shall call the western Esquimaux, in order to distinguish them from the tribes inhabiting Hudson's Bay, Greenland, Igloolik, and indeed from all the places eastward of Point Barrow. This line ought properly to be drawn at M'Kenzie River, in consequence of certain peculiarities connecting the people seen near that spot with the tribe to the westward; but it will be more convenient to confine it within the above-mentioned limits.

These people inhabit the north-west coast of Ame-

CHAP. VII. Oct. 1827. rica, from 60° 34′ N. to 71° 24′ N., and are a nation of fishermen dwelling upon or near the sea shore, from which they derive almost exclusively their subsistence. They construct yourts or winter residences upon those parts of the shore which are adapted to their convenience, such as the mouths of rivers, the entrances of inlets, or jutting points of land, but always upon low ground. They form themselves into communities, which seldom exceed a hundred persons; though in some few instances they have amounted to upwards of two hundred. Between the above-mentioned limits we noticed nineteen of these villages, some of which were very small, and consisted of only a few huts, and others appeared to have been deserted a long time; but allowing them all to be inhabited in the winter, the whole population, I should think, including Kow-ee-rock, would not amount to more than 2500 persons. I do not pretend to say that this estimate is accurate, as from the manner in which the people are dispersed along the coast in the summer, it is quite impossible that it should be so; but it may serve to show that the tribe is not very numerous.

As we landed upon every part of the coast, to which these villages appear to be confined, it is not likely that many escaped our observation; neither is it probable that there are many inland or far up the rivers, as frequent access to the sea is essential to the habits of the people. Besides this may further be inferred, from the circumstance of no Esquimaux villages being found up either the M'Kenzie or Coppermine rivers, and from the swampy nature of the country in general, and the well-known hostile disposition of the Indians towards the Esquimaux.

Their yourts or winter residences are partly excavated in the earth, and partly covered with moss laid upon poles of driftwood. There are, however, several kinds of habitations, which seem to vary in their construction according to the nature of the ground and the taste of the inhabitants. Some are wholly above ground, others have their roof scarcely raised above it; some resemble those of the Tschutschi, and others those of the natives near Prince William Sound; but they all agree in being constructed with driftwood covered with peat, and in having the light admitted through a hole in the roof covered with the intestines of sea animals. The natives reside in these abodes during the winter, and when the season approaches at which they commence their wanderings, they launch their baidars, and taking their families with them, spread along the coast in quest of food and clothing for the ensuing winter. An experienced fisherman knows the places which are most abundant in fish and seals, and resorts thither in the hope of being the first occupier of the station. Thus almost every point of land and the mouths of all the rivers are taken possession of by the tribe. Here they remain, and pass their time, no doubt, very happily, in the constant occupation of taking salmon, seals, walrusses, and reindeer, and collecting peltry, of which the beaver-skins are of very superior quality, or whatever else they can procure, which may prove useful as winter store.

During their absence the villages are left in charge of a few elderly women and children, with a youth or two to assist them, who, besides preventing depredations, are deputed to cleanse and prepare the yourts for the reception of the absentees at the approach of winter. As long as the fine weather lasts they live under tents made of deer-skins laced upon poles; but about the middle of September, they break up these

CHAP. VII. Oct. 1827.

establishments, load their baidars with the produce of their labour, and track them along the coast with dogs towards their yourts, in which they take up their winter station as before, and regale themselves after their success by dancing, singing, and banqueting, as appears to be the custom with the Eastern Esquimaux, and from their having large rooms appropriated to such diversions.

These winter stations may always be known at a distance by trunks of trees, and frames erected near them; some supporting sledges and skins of oil, and others the scantling of boats, caiacs, fishing implements, &c.

We had no opportunity of witnessing their occupations in the winter, which must consist in the construction of implements for the forthcoming season of activity, in making clothes, and carving and ornamenting their property, for almost every article made of bone is covered with devices. They appear to have no king or governor, but, like the patriarchal tribes, to venerate and obey the aged. They have sometimes a great fear of the old women who pretend to witchcraft.

It seems probable that their religion is the same as that of the Eastern Esquimaux, and that they have similar conjurers and sorcerers. We may infer that they have an idea of a future state, from the fact of their placing near the graves of their departed friends the necessary implements for procuring a subsistence in this world, such as harpoons, bows, and arrows, caiacs, &c. and by clothing the body decently; and from the circumstance of musical instruments being suspended to the poles of the sepulchres, it would seem that they consider such state not to be devoid of enjoyments. Their mode of burial differs from that of the Eastern Esquimaux, who inter their dead; whereas these people dispose the corpse upon a platform of wood, and raise a pile over it with young trees. The position in which the bodies are laid also differs; the head being placed to the westward by this nation, while in the eastern tribes it lies to the north-east.

They are taller in stature than the Eastern Esquimaux, their average height being about five feet seven and a half inches. They are also a better looking race, if I may judge from the natives I saw in Baffin's Bay, and from the portraits of others that have been published. At a comparatively early age, however, they (the women in particular) soon lose this comeliness, and old age is attended with a haggard and care-worn countenance, rendered more unbecoming by sore eyes, and by teeth worn to the gums by frequent mastication of hard substances.

They differ widely in disposition from the inhabitants of Igloolik and Greenland, being more continent, industrious, and provident, and rather partaking of the warlike, irascible, and uncourteous temper of the Tschutschi. Neither do they appear by any means so deficient in filial affection as the natives of Ingloolik, who as soon as they commenced their summer excursions left their aged and infirm to perish in the villages; of whom it will be recollected that one old man, in particular, must have fallen a victim to this unnatural neglect, had not his horrible fate been arrested by the timely humanity of the commander of the polar expedition.

With the Western Esquimaux, as indeed with almost all uncivilized tribes, hospitality seems to form one characteristic feature of the disposition; as if Nature, by the gift of this virtue, had intended to check, in some measure, that ferocity which is otherwise so predominant.

CHAP. VII. Oct. 1827.

Smoking is their favourite habit, in which they indulge as long as their tobacco lasts. Parties assemble to enjoy the fumes of this narcotic, and the pipe passes round like the calumet of the Indians, but apparently without the ceremony being binding. Their pipes are short, and the bowls of some contain no more tobacco than can be consumed in a long whiff; indeed, the great pleasure of the party often consists in individuals endeavouring to excel each other in exhausting the contents of the bowl at one breath, and many a laugh is indulged at the expense of him who fails, or who, as is very frequently the case, is thrown into a fit of coughing by the smoke getting into his lungs.

They seldom use tobacco in any other way than this, though some natives whom we saw to the southward of Beering's Strait were not averse to chewing it, and the St. Lawrence islanders indulged in snuff. Their predilection for tobacco is no doubt derived from the Tschutschi, who are so passionately fond of it, that they are said, by Captain Cochrane, to snuff, chew, and smoke, all at the same time. The practice of adulterating tobacco is common with the Tschutschi, and has, no doubt, passed from them to the Esquimaux, who often adopt it from choice. That which finds its way to the N.W. coast of America is of very inferior quality, and often has dried wood chopped up with it.

The ornaments worn in the lip, described in the course of this narrative, are peculiar to the males of the Western Esquimaux, and are in use only from Norton Sound, where they were seen by Captain King, to the Mackenzie River, where they were worn by the party which attacked Captain Franklin. The practice is by no means modern, as Deschnew, as far back as 1648, describes the inhabitants of the islands opposite Tsčhutskoi Noss as having pieces of sea-horse tusk thrust into holes in their lips. No lip ornaments similar to these have been seen to the eastward of the Mackenzie River; and indeed we know of no other tribe which has adopted this singular custom of disfiguring the face, except that inhabiting the coast near Prince William Sound, and even there the arrangement differs. It is remarkable that the practice with them is confined to the women, while in the tribe to the northward it is limited to the men. It is also singular, that this barbarous custom of the males is confined to so small a portion of the coast, while that by which the females are distinguished extends from Greenland, along the northern and western shores of America, down to California.

Nasal ornaments, so common with the tribes to the southward of Oonalaska, were seen by us in one instance only, and were then worn by the females of a party whose dialect differed from that in general use with the tribe to the westward of Point Barrow. The custom disappears to the northward of Alaska, and occurs again in the tribe near the Mackenzie River. A similar break in the link of fashion in the same nation may be traced in the practice of shaving the crown of the head, which is general with the Western Esquimaux, ceases at the Mackenzie River, and appears again in Hudson's Bay, and among a tribe of Greenlanders, who, when they were discovered by Captain Ross, had been so long excluded from intercourse with any other people, that they imagined themselves the only living human beings upon the face of the globe.*

* See a letter from Captain Edward Sabine, Journal of Science, vol. vii.

CHAP. VII. Oct. 1827.

It was remarked that the inhabitants of Point Barrow had copper kettles, and were in several respects better supplied with European articles than the people who resided to the southward. Captain Franklin found among the Esquimaux near the Mackenzie several of these kettles, and other manufactures, which were so unlike those supplied by the North west Company, as to leave no doubt of their being obtained from the westward. Connecting these facts with the behaviour of the natives who visited us off Wainwright Inlet, and the information obtained by Augustus, the interpreter, it is very probable that between the Mackenzie River and Point Barrow there is an agent who receives these articles from the Asiatic coast, and parts with them in exchange for furs. Augustus learned from the Esquimaux that the people from whom these articles were procured resided up a river to the westward of Return Reef. The copper kettles, in all probability, come from the Russians, as the Tschutschi have such an aversion to utensils made of that metal, that they will not even use one when lined with tin.* From the cautious manner in which the whole tribe dispose of their furs, reserving the most valuable for larger prices than we felt inclined to give, and sometimes producing only the inferior ones, we were induced to suspect that there were several Esquimaux acting as agents upon the coast, properly instructed by their employers in Kamschatka, who, having collected the best furs from the natives, crossed over with them to the Asiatic coast, and returned with the necessary articles for the purchase of others.

I regret that we never had an opportunity of seeing

* Captain Cochrane's Journey in *Siberia.*

the Esquimaux in pursuit of their game, or in any way actively employed, except in transporting their goods along the coast. One cause for this is, that they relinquished all occupation on our appearance, to obtain some of the riches that were on board the ship. It may, however, be inferred, from the carvings upon their ivory implements, that their employments are numerous, and very similar to those practised by the Greenlanders. Of these, rein-deer hunting appears to be the most common. If we may credit the sculptured instruments, they shoot these animals with bows and arrows, which, from the shyness of the deer, must require great skill and artifice to effect. The degree of skill may be inferred from the distance at which some of the parties are drawn shooting their arrows, and the artifice is shown by a device of a deer's head and horns placed upon the shoulders of a person creeping on all-fours towards the animal, after the manner of the Californian Indians, and of some of the inland tribes of North America. We found the flint head of an arrow which had been used for this purpose broken in a haunch of venison that was purchased from the inhabitants near Icy Cape. In some of the representations the deer are seen swimming in the water, and the Esquimaux harpooning them from their caiacs, in the manner represented in the plate in Captain Parry's Second Voyage, p. 508.

As an instance of their method of killing whales, we found a harpoon in one that was dead, with a drag attached to it made of an inflated seal-skin. It must be extremely difficult for these people, with their slender means, to capture these enormous animals, and it must require considerable perseverance. The occupation, however, appears to be less hazardous

CHAP. VII. Oct. 1827.

than that of killing walrusses, which, by the devices upon the instruments, occasionally attack the caiacs. The implements for taking these animals are the same as described by Captain Parry. Seals are also captured in the manner described by him. Upon some of the bone implements there are correct representations of persons creeping along the ice towards their prey, which appears to have been decoyed by an inflated seal-skin placed near the edge of the ice; an artifice frequently practised by the eastern tribes. These animals are also taken in very strong nets made of walrus-hide; and another mode is by harpooning them with a dart about five feet in length, furnished with a barb, which is disengaged from its socket when it strikes the animal, and being fastened by a line to the centre of the staff, the harpoon acts as a drag. This instrument is discharged with a throwing board, which is easily used, and gives very great additional force to the dart, and in the hands of a skilful person will send a dart to a considerable distance. The throwing board is mentioned also by Captain Parry, by Crantz, and others, and corresponds with the *womoru* of New Zealand.

We noticed in the possession of a party to the northward of Kotzebue Sound a small ivory instrument, similar to the *keipkuttuk* of the Igloolik tribe.

Birds are likewise struck with darts which resemble the *nuguit* of Greenland; they are also caught in whalebone snares, and by having their flight arrested by a number of balls attached to thongs about two feet in length: they are sometimes shot with arrows purposely constructed with blunt heads.

The practice of firing at a mark appears to be one of the amusements of the Esquimaux; and judging from what we saw at Chamisso Island, there are some extraordinary performers in this way among the tribe. One day a diver was swimming at the distance of thirty yards from the beach, and a native was offered a reward if he would shoot it: he fired, but the bird evaded the arrow by diving. The Esquimaux watched its coming to the surface, and the instant his head appeared he transfixed both eyes with his arrow. He was rewarded for his skilfulness, and the skin was preserved as a specimen of ornithology and of Indian archery. Generally speaking, however, I do not think they are expert marksmen.

Their bows are shaped differently to those of Igloolik, and are superior to any on the eastern coast of America; they are, however, made upon the same principle, with sinews and wedges at the back of the wood. On the western coast driftwood is so abundant that the inhabitants have their choice of several trees, and are never obliged to piece their implements. It requires some care to bring a bow to the form which they consider best; and for this purpose they wrap it in shavings soaked in water, and hold it over a fire; it is then pegged down upon the earth in the form required. If not attended to when used, the bows are apt to get out of order, and the string to slip out of its place, by which the bow bends the wrong way, and is easily broken.

In these bows the string is in contact with about a foot of the wood at each end, and when used makes a report which would be fatal to secrecy. The Californians, accustomed to fight in ambush, are very careful to have that part of the string muffled with fur, but I never saw any precaution of the kind used by the Esquimaux. To protect the wrist from the abrasion which would ensue from frequent firing, the

Esquimaux buckle on a piece of ivory, called *mun-era*, about three or four inches long, hollowed out to the wrist, or a guard made of several pieces of ivory or wood fastened together like an ironholder.

Fishing implements are more numerous and varied with the Western Esquimaux than with the others, and some are constructed with much neatness and ingenuity; but I do not know that any of them require description, except a landing net, and that only because it is not mentioned by Captain Parry. This consists of a circular frame of wood or bone, about eight inches in diameter, worked across with whalebone like the bottoms of cane chairs, and fixed upon a long wooden handle.

Of all their manufactures, that of ivory chains is the most ingenious. These are cut out of solid pieces of ivory, each link being separately relieved, and are sometimes twenty-six inches in length. For what purpose they are used I know not: but part of the last link is frequently left solid, and formed in imitation of a whale; and these chains being strong, they may in some way or other be appropriated to the capture of that animal.

Among a great many singularly shaped tools in the possession of these people, we noticed several that are not in Captain Parry's catalogue, such as instruments for breaking wood short off; small hand chisels, consisting of pieces of hard stone fixed in bone handles adapted to the palm of the hand; meshes for making nets; an instrument made with the claws of a seal, for cleansing skins of their fat, &c. Though I never saw the screw in use among this tribe, yet I found a worm properly cut upon the end of one of their fishing implements. The *panna*, or double-edged knife, is also in use with these people; some of them were inlaid with brass, and undoubtedly came from the Tschutschi.

The language of the Western Esquimaux so nearly resembles that of the tribes to the eastward, as scarcely to need any further mention, particularly after the fact of Augustus, who was a native of Hudson's Bay, being able to converse with the Esquimaux whom he met at the mouth of the Mackenzie River. It may, however, be useful to show, by means of a vocabulary compiled from the people we visited, how nearly it coincides with that given by Captain Parry; some allowances being made for the errors to which all collectors are liable, who can only make themselves understood by signs, and who collate from small parties, residing perhaps at a distance from each other, and who, though they speak the same language, may make use of a different dialect. It does not appear that this language extends much beyond Norton Sound, certainly not down to Oonalashka; for the natives of that island, who are sometimes employed by the Russians as interpreters, are of no use on the American coast, near Beering's Strait. The language, notwithstanding, has a great affinity, and may be radically the same.

It is unnecessary to pursue further the peculiarities of these people, which are so similar to those of the eastern tribes, as to leave no doubt of both people being descended from the same stock; and though the inhabitants of Melville Peninsula declared they knew of no people to the westward of Akoolee, there is much reason to believe, from the articles of Asiatic manufacture found in their possession, that there is

CHAP. VII.

an occasional communication between all the tribes on the north coast of America.

Oct. 1827.

The subject of currents in Beering's strait has lost much of its interest by the removal of the doubt regarding the separation of the continents of Asia and America; and it is now of importance only to the navigator, and to the natural philosopher.

It does not appear, from our passages across the sea of Kamschatka, that any great body of water flows towards Beering's Strait. In one year the whole amount of current from Petropaulski to St. Lawrence Island was S. 54° W. thirty-one miles, and in the next N. 50° W. fifty-one miles, and from Kotzebue Sound to Oonemak N. 79°. W. seventy-nine miles. Approaching Beering's Strait, the first year, with light southerly winds, it ran north sixteen miles per day; and in the next, with strong S. W. winds, north five miles; and with a strong N. E. wind, N. 34° W. twenty-three miles. Returning three different times with gales at N. W. there was no perceptible current.

By these observations it appears that near the strait with southerly and easterly winds there is a current to the northward; but with northerly and north-westerly winds there is none to the southward, and consequently that the preponderance is in favour of the former, and of the generally received opinion of all persons who have navigated these seas. I prefer this method of arriving at the set of the current to giving experiments made occasionally with boats, as they would lead to a result, which would err according to the time of the tide at which they were made.

To the northward of Beering's Strait, the nature of the service we were employed upon confined us within a few miles of the coast; there the northerly current was more apparent. We first detected it off Schismareff Inlet; it increased to between one and two miles an hour off Cape Krusenstern, and arrived at its maximum, three miles an hour, off point Hope: this was with the flood tide; the ebb ran W. S. W. half a mile an hour. Here the current was turned off to the north-west by the point, and very little was afterwards felt to the northward. The point is bold and shingly, and shows every indication of the current being prevalent and rapid.

This current, as I have before remarked, was confined nearly to the surface and within a few miles of the land; at the depth of nine feet its velocity was evidently diminished, and at three and five fathoms there was none. The upper stratum, it should be observed, was much fresher than sea water; and there is no doubt that this current was greatly accelerated, if not wholly occasioned, by rivers; but why it took a northerly course is a question I am not prepared to answer.

To the northward and eastward of Cape Lisburn we found little or no current until we arrived at Icy Cape. Off this projection it ran strong, but in opposite directions, and seemed to be influenced entirely by the winds. Near Point Barrow, with a southwesterly gale, it ran at the rate of three miles an hour and upwards to the N. E., and did not subside immediately with the wind; but the current must here have been increased by the channel between the land and the ice becoming momentarily narrowed by the *pack* closing the beach; and it must not be imagined that the whole body of water in the Polar Sea was going at the rapid rate above mentioned, which would be con-

CHAP. VII. Oct. 1827.

trary to our experiments in the offing. Another cause of this may be a bank lying to the westward of Icy Cape, upon one part of which the water shoals from thirty-two fathoms to nineteen, and the bottom is changed from mud to stones.

It is evident, from the above-mentioned facts, that a current prevails in a northerly direction, although we are unable to state with precision its amount, which cannot under any circumstances be great, nor, I should think, exceed a mile an hour on the average. To be able to speak positively on this subject would require a vast number of trials to be made in the same place, and at a distance from the land, out of the influence of rivers. We may however presume, that the above-mentioned direction is that of the prevailing current throughout the year; for, upon examining the shoals off the principal headlands, we find them all to extend to the north-west, as may be seen on referring to St. Lawrence Island, Capes Prince of Wales, Krusenstern, and Lisburn, and also to Point Hope. This I conceive to be the most certain mode of deciding the question, without purposely stationing a vessel in the strait, and it is satisfactory to find that the result fully coincides with the experiments made near the shore by the Blossom and her boats.* Our observations, of course, apply to one season of the year only, as no experiments have as yet been made in the winter.

* I was in hopes that I had expressed myself clearly on this subject in the preceding edition of my work; but I find that I have been misunderstood, and even supposed in one place to have contradicted my statement in another. This apparent disagreement has arisen partly, if not wholly, from an oversight in some of my readers, who have compared observations, made at the *surface* of the sea at one place, with those at *five fathoms below* it at another

The course of this current, after it passes Cape Lisburn, is somewhat doubtful; we should expect it to diverge, and one part to sweep round Icy Cape and Point Barrow; but the shoals off the former place, like the currents themselves, do not furnish any satisfactory inference. These shoals lie parallel with the shore, and may be occasioned by ice grounded off the point. It may be observed here, that voyagers have frequently mentioned westerly currents along the northern coast of Asia and Nova Zembla, and we know from experience, that, in the summer, at least, there is a strong westerly current between Spitzbergen and Greenland. In the opposite direction, we find only a weak stream passing through the narrow strait of Hecla and Fury, and none through Barrow Strait. It seems, therefore, probable, that the principal part of the water which flows into the Polar Sea, from the Pacific, finds its way to the westward.

By many experiments made on shore at Icy Cape by Lieutenant Belcher, it appeared that southerly and westerly winds occasioned high tides, and northerly and easterly winds very low ebbs. It would seem, from this fact, that the water finds some obstruction to the northward, and I think it probable that the be-

nearly 200 miles distant. If the reader will have the candour to compare the observations made at the *surface at both places*, he will find them to agree, with the exception that the current at one place ran faster than that at the other, the reason of which I have endeavoured to account for in page 313 of this volume. I should observe here, that, although I have not encumbered my narrative with a notice of every time the current was tried, such observations were made repeatedly, whenever the nature of the service I was employed upon would admit of it; but I wish it to be borne in mind, that the situation of the ship, necessarily close in shore, was highly unfavourable to the determination of the question under discussion.

fore mentioned shoal, which closes the land toward Point Barrow, may extend to the northward; nay, it may even lie off the coast of some polar lands, too low and too far off to be seen from the margin of the ice; and which can only be ascertained by journeys over the ice, in a similar manner to that in which the mountains to the northward of Shelatskoi Noss were discovered by the Russians. It was this shoaling of the water to the northward of Cape Lisburn that induced the late Captain Burney to believe the continents of Asia and America were connected.

To the northward of Beering's Strait the tide rises about two feet six inches at full and change, and the flood comes from the southward.

The quantity of drift wood found upon the shores of Beering's Strait has occasioned various conjectures as to the source from which it proceeds; some imagining it to be brought down the rivers; others to be drifted from the southward.

We found some at almost every place where we landed, and occasionally in great quantities. There was more at Point Rodney than in any other part; a great deal upon Point Spencer; some upon Cape Espenburg, but more in Kotzebue Sound. Between Cape Krusenstern and Cape Lisburn there was very little, and in the bay to the eastward of the Cape scarcely any; but when the coast turned to the northward it became more plentiful, and it was afterwards tolerably abundant, and continued so all the way to Point Barrow. In addition to this, it should be remembered, that a great deal is used by the Esquimaux for boats, implements of all sort, houses, and fuel.

These trees are principally, if not all, either pine or birch; all that we examined were of these two species, and we lost no opportunity of making inquiry on this subject. The wood is often tough and good; indeed some that was taken from Choris Peninsula was superior to the pine we procured at Monterey; but from this stage of preservation it may be traced to old trunks crumbling to dust. Some trees still retained their bark, and appeared to have been recently uprooted; and comparatively few showed marks of having been at sea.

Some circumstances favour an opinion, by no means uncommon, that this wood is drifted from the southward; such as its being found in large quantities on Point Rodney, the many floating trees met with at sea to the southward of Kamschatka, &c.; but the quantity of this material found by Captain Franklin and Dr. Richardson at the mouths of the rivers on the northern coast of America, and some being found by us high up Kotzebue Sound, in Port Clarence, and other places, where it is hardly possible for it to be drifted, considering the outset of fresh water, renders it more probable that it is brought down from the interior of America. Rivers quite sufficient for this purpose will be found on an inspection of the chart, but without this we need only advert to the before-mentioned rapid current of nearly fresh water to prove their existence. Did the wood come by sea from the southward, we could scarcely have failed seeing some of it in our passage from Petropaulski, and during our cruises to the northward of Beering's Strait; but scarcely any was observed between Kamschatka and St. Lawrence Island; none between that place and Beering's Strait; and only six or seven pieces of short wood to the northward, notwithstanding the coast was closely navigated in both years by the ship and the barge. Besides,

CHAP. VII. Oct. 1827. the westerly current, which is prevalent to the southward of Beering's Strait, is very much against the probability of its being drifted from the southward.

We passed the Aleutian Islands on the night of the 14th, and as in the preceding year entered a region of fine clear weather. The volcano on Oonemak was still emitting flashes, which were visible at a very considerable distance. It being my intention now to make the best of my way to England, I directed the course towards California, for the purpose of refitting the ship, and of recruiting the health of the ship's company. In this passage nothing remarkable occurred until the 20th, on which day the sun was eclipsed, when we were overtaken by a violent storm, beginning at S.E. and going round the compass in a similar manner to the typhoons in the China Sea. As the gale increased, our sails were gradually reduced, until a small storm staysail was the only canvass we could spread. The sea had the appearance of breakers, and the birds actually threw themselves into the water, apparently to escape the fury of the wind. About four in the afternoon, just before the gale was at its highest, the wind shifted suddenly eight points, and brought the ship's head to the sea, which made a clear breach over the forecastle. Anticipating a change of this nature, we fortunately wore round a few hours before it occurred, and escaped the consequences which must have attended the stern of the ship being opposed to such breakers. The barometer during this gale fell an inch in eleven hours, and rose the same quantity in five hours, standing at 28·4 when at its lowest altitude. The temperature of the air rose nine degrees from eight in the morning to noon, and fell again to its former altitude at eight at night.

On the 24th, we were concerned to find several of the seamen afflicted with scurvy. Had this disease appeared the preceding year, in which they had been a very long time upon half allowance of salt provisions, and without any vegetable diet, it would not have been extraordinary; but in this year the seamen had been on full allowance of the best kind of provision, and had been living upon fresh beef in China, turtle and fish in the Arzobispo Islands and Petropaulski, besides the full allowance of lemon juice, pickled cabbages, and other anti-scorbutics. The season to the northward, it is true, had been more severe than that of the preceding year, and the duty in consequence more harassing; but this is not sufficient in my opinion to occasion the difference, and I cannot but think that the indulgence in turtle, after leaving the Arzobispo Islands, which was thought so beneficial at the moment, induced a predisposition to the complaint. The disease assumed an unusual character, by scarcely affecting the gums, while patients were otherwise so ill that a disposition to syncope attended the exertion of walking. Our cases fortunately were not numerous, being confined to six, and, after a few days' fresh provisions in California, were entirely cured.

On the 29th we were apprised of our approach to the coast of California by some large white pelicans, which were fishing a few miles to the westward of Point Pinos. We soon afterwards saw the land, and at eight at night moored in the Bay of Monterey. Early the following morning I waited upon the governor, and despatched messengers to the missions of

CHAP. VII. Nov. 1827.

St. Carlos and St. Cruz for vegetables, which were afterwards served daily in double the usual proportion to the ship's company, who benefited so much by the diet that, with one exception, they very soon recovered from all indisposition.

By some English newspapers, which were found in this remote part of the world, we learned the melancholy news of the death of His Royal Highness the Duke of York, and put the ship in mourning, by hoisting the flag half-mast during the time she remained in the port.

In my former visit to this country I remarked that the padres were much mortified at being desired to liberate from the missions all the Indians who bore good characters, and who were acquainted with the art of tilling the ground. In consequence of their remonstrances the governor modified the order, and consented to make the experiment upon a few only at first, and desired that a certain number might be settled in the proposed manner. After a few months' trial, much to his surprise, he found that these people, who had always been accustomed to the care and discipline of schoolboys, finding themselves their own masters, indulged freely in all those excesses which it had been the endeavour of their tutors to repress, and that many having gambled away their clothes, implements, and even their land, were compelled to beg or to plunder in order to support life. They at length became so obnoxious to the peaceable inhabitants, that the padres were requested to take some of them back to the missions, while others who had been guilty of misdemeanors were loaded with shackles and put to hard work, and when we arrived were employed transporting enormous stones to the beach to improve the landing-place.

The padres, conscious that the government were now sensible of the importance of the missions, made better terms for themselves than they had been offered by the Republican government. They were allowed to retain their places, and had their former salary of four hundred dollars a year restored to them, besides a promise of payment of arrears. In return for this a pledge was exacted from the padres, binding them to conform to the existing laws of the country, and in every way to consider themselves amenable to them. Thus stood the missionary cause in California when we quitted that country.

CHAP. VII. Nov. 1827.

We remained in Monterey until the 17th, and then sailed for St. Francisco to complete our water, which at the former place, besides being so scarce that we could hardly procure sufficient for our daily consumption, was very unwholesome, being brackish and mingled with the soapsuds of all the washerwomen in the place, and with streams from the bathing places of the Indians, into which they were in the habit of plunging immediately on coming out of the Temeschal.

Sán Francisco had undergone no visible change since 1826, except that the presidio had suffered from the shock of an earthquake on the 22d of April, which had greatly alarmed its inhabitants.

We had here the misfortune to lose James Bailey, one of our marines, who had long been an invalid.

Dec.

The third of December we left the harbour of St. Francisco, the shores of which, being newly clothed with snow, had a very wintry appearance; and on the 13th saw Cape St. Lucas. The next day we were off the Tres Marias, three high islands, situated seventy-five miles to the westward of San Blas, and well known by the frequent mention of them in the history of the

CHAP. VII. Dec. 1827.

Buccaneers, and by other early navigators in these seas. In consequence of a current setting out of the Gulf of California we were more to leeward than we were aware, and, with a view of saving time, passed through the channel between the two northernmost islands. In doing this we were becalmed several hours, and fully verified the old proverb, that the longest way round is often the shortest way home.

This channel appears to be quite safe; and in the narrowest part has from sixteen to twenty-four fathoms water; but the ground in other places is very steep, and at two miles distance from the shore to the westward there is no bottom at a hundred fathoms. When the wind is from the northward it is calm in this channel, and a current sometimes sets to the southward, which renders it advisable, on leaving the channel, to take advantage of the eddy winds which intervene between the calm and the true breeze to keep to the northward, to avoid being set down upon St George's Island. We found these islands twenty miles further from San Blas than they were placed on the charts.

The next morning the mountains on the mainland were seen towering above the white vapour which hangs over every habitable part of the land near San Blas. The highest of these, San Juan, 6,230 feet above the sea, by trigonometrical measurement, is the best guide to the road of San Blas, as it may be seen at a great distance, and is seldom obscured by fogs, while the low lands are almost always so. In my chart of this part of Mexico I have given its exact position. When the Piedra de Mer can be seen, it is an equally certain guide. This is a rock about ten miles west of the anchorage, a hundred and thirty feet high, with twelve fathoms water all round it.

The afternoon was well advanced before we anchored in the Road of San Blas, and the refreshing sea-breeze, sweeping the shores of the bay, had already dispersed the mist, which until then steamed from the hot swampy savannahs that for many miles surround the little isolated rock upon which the town is built. The inhabitants had not yet returned from Tepic, to which place they migrate during the *tiempo de las aguas*; the rainy season, so called from the manner in which the country is deluged with rain in the summer time.

At the time of our arrival in Mexico political affairs were very unsettled, and the property of British merchants was so much endangered, that I was compelled to accede to a request of the merchants, made through the vice-consul of San Blas, that I would delay my return to England, and remain until they could collect their funds, and that I would receive them on board for conveyance to Europe. As it would require several weeks before this specie could be got together, I proposed to visit Guaymas, and to examine the eastern coast of the Gulf of California; but this was frustrated by the revolt of Bravo, the vice-president of Mexico, and by the affairs of the state becoming so disorganized that the merchants further requested me not to quit the anchorage until they assumed a less dangerous aspect.

Shortly after our arrival we began to feel the effects of the unhealthy climate of San Blas, by several of the seamen being affected with intermittent fevers and agues, the common complaints of the place, particularly with persons who reside upon low ground, or who are

CHAP. VII. Jan. 1828.

exposed to the night air; and I regret to add that we here lost Thomas Moore, one of our most active seaman.

On the 27th of January, 1828, the agitation occasioned by the revolt had subsided, but unfortunately too late for me to proceed to Guaymas. However, as the principal part of the specie was to be shipped at Mazatlan, we put to sea a few days earlier than necessary for that purpose, that we might examine the Tres Marias and Isabella Islands, of which an account will be found in the Appendix. On the 3d February we reached Mazatlan, a very exposed anchorage, in which ships are obliged to lie so close to the shore that there would be very great difficulty in putting to sea with the wind from the W.S.W. to S.E. In the course of our survey, a rock having only eleven feet water upon it was discovered nearly in the centre of the anchorage, and occasioned no little surprise that of the many vessels which had put into the port all should have escaped being damaged upon it. Mazatlan is more healthy than San Blas, and our people here began to recover from the disorders they had contracted at that place.

Feb.

February 7th.—Having embarked the specie on the 24th, we put to sea on our return to San Blas, and ran along the shore with a northerly wind which is here prevalent from November to June. Lieutenant Belcher, in the cutter, kept in shore of the ship, and filled in those parts of the coast which could not be seen by her; and we thus completed a survey of the coast from Mazatlan to several miles South of San Blas. Between these two ports the water shoals so gradually that there is no danger whatever.

In my former visit to this place I found it necessary to proceed to Tepic to meet the merchants in consultation, and on that occasion I carried with me the necessary instruments for determining its position; by which it appears that it is only twenty-two miles direct from the port, though by the road it is fifty-two. It is in latitude 21° 30′ 42″ N., and its height above the sea 2,900 feet. By a register kept there during our stay, its mean temperature was 8°.1 below that of San Blas, and the range 2°.8 greater.

Tepic is the second town in importance in Xalisco, now called Guadalaxara, and contains 8000 inhabitants; but this population is augmented to about 11,000 in the unhealthy season upon the coast, at which time the people resort to Tepic. The town stands in the lowest part of a plain nearly surrounded by mountains, and not far from a large lake which exhales a malaria fatal to those who attempt to live upon its banks. On hot sunny days, of which there are many, the clouds as they pass often envelope the town, and strike a chill which proves fatal to hundreds of persons in the course of the year; and immediately the sun has set behind the mountains a cold deposit takes place, which is so great that it soon wets a person through. Under these circumstances Tepic is itself scarcely more healthy than the sea coast, and by the records of the Church it appears that the deaths exceed the births.

About a league and a half from Tepic, at the foot of Mount San Juan, stands Xalisco, near the site of the ancient town of that name. This town, though so close to Tepic, is very salubrious. I had the curiosity to examine the parish books here, in order to compare them with those at Tepic, and found the births to exceed the deaths in the proportion eighty-four to nineteen. In a population of only 3000, there were

CHAP. VII. March, 1828.

several persons upwards of a hundred years of age, while in Tepic there are very few above seventy-two. The Spaniards are fully aware of this difference of climate, and often send invalids from Tepic to Xalisco to recover their health; yet they continue to reside, and even to build new houses in the unhealthy spot their ancestors have chosen.

I had the good fortune to procure at this place, through the kindness of a gentleman who was residing at Tepic, a curious hive, constructed by bees, which had never been described, and of which an account will be found in the Appendix by Mr. Edward Bennet, to whom I am also indebted for his remarks upon the fishes we collected, which will appear in the natural history of the voyage.

The 1st of March was the day appointed for the embarkation of the specie at San Blas; but it was the 6th before it arrived, and the 8th before we could put to sea. On my way to the southward it became necessary to call at Acapulco for the purpose of securing the bowsprit previous to the passage round Cape Horn, as this could not be done conveniently in the open road of San Blas. While we were at anchor we received very distressing accounts of the state of affairs at Acapulco, and several vessels arrived from that place with passengers, who had been obliged to seek their safety by flight. It appeared that shortly after the revolt of Bravo, the Spaniards, with certain exceptions, were expelled from the Mexican teritory; and that Montesdeoca, a republican general, who was deeply indebted to some Spaniards at Acapulco, took advantage of this proclamation to liquidate his debt by marching against the town with a lawless troop of half-cast Mexicans, and by obliging the Spaniards to take refuge on board the vessels in the harbour, or to secrete themselves in the woods.

On putting to sea from San Blas, we kept along the land; the next day we determined the position of Cape Corrientes, a remarkable promontory on this coast, and on the 10th were within sight of the volcano of Colima. This mountain, by our measurement from a base of forty-eight miles, is 12,003 feet above the sea; and is situated in latitude 19° 25′ 24″ N. and longitude 1° 41′ 42″ E. of the arsenal at San Blas. On the 11th, in latitude 17° 16′ N., our temperature underwent a sensible change; previous to this date the thermometer had ranged between 71° and 73°, but on this day it rose to 82°, and did not fall again below 80° until after we quitted Acapulco. I notice the circumstance in consequence of Captain Hall having experienced precisely the same change in the same situation.*

Early in the morning of the 12th March we came within view of the Tetas de Coyuca, two peaked hills, which are considered by seamen the best guide to the port of Acapulco, and the next morning came to anchor in the most perfect harbour of its size that can be imagined.

The town of Acapulco was now tranquil, two Spaniards only being left in the place, and Montesdeoca having retired to Tulincinga, and disbanded his troops by order of the congress. The government of Acapulco was administered by Don Jose Manuella, a tool of Montesdeoca, who received me in his shirt, seated upon a Guyaquil hammock, in which he was swinging from side to side of the apartment.

Having effected our purpose in putting into the

* Hall's South America, p. 182.

CHAP. VII. March, 1828. port, and taken on board a supply of turkeys and fruit, which are finer here than in any other part of the world with which I am acquainted, we put to sea on the 18th. On the 29th March we crossed the equator in 99° 40′ W., and arrived at Valparaiso on the 29th of April, where we had the gratification to find, that his Royal Highness the Lord High Admiral had been pleased to mark his approbation of our proceedings on our voyage to the northward in 1826, by honouring the Blossom with the first commissions for promotion which had been issued under his Royal Highness's auspices. Here also I found orders awaiting my arrival to convey to Europe the remittances of May. specie, part of which arrived on the 19th May, and on the 20th we proceeded to Coquimbo to take on board the remainder.

On the 23d, when seven leagues S.W. ½ W. of this port, we were surprised by the shock of an earthquake, which shook the ship so forcibly, that some of the seamen imagined the anchor had been let go by accident, and was dragging the chain-cable with it to the bottom; while others supposed the ship had struck upon a shoal. An hour afterwards we felt a second shock, but much lighter. On our arrival in Coquimbo we found that these shocks had been felt by the inhabitants, and that there had been one the preceding night, which made the churches totter until the bells rang. Several slight shocks were afterwards felt by the inhabitants, who are very sensible to these subterraneous convulsions.

We remained several days in this port, which enjoys one of the most delightful climates imaginable, where gales of wind are scarcely ever felt, and in which rain is a very rare occurrence. Situated between the ports of Valparaiso and of Callao, where the dews alone irrigate the ground, it seems to partake of the advantages of the climates of each, without the inconveniences of the rainy season of the one, or of the heat and enervating qualities of the other.

On the 3d June all the specie was embarked, and we put to sea on our way to Brazil; passed the meridian of Cape Horn on the 30th, in very thick snow-showers, and after much bad weather arrived at Rio Janeiro on the 21st July. Here we received on board the Right Hon. Robert Gordon, ambassador to the court of Brazil, and after a passage of forty-nine days arrived at Spithead, and on the 12th October paid the ship off at Woolwich.

In this voyage, which occupied three years and a half, we sailed seventy-three thousand miles, and experienced every vicissitude of climate. It cannot be supposed that a service of such duration, and of such an arduous nature, has been performed without the loss of lives, particularly as our ship's company was, from the commencement, far from robust; and I have to lament the loss of eight by sickness, of four by shipwreck, of one missing, of one drowned in a lake, and of one by falling overboard in a gale of wind; in all fifteen persons. To individuals nothing probably can compensate for the selosses; but to the community, considering the uncertainty of life under the most ordinary circumstances, the mortality which has attended the present undertaking will, I hope, be considered compensated by the services which have been performed by the expedition.

In closing this narrative I feel it my duty to the

officers employed under my command, particularly to those whose immediate assistance I have acknowledged in my introduction, briefly to enumerate these services, as they are of such a nature that they cannot appear in a narrative, and as my professional habits have unqualified me for executing, with justice to to them, or with satisfaction to myself, the task of authorship which has devolved upon me as commander of the expedition, and which I should not have undertaken had I not felt confident that the candid public would look more to what has been actually done, than to the mode in which the proceedings have been detailed. In the Appendix to the quarto edition I have collected as much information as the nature of the work would admit. Besides the interesting matter which it will be found to contain, the expedition has surveyed almost every place it touched at, and executed plans of fourteen harbours, of which two are new; of upwards of forty islands, of which six are discoveries; and of at least six hundred miles of coast, one-fifth of which has not before been delineated. There have also been executed drawings and views of headlands, too numerous to appear in one work; and I hope shortly to be able to lay before the public two volumes of natural history.

In taking my leave, it is with the greatest pleasure I reflect that the Board of Admiralty again marked the sense they entertained of our exertions, by a further liberal promotion at the close of the expedition.

END OF THE NARRATIVE.

APPENDIX.

APPENDIX.

On the occurrence of the remains of Elephants, and other Quadrupeds, in the cliffs of frozen mud, in Eschscholtz Bay, within Beering's Strait, and in other distant parts of the shores of the Arctic seas.

BY THE REV. WM. BUCKLAND, D.D., F.R.S., F.L.S., F.G.S., AND PROFESSOR OF GEOLOGY AND MINERALOGY IN THE UNIVERSITY OF OXFORD.

HAVING been requested, at the time of Captain Beechey's return to England in October, 1828, to examine the collection of animal remains which he brought home from the shores of Eschscholtz Bay, and to prepare a description of them for the present publication, I attended at the Admiralty to assist at the opening and distribution of these specimens. The most perfect series, including all the specimens engraved in plates 1, 2, 3, (fossils), was selected for the British Museum; another series, including some of the largest tusks of elephants, was sent to the Museum of the College at Edinburgh, and other tusks to the Museum of the Geological Society of London. To the plates of these fossils, I have added a map of the bay in which they were collected, on the same spot where similar remains were first discovered by Lieutenant Kotzebue and Dr. Eschscholtz, on the 8th of August, 1816. Captain Beechey, in the course of his Narrative (p. 352 and 444, Vol. I.), has given a general description of the circumstances attending the examination of the locality in which the existence of these bones had been indicated by Lieutenant Kotzebue, and before I proceed to offer any observations of my own on these remarkable organic remains, or on the causes that may have collected them in such abundance on the spots where they are now found, I shall extract a further and more detailed account of the place and circumstances in which they were discovered, from the journal of Mr. Collie (surgeon to the English Expedition), by whom the bones were principally collected, and the chief observations and experiments made, on which Captain Beechey has founded his opinion, in which his officers, Lieutenant Belcher and Mr. Collie, entirely coincide with him, that the cliffs containing bones, which have been described by Kotzebue and Eschscholtz as icebergs covered with moss and grass, are not composed of pure ice, but are merely one of the ordinary deposits of mud and gravel, that occur on many parts of the shores of the Polar Sea, being identical in age and character with diluvial deposits of the same kind which are known to be dispersed over the whole of Europe, and over a large part of Northern Asia and North America; and presenting no other peculiarities in the frozen regions of the North, than that which results from the present temperature of these regions, causing the water which percolates this mud and gravel to be congealed into ice.

The question of fact, whether the cliffs containing these bones of elephants, and other land quadrupeds, are composed of "masses of the purest ice, a hundred feet high, and covered on their surface with vegetation," as stated in the voyage of Lieutenant Kotzebue, (p. 219, English translation), or are simply composed, as Captain Beechey thinks them to be, of ordinary diluvium, having its interstices filled up with frozen water, is important, as it affects materially the consideration of the further question, as to what was the state of the climate of the arctic regions at the time when they were thickly inhabited by genera of the largest quadrupeds, such as at present exist only in our warmest latitudes; this being a point of much interest and curiosity, in relation to the history of the physical revolutions that have affected our planet, and on which there still exists a difference of opinion among those individuals who have paid the greatest attention to the subject.

Before I proceed to Mr. Collie's observations on the spot in which they were found, I shall extract from his journal a list of the total number of animal remains collected during the short time he was with Captain Beechey in Eschscholtz Bay, and add my own list and description of the most perfect of these specimens, which I have selected to be engraved.

List, showing the total number of animal remains collected in Eschscholtz Bay, taken from the Journal of Mr. Collie.

ELEPHANT.

1 Lower jaw, nearly complete.
7 Molar teeth.
9 Tusks. Five of them large, and weighing from one hundred to one hundred and sixty pounds each. Four small; one of these was found in the debris of the cliff half way up; the circumference of the largest tusk at its root is twenty inches, and at three feet above the root twenty-one inches and a half: another tusk, in which part of the tip is wanting, measures nine feet two inches along the curve from the root to the tip, and five feet two inches across the chord of its curve.
4 Fragments of tusks.
3 Dorsal vertebræ, five inches and a half in diameter.
1 Atlas.
1 Os innominatum, nearly perfect.
1 Ilium, imperfect.
1 Os pubis, imperfeet.
4 Fragments of scapulæ, one of them tolerably complete.
1 Portion of humerus.
5 Femora, one of them almost complete.
4 Fragments of femora.
2 Tibiæ, one of them nearly complete.
1 Tarsal bone.
1 Os calcis, entire, taken out of the cliff.
1 Cuboides, nearly entire.
1 Cuneiform.
1 Phalangal bone.

URUS.

1 Skull, incomplete.
3 Fragments of horns.
1 Femur.
3 Tibiæ.
1 Dorsal vertebra.
1 Sacrum.

MUSK-OX.

1 Skull, with horns attached, incomplete and very modern.

DEER.

1 Fragment of antler.
4 Tibiæ, entire.
3 Metatarsal bones.
1 Os calcis.

Some of these are probably casual and modern, and derived from rein-deer that now frequent this part of America.

HORSE.

1 Astragalus.
1 Metacarpus.
1 Metatarsus.

Description of the most perfect specimens of animal remains brought home by Captain Beechey from Eschscholtz Bay, and selected by Dr. Buckland to be engraved in pl. 1, 2, 3, *(fossils). All these specimens are deposited in the British Museum.*

PLATE I.—(Fossils.)

Fig. 1. Lower jaw of extinct elephant, containing two molar teeth.
2. Profile of No. 1, on the left side.
3. Molar tooth of elephant.

If we compare this jaw and the teeth with the fossil jaws and teeth described by Cuvier, we shall find them to exhibit all the leading characters pointed out by that great naturalist, as distinguishing the fossil elephant from any existing species.

First. The teeth possess that broadness of surface which is more constant in the fossil teeth than

Plate 1. (fossils)

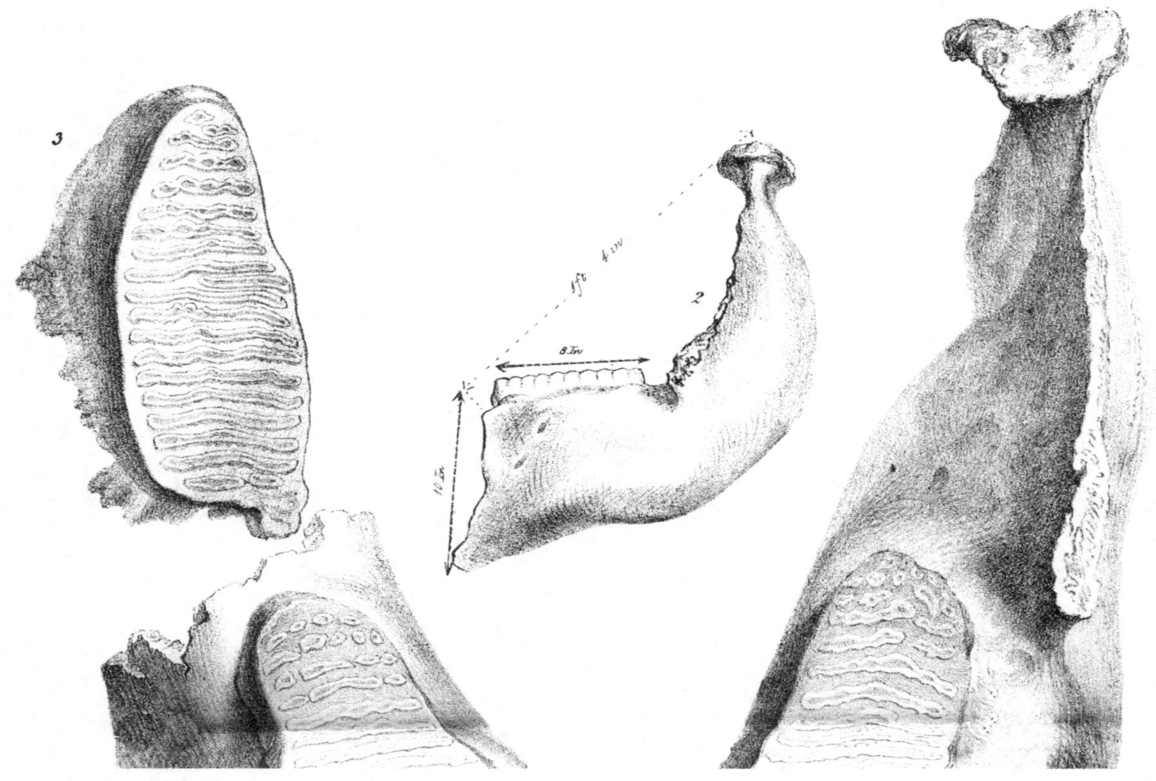

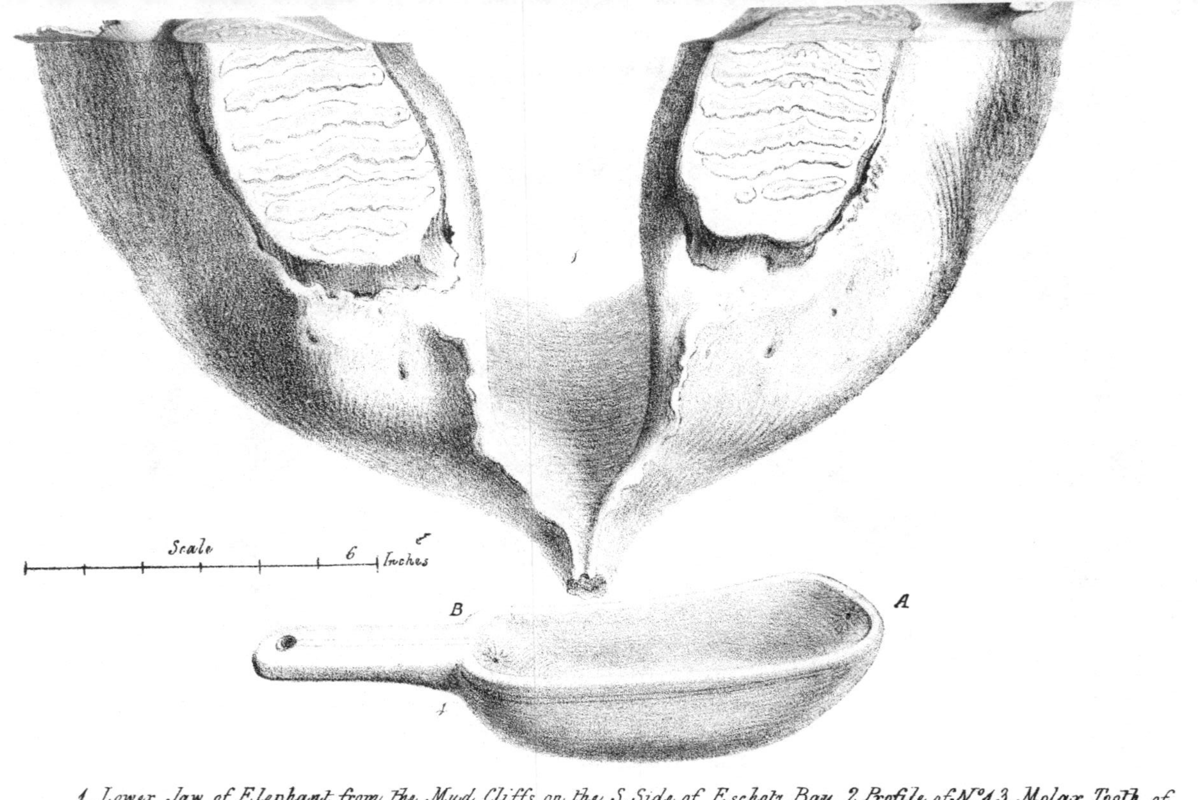

1. Lower Jaw of Elephant from the Mud Cliffs on the S. Side of Escholz Bay, 2 Profile of No. 1. 3. Molar Tooth of Elephant, from the same place 4 Ivory scoop made by the Esquimaux near Escholtz Bay from a fossil Tusk

Plate 2 (fossils)

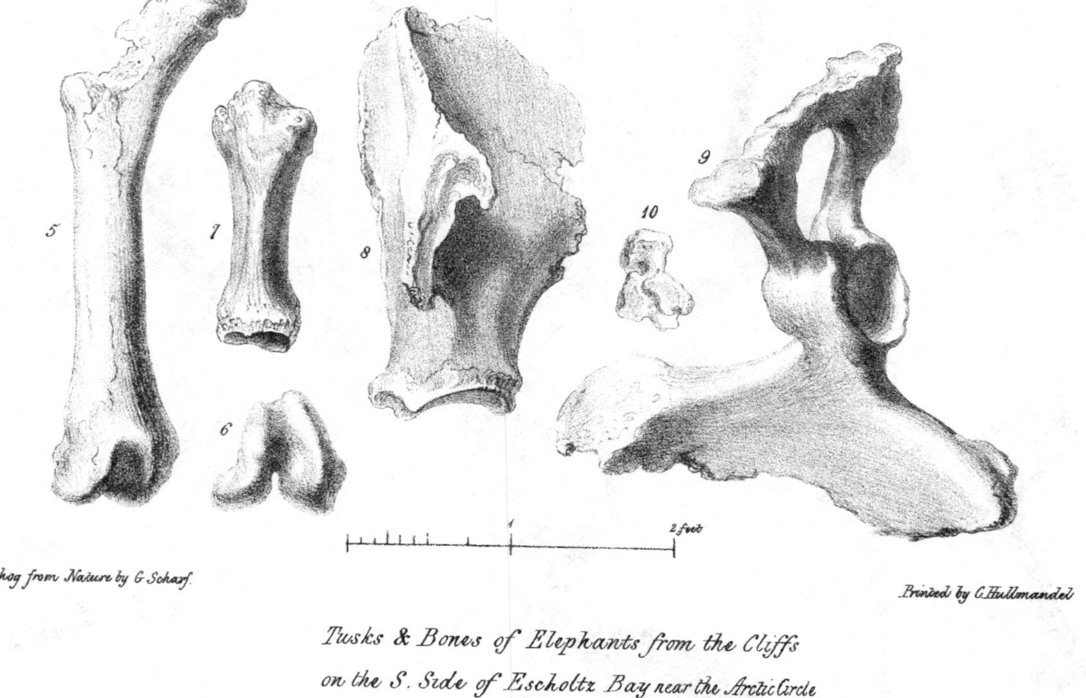

Lithog from Nature by G. Scharf.

Printed by C. Hullmandel

Tusks & Bones of Elephants from the Cliffs on the S. Side of Escholtz Bay near the Arctic Circle

Plate 3 (fossils)

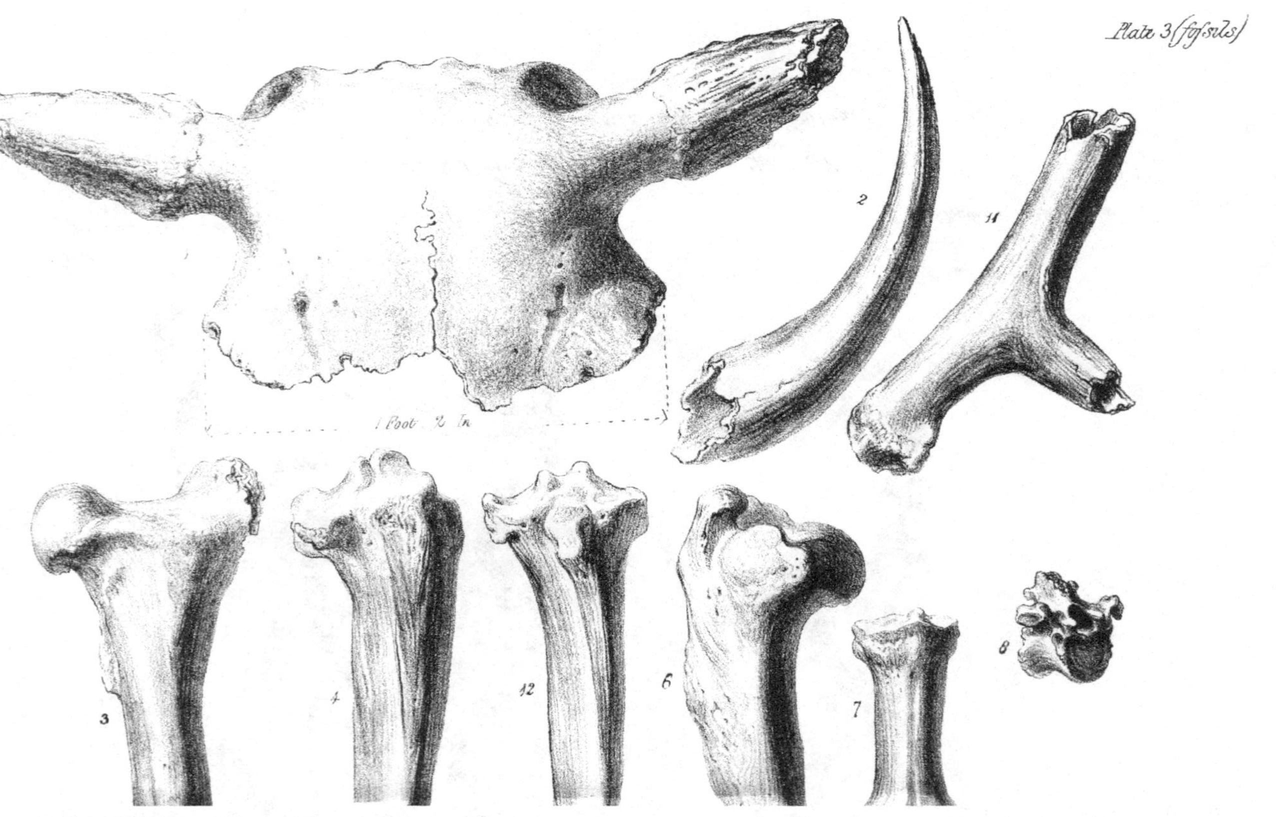

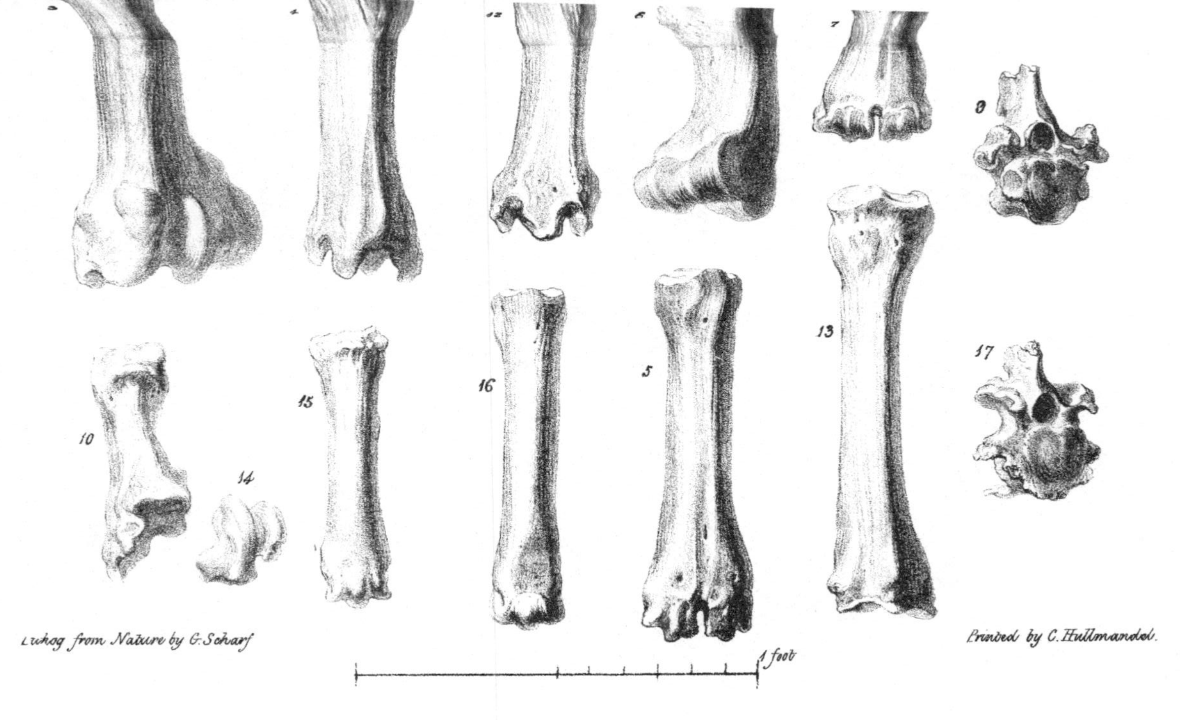

Lithog from Nature by G. Scharf

Printed by C. Hullmandel.

Bones & Horns of Oxen & Deer, & Bones of Horses, from the Cliffs, on the S Side of Escholtz Bay near the Arctic Circle

either the greater number or greater thinness of the component laminæ.

Secondly. The position of the teeth in the jaw is at a less acute angle, and more nearly parallel than in the recent species.

Thirdly. The channel within the chin at the junction of the two sides of the jaw is broader in proportion to its length; the exterior projecting point of the chin, also at the apex of the jaw, is not so prominent as in recent elephants, but truncated as in the fossil species. Compare this jaw with those of fossil elephants engraved in Cuvier's Ossemens Fossiles, vol I. pl. II. fig. 1, 4, 5. Pl. V. fig. 4, 5. Pl. VIII. fig. 1. Pl. IX. fig. 8, 10. Pl. XI. fig. 2.

4. An ivory scoop, purchased by Captain Beechey from the Esquimaux, and made of a portion of a very large fossil tusk; it shows at the extremities of the excavated part at A and B a point that indicates the axis of the tusk; this ivory is firm and solid, and in nearly the same high state of preservation as the entire tusks from Eschscholtz Bay.

Plate II.—(Fossils.)

Fig. 1. Entire tusk of an elephant, measuring ten feet in the curve and six inches in diameter at the largest part, and weighing one hundred and sixty pounds.

2. Another tusk of an elephant, nine feet six inches in the curve.

Both these tusks are nearly perfect; two other tusks of nearly the same size have marks of having been chopped with some cutting instrument; this has probably been done by the Esquimaux to ascertain their solidity and fitness for making their utensils: the large scoop made of fossil ivory—see Plate I. (Fossils) fig. 4—shows that these people apply the fossil tusks to such purposes. The tusks which are thus chopped appear to have been left on the shore as unfit for use, on account of the shattered condition of their interior.

3, 4. Longitudinal view of the tusks represented laterally in figs. 1, 2. They both possess the same double curvature as the tusks of the great fossil elephant in the Museum at Petersburgh from the icy cliff at the mouth of the Lena, in Siberia.

This curvature is very frequent but not constant in fossil tusks; it occurs also sometimes in the tusks of recent elephants: there is a similar double curvature in the recent tusk of a small elephant from Ceylon in the possession of the Earl of Carnarvon, at Highclere, Hants.

Fig. 5. Femur of elephant.

6. Epiphysis from the lower extremity of another femur of elephant.

7. Tibia of elephant.

8. Scapula of elephant.

9. Os innominatum of elephant.

10. Os calcis of elephant.

PLATE III.—(Fossils.)

Fig. 1. Head of a boss urus, in precisely the same condition with the fossil bones of elephants, and very different from the state of the head of a musk-ox with the external case of the horns still attached to it, which was brought home with the fossil bones, and was found with them on the beach at the bottom of the mud cliff in Eschscholtz Bay, but is so slightly decayed that it seems to have been derived from a carcass that has not long since been stranded by the waves. This head of a musk-ox is not engraved, as it cannot be considered fossil.

2. External horny case detached from the bony core of the horn of an ox: it is in a state equally fresh with the head of the musk-ox just mentioned; and, like it, appears to be derived from an animal recently cast on shore.

3. Femur of an ox.

4. Tibia of an ox.

5. Metatarsus of an ox.

6. Humerus of an ox.
7. Metacarpus of an ox.
8. Dorsal vertebra of an ox.
9. Dorsal vertebra of an ox.
10. Os calcis of an ox.
11. Base of the horn of a deer, similar to horns that occur in the diluvium of England, and somewhat resembling the horn of a rein-deer.
12. Tibia of a large deer.
13. Radius of a large deer.
14. Astragalus of a horse.
15. Metacarpus of a horse.
16. Metatarsus of a horse.
17. Cervical vertebra of an unknown animal. It has been compared with all the skeletons in the collection at Paris, by Mr. Pentland, without finding any to which it can be referred: he thinks the nature of the articulation more resembles that in the sloth and ant-eaters than in any other animal; but the bone differs from them in other respects, and approaches to the character of the Pachydermata. The animal, whatever it was, seems to have differed essentially from any that now inhabit the Polar Regions of the Northern Hemisphere.

NOTES EXTRACTED FROM THE JOURNAL OF MR. COLLIE.

" The attention of the world has been called to the remarkable cliff in which fossil bones were found by Dr. Eschscholtz in August, 1816. On my first visit to it in the month of July, 1826, time did not permit me to do more than take a view of the most eastern part, and examine the nature of the icy fronting which it presented. At that time I saw no traces of fossils; this cliff faces to the N., and extends in nearly a right line, with few interruptions, for two miles and a half, and is in general about ninety feet high. It is composed of clay and very fine quartzy and micaceous sand, assuming a grayish appearance when dry. The land hehind rises gradually to an additional height of one hundred feet, and is covered with a black boggy soil nourishing a brown and grey lichen, moss, several species of ericæ, graminiæ and other herbaceous plants, and is intersected with a few valleys containing small streams, and having their more protected declivities adorned with shrubs of willow and dwarf betula *(betula incana)*.

A continual waste of the cliff is produced at the upper part by its falling down in considerable lumps to the bottom, where the debris remains for a longer or shorter time, and covers the front to a greater or less height; in some places, almost to the very top. Large masses are sometimes seen rent off and standing out from the body of the cliff ready to have their last slight hold washed away by the next shower, or by a little more thawing and separation of the frozen earth that serves them for attachment. The lumps of soil that fall are still covered with the herbaceous and shrubby verdure that grew upon them. The perpendicular front of the cliff of frozen mud and sand is every summer gradually decreasing by the melting of the ice between its particles into water, which trickles down and carries with it loose particles of earth. In some portions of the cliff the earthy surface is protected with ice, partly the effect of snow driven into the hollows and fissures, and partly from the congelation of water, which may have collected in chinks or cavities: these masses of ice dissolve in summer, and the water running from them carries with it any earth that lies in its way, and mixes itself with, and moves forward, the mass of debris below. By this gradual thawing and falling of the cliff, the black boggy soil at the surface becomes undermined, and assumes the projecting and overhanging appearance which is so remarkable. At the base of the greater part of the cliff the debris is washed by the sea at full tide, and being gradually carried away by the retiring waters, is spread out into an extensive shoal along the coast. It was in this shoal, where it is left dry by the ebbing tide, to the distance of fifty or a hundred yards from the cliff that the greater number of the fossil bones and teeth were discovered, many of them so concealed as only to leave a small end or knob sticking up; they were dispersed very irregularly. Remains of the musk-oak were found on this shoal, along with those of elephants.

The few specimens taken out of the cliff, or more properly from the debris, on the front of it (for none, I believe, were taken out of the very cliff), were in a better state of preservation than those which had been alternately covered and left exposed by the flux and reflux of the tide, or imbedded in the mud and clay of the shoal.

A very strong odour, like that of heated bones, was exhaled wherever the fossils abounded. Quantities of rolled stones, mostly of a brownish sandstone, lay upon the shoal, left dry by the receding sea. With these were also porphyritic pebbles.

Parts of some of the tusks, where they had been imbedded in the clay and sand, were coloured blue by phosphate of iron, and many of the teeth were stained in the same manner. The circular layers of the tusks in the more decayed specimens were distinctly separated by a thin vein of fibrous gypsum.

In those parts of the bay where there are no cliffs, the waves are kept at a distance from the land by a gravelly beach, which they have thrown up for a considerable extent round the entrance of the streams which come down the valleys. These beaches have formed rounded flats containing marshes or lakes: not unfrequently rather a luxuriant herbage covers their surface. The land behind them rises by a gentle slope. Great part of the shore of Kotzebue Sound is made up of a diluvial formation, similar to that on the south shore of Eschscholtz Bay. From Hut Peak to Hotham Inlet it exhibits many cliffs similar to those just described, and also others with an uniform and steep slope, partly covered with verdure, and partly exposing the dry sand and clay which compose them. The most elevated cliffs form the projecting head-land of Cape Blossom, and abound in ice, notwithstanding their southern aspect, particularly at Mosquito station and Cape Blossom. In their neighbourhood I observed the natives had recently formed coarse ivory spoons from the external layer of a fossil elephant's tusk. The ice here in the end of September showed itself more abundantly than it did in the middle of the same month on the cliffs of Escholtz Bay which have a northern aspect."

Mr. Collie then proceeds to explain still further his ideas of the manner in which masses and sheets of pure ice may have been collected in hollows and fissures on and near the front of the cliff in Eschscholtz Bay.

1st. By the accumulation of snow drifted into hollows subjacent to the overhanging stratum of black boggy soil that forms the brink of the cliff, and subsequently converted into ice by successive thawing and freezing in spring and summer.

2dly. They may have been formed from water collected in deep fissures and cavities that intersect the falling cliff near its margin. The inclined position of the land immediately above this margin of peat, and the annual undermining which is produced by the thawing of the frozen mud beneath it, produce occasional land slips, and movements of the edge of the cliff towards the sea; these cause cracks and fissures of the soil in various directions, but chiefly parallel to the external face of the cliff. When these fissures descend through the black boggy soil of the surface into the frozen mud below, they become receptacles for the formation of ice, since the water that oozes into them is congealed upon their sides until it entirely fills them with a wall or dyke of solid ice. The fall of a mass of mud from the outer side of one of these walls would expose this ice, forming a case over the inner side of the fissure in which it was accumulated.

3dly. The manner in which an extensive facing of pure ice may be formed on these cliffs, by water during the summer trickling down their frozen surface from the soil above, and becoming converted to ice in the course of its descent, has been described by Captain Beechey (pages 353 and 454, Vol. I.)

Lieutenant Belcher, in his notes, proposes another theory to explain the occurrence of masses of pure ice immediately below the margin of peat on the top of the cliff on the southern shore of Eschscholtz Bay. He conceives that between the superficial bed of spongy peat, and the mass of frozen mud which forms the body and substance of this cliff, the water oozing downwards through the peat, during the thaw of each successive summer, is stopped at the point where it comes into contact with the perpetually frozen earth below, and there accumulates into a thick horizontal sheet of pure transparent ice, and that it is the broken edge of this icy stratum which becomes exposed in the margin of the cliff during the process of slow and gradual destruction which it is continually undergoing.

This opinion, however, is I believe peculiar to Lieutenant Belcher. The experiment made by Mr. Collie, in boring horizontally into the cliff through a vertical face of ice, until he penetrated the frozen mud behind it, shows, that in this case the ice was merely a superficial facing of frozen water, consolidated as it descended the front of the cliff; and his further experiments in digging vertically downwards, in two places, through the peat into frozen mud, and finding no traces of any intermediate bed of ice, appear unfavourable to any hypothesis as to the formation of a stratum of pure ice between the superficial peat and subjacent mud.

It has just been stated that Captain Beechey and Mr. Collie propose three different solutions to explain the origin of these hanging masses of ice near the upper margin of the vertical cliffs: 1st, That they may have been formed from snow drifted into hollows of the cliffs, and subsequently converted into ice; 2dly, From water consolidated into ice within fissures and cavities, caused by the subsidence and falling forwards of the frozen mud; 3dly, From water trickling down the external surface of the cliff, and freezing as it descended. To these the theory of Lieutenant Belcher would add a fourth process, by which a horizontal bed of ice is formed between a superficial bed of peat and the subjacent mud. These hanging masses of ice, whatever may be their origin, appear to have been so abundant at the time of the Russian expedition to this coast, as to have made Kotzebue and Eschscholtz imagine the entire cliff behind them to be an iceberg; an opinion which all the English officers agree in considering to be erroneous, since the view and descriptions of the cliff on the south shore of Escholtz Bay, given at p. 219 of the English translation of Kotzebue's Voyage, do not correspond with the state of this coast when it was subsequently visited by the crew of the Blossom.

The following are Captain Kotzebue's observations respecting it: * "We had climbed much about, without discovering that we were on real icebergs. Dr. Eschscholtz found part of the bank broken down, and saw, to his astonishment, that the interior of the mountain consisted of pure ice. At this news we all went, provided with shovels and crows, to examine these phenomena more closely, and soon arrived at a place where the bank rises almost perpendicularly out of the sea to the height of a hundred feet, and then runs off, rising still higher: we saw masses of the purest ice, of the height of a hundred feet, which are under a cover of moss and grass, and could not have been produced but by some terrible revolution. The place, which by some accident had fallen in, and is now exposed to the sun and air, melts away, and a good deal of water flows into the sea. An indisputable proof that what we saw was real ice is the quantity of mammoth's teeth and bones which were exposed to view by the melting, and among which I myself found a very fine tooth. We could not assign any reason for a strong smell, like that of burnt horn, which we perceived in this place. The covering of these mountains, on which the most luxuriant grass grows to a certain height, is only half a foot thick, and consists of a mixture of clay, sand, and earth; below which the ice gradually melts away, the green cover sinks with it, and continues to grow."

Mr. Collie's experiments, which I have before alluded to, in digging both horizontally and vertically through the ice and peat into frozen mud, show that, at the points where they were made, the cliff formed no part of any iceberg. Still more decisive is the important fact, that on the two occasions when it was visited by the English expedition, the patches of ice upon the cliff in question were very few in number, and variable from one year to another; that the "masses of the purest ice of the height of a hundred feet," which were seen by the Russian officers, had entirely vanished; and that nearly the whole front of the cliff, from the sea at its base to the peat that grew on its summit, presented a continuous mass of indurated mud and sand, or of under-cliffs formed by the subsidence of these materials.

* Kotzebue's Voyage of Discovery, Vol. I. p. 220.

It seems quite certain therefore that there must have been a material change in the quantity of ice on the cliff in Eschscholtz Bay in the interval between the visits of Lieutenant Kotzebue and Captain Beechey; and if we suppose that, during this interval, there was an extensive thawing of the icy front that was seen by Kotzebue, but which existed not at the time of Beechey's visit, we find in this hypothesis a solution of the discrepancy between these officers; since what to the first would appear a solid iceberg, when it was glazed over with a case of ice, would, after the melting of that ice, exhibit to the latter a continuous cliff of frozen diluvial mud. Whilst the ice prevailed all over the front of the cliff, any bones that had fallen from it before the formation of this ice, and which lay on the under cliffs or upon the shore, must, by an error almost inevitable, have been presumed to fall from the imaginary iceberg.

This circumstance seems to suggest to us that it is worthy of consideration whether or not there may have existed any similar cause of error in the case of the celebrated carcass of an elephant in Siberia, which is said to have fallen entire from an iceberg in the cliffs near the Lena. The Tungusian, who discovered this carcass suspended in what he called an iceberg, may possibly have made no very accurate distinction between a pure iceberg and a cliff of frozen mud.

It is stated by Lieutenant Belcher, that at a spot he visited on the S. E. shore of Eschscholtz Bay, on ascending what appeared at first to be a solid hillock, he found a heap of loose materials, unsafe to walk on, and having streams of liquid mud oozing from it on all sides through coarse grass; that as the melting subsoil of this hillock sinks gradually down, the incumbent peat subsides with it; so that at no very distant period the entire hillock will disappear. In other mud cliffs, also, he observed similar streams of liquid mud, accompanied by a depression of the surface immediately above them. Thus, from the month of June to October these cliffs are constantly thawing, and throwing down small avalanches of mud, which, between Cape Blossom and Cape Kruzenstern, are so numerous, that you can scarce stand there an hour without witnessing the downfall of some portion of the thawing cliffs. Hence originates a succession of ravines and gullies, which do not run far inland, and afford no sections, being covered with the debris of the superficial peat that falls into them. Small streams of muddy water, of the consistence of cream, ooze from the sides of these ravines, the water being supplied by the melting of the particles of ice which pervade the substance of the frozen mud and peat.

There remain, then, three important points, on which all the English officers concur in the same opinion: 1st, That the bones and tusks of elephants at Eschscholtz Bay are not derived from the superficial peat; 2dly, That they are not derived from any masses of pure ice; 3dly, That, although collected chiefly on the shore at the base of the falling cliff, they are derived only from the mud and sand of which this cliff is composed.

The occurrence of cliffs composed of diluvial mud is by no means peculiar to the south shore of Eschscholtz Bay. It will be seen by reference to the map (plate I. Geology), that they are more extensive, but at a less elevation along the north shore of this same bay, and also on the south-west of it, at Shallow Inlet, in Spafarief Bay. Indeed, in following the line of coast north-eastwards, from the Arctic Circle, near Beering's Strait, to lat. 71° N., wherever the coast is low, there is a long succession of cliffs of mud, in the following order: 1. Schischmareff Inlet. 2. Bay of Good Hope, on the south of Kotzebue's Sound. 3. Spafarief Bay, at the south-east extremity of Kotzebue's Sound. 4. Elephant Point, in Eschscholtz Bay. 5. At the mouth of the Buckland River, at the head of Eschscholtz Bay. 6. The north coast of Eschscholtz Bay. 7. Cape Blossom. 8. Point Hope. 9. From Cape Beaufort to twenty miles east of Icy Cape. 10. Lunar Station, near lat. 71°.—At the base of the mud cliff, fifteen feet high, in the Bay of Good Hope, a small piece of a tusk of an elephant was found upon the shore. At Shallow Inlet, the mud cliff was fifteen feet high, without any facings of ice, or appearance of bones; yet there was the same smell at low water as in the cliffs near Elephant Point, that abound so much in bones. At Icy Cape the cliffs of mud behind the islands were about twenty feet high, but were not examined. Patches of pure ice were observed hanging on the mud cliffs in many places along this coast, but only where there was peat at the top; hence it may be inferred, that the ice, in such cases, is formed by water oozing from the peat. At High Cape, near Hotham Inlet, is a cliff of mud, a hundred feet high, covered at the top with peat, and having patches of ice upon its surface; but no bones were found here. In those parts of the coast where the cliffs are rocky there were no facings of ice.

Having thus far stated the evidence we possess respecting the facts connected with the discovery of these bones in Eschscholtz Bay, I will proceed to offer a few remarks in illustration and explanation of them, and to consider how far they tend to throw light on the curious and perplexing question, as to what was the climate of this portion of the world at the time when it was inhabited by animals now so foreign to it as the elephant and rhinoceros, and as to the manner in which, not only their teeth and tusks and dislocated portions of their skeletons, but, in some remarkable instances, the entire carcasses of these beasts, with their flesh and skin still perfect, became entombed in ice, or in frozen mud and gravel, over such extensive and distant regions of the northern hemisphere.

The bones from Eschscholtz Bay, like most of those we find in diluvial deposits, are no way mineralized: they are much altered in colour, being almost black, and are to a certain degree decomposed and weakened; yet they retain so much animal matter, that not only a strong odour like that of burnt horn is emitted from them on the application of heated iron, but a musty and slightly ammoniacal smell is perceptible on gently rubbing their surface.

It must not, however, be inferred that this high state of preservation can exist only in bones that have been imbedded in frozen mud or frozen gravel, since dense clay impermeable to water has been equally effective in preserving the remains of the same extinct species of animals in the milder climate of England. There are, in the Oxford Museum, bones of the elephant and rhinoceros, from diluvial clay in Warwickshire and Norfolk, that are scarcely at all more decomposed than those brought by Captain Beechey from Eschscholtz Bay, and are nearly of the same colour and consistence with them. I have also a fragment of the tusk of an elephant from the coast of Yorkshire, near Bridlington, of which great part had been made into boxes by a turner of ivory before the remainder came into my possession; and on comparing the state of the residuary portion of this tusk from Yorkshire with that of the scoop made of a fossil tusk by the Esquimaux in Eschscholtz Bay, I find the difference scarcely appreciable.

It is mentioned, both by the Russian and English officers, that a strong odour like that of burnt bones is emitted from the mud of the cliffs in which they discovered these animal remains in Eschscholtz Bay: other observers have stated the same thing of the mud cliffs in Siberia, near the mouth of the Lena, which contain similar organic remains. But it is also stated by Mr. Collie that a like odour was perceived at the base of another mud cliff in Shallow Inlet, near Eschscholtz Bay, where there were no bones; and as in this latter case we must attribute it to some cause unconnected with the bones, and probably to gaseous exhalations from the mud itself, we may, I think, draw the same inference as to the origin of the odour in all the other cases also; thus in Eschscholtz Bay, where nearly all the bones were collected at the base of the cliff on the beach below high water, how can the presence of two or three bones only, lying half way up the cliff, account for the odour which is emitted over a distance of more than a mile along this shore? How inadequate is a cause so partial to so general an effect! since, however numerous may be the animal remains that are buried in the interior of the cliff, no exhalations from them can escape through their impenetrable matrix of frozen mud; and even if that fallen portion of mud which constitutes the under-cliff be ever so abundantly loaded with fossil bones, it is scarcely possible that these should undergo such rapid decomposition as to transmit strong exhalations to the surface through so dense a substance as saturated clay; in fact, their high degree of preservation shows that no such rapid decomposition has taken place.

With respect to the matrix of frozen mud, from which these remains are said to be derived, it appears, from specimens of it adhering to the bones, that it consists of micaceous sand and quartzose sand, intermixed with fine blue clay. In the hollow of one of the tusks I found a quantity of this compound, and some fragments of mica slate. All these ingredients may have been derived from the detritus of primitive micaceous slates, such as constitute a large part of the fundamental rocks of the neighbourhood of Eschscholtz Bay.

Pebbles of porphyry also are said to occur in the cliff, and also on the beach below it, mixed, in the latter case, with pebbles of basalt and sandstone, and a few large blocks of basalt. No rock was noticed in this district from which these rolled stones could have been derived; some of those upon the beach may possibly have been drifted thither on floating icebergs. The tranquil state and retired position of the bay render it improbable that these pebbles have been brought to their present place by the influence of any existing submarine currents.

It is important to clear from confusion two facts mentioned by Captain Beechey, viz. the occurrence of remains of the rein-deer, and of the musk-ox, along with bones of the elephant in Eschscholtz Bay. Had the bones of either of these arctic animals been found unequivocally mixed with the bones of elephants in any undisturbed part of the high cliff, it would have followed that the rein-deer and the musk-ox must have been coeval with the fossil elephant; and this fact would have been nearly decisive of the question as to the climate of this region at the time when it was inhabited by these three species of animals. But as all the fossil remains collected in Eschscholtz Bay, with the exception of a very few bones and the tusk of an elephant that lay high up in the under cliff, were collected on the beach between high and low water mark, nothing is more probable than that the bones of modern animals should become mixed with these fossils after they had fallen upon the beach in the recesses of a quiet bay.

Kotzebue (vol. I. p. 218) says he saw many horns of reindeer lying on the shore in Eschscholtz Bay, and conjectures that the Americans, who frequent these coasts occasionally in the hunting season, may have brought with them the rein-deer from which these horns had been derived. This hypothesis may explain the presence of such horns in a spot which no wild rein-deer are known to frequent at present; but as Kotzebue (p. 219) mentions also the abundance of drift-wood upon the shores of this bay, it is probable that the same currents which brought the wood may have also brought the carcasses of rein-deer, and have stranded them on the shores where their horns were found.

The agency of the same currents, to which I have referred the drifting of the carcasses of rein-deer into Eschcholtz Bay, will also equally explain the presence of recent bones of the musk-ox in this bay, on the same shoal, with the bones of elephants that had fallen from the cliff. I have already stated that the condition of the skull and horns of a musk-ox, which were brought home with the fossil bones, is so very recent, and differs so essentially from the condition of all the bones of elephants from this place, that it is impossible it can have been buried in the same matrix with them; for, in such case, all would have been nearly in the same state, either of preservation or decay.

It is stated by Cuvier (Ossemens Fossiles, second edition, vol. iv. p. 165), that a similar doubt is attached to the heads of musk-oxen described by Pallas and Ozeretzkovsky, as found near the mouth of the Ob, and at the embouchure of the Yana, and that there is yet no sufficient proof of the existence of any fossil species of musk-ox that may be considered of the same age with the fossil elephant, or which can be brought in evidence as to the question of the climate of the polar regions when these elephants were living. Of the very few remains of musk-oxen which have yet been found, it does not appear that any have been buried at a great depth.

There is nothing peculiar to Eschscholtz Bay in the occurrence of bones of horses with those of elephants: from the number of localities in which their teeth and bones have been found together, in diluvial deposits, it appears that more than one species of horse was co-extensive with the fossil elephant in its occupation of the ancient surface of the earth. Wild horses are at present almost unknown, except in warm or temperate latitudes.

We may now consider how far the facts we have collected respecting the bones in Eschscholtz Bay are in accordance with similar occurrences, either in the adjoining regions of the north, or in other still more distant parts of the earth, and in different latitudes.

It is stated by Pallas, in the 17th volume of the New Commentaries of the Academy of Petersburgh, 1772, that throughout the whole of northern Asia, from the Don to the extreme point nearest America, there is scarce any great river in whose banks they do not find the bones of elephants and other large animals, which cannot now endure the climate of this district, and that all the fossil ivory which is collected for sale throughout Siberia is extracted from the lofty, precipitous, and sandy banks of the rivers of that country; that in every climate and latitude, from the zone of mountains in central Asia to the frozen coasts of the Arctic Ocean, all Siberia abounds with these bones, but that the best fossil ivory is found in the frozen lands adjacent to the arctic circle; that the bones of large and small animals lie in some places piled together in great heaps, but in general they are scattered separately, as if they had been agitated by waters, and buried in mud and gravel.

The term mammoth has been applied indiscriminately to all the largest species of fossil animals, and is a word of Tartar origin, meaning simply "animal of the earth." It is now appropriated exclusively to the fossil elephant, of which one species only has been yet established, differing materially from the two existing species, which are limited, one to Asia the other to Africa.

Of all the fossil animals that have been ever discovered, the most remarkable is the entire carcass of a mammoth, with its flesh, skin, and hair still fresh and well preserved, which in the year 1803 fell from the frozen cliff of a peninsula in Siberia, near the mouth of the Lena*. Nearly five years elapsed between

* The details of this case were published by Dr. Tilesius in the fifth vol. of the Memoirs of the Academy of Petersburg, and also by Mr. Adams in the Journal du Nord, printed at Petersburg in 1807,

the period when this carcass was first observed by a Tungusian in the thawing cliff in 1799, and the moment when it became entirely disengaged, and fell down upon the strand, between the shore and the base of the cliff. Here it lay two more years, till great part of the flesh was devoured by wolves and bears: the skeleton was then collected by Mr. Adams and sent to Petersburgh. Many of the ligaments were perfect, and also the head, with its integuments, weighing four hundred and fourteen pounds without the tusks, whose weight together was three hundred and sixty pounds. Great part of the skin of the body was preserved, and was covered with reddish wool and black hairs; about thirty-six pounds of hair were collected from the sand, into which it had been trampled by the bears.

The following description, by Mr. Adams, of the place in which this mammoth was found will form an interesting subject of comparison with Captain Beechey's account of the cliff in Eschscholtz Bay: " The place were I found the mammoth is about sixty paces distant from the shore, and nearly a hundred paces from the escarpment of the ice from which it had fallen. This escarpment occupies exactly the middle between the two points of the peninsula, and is two miles long: and in the place where the mammoth was found, this *rock* has a perpendicular elevation of thirty or forty toises. Its substance is a clear pure ice; it inclines towards the sea; its top is covered with a layer of moss and friable earth fourteen inches in thickness. During the heat of the month of July a part of this crust is melted, but the rest remains frozen. Curiosity induced me to ascend two other hills at some distance from the sea; they were of the same substance, and less covered with moss. In various places were seen enormous pieces of wood of all the kinds produced in Siberia: and also mammoth's horns, in great numbers, appeared between the hollows of the rocks; they all were of astonishing freshness. The escarpment of ice was from thirty-five to forty toises high: and according to the report of the Tungusians, the animal was, when they first saw it, seven toises below the surface of the ice," &c.

I have to observe on this passage, that it contains no decisive evidence to show that the ice seen by Mr. Adams on the front of the cliff from which the elephant had fallen, was any thing more than a superficial facing, similar to that found by Captain Beechey on parts of the front of the earthy cliff in Eschscholtz Bay; the same cliff which, a few years before, when visited by Kotzebue, seems to have been so completely incased with a false fronting of ice as to induce him to consider the entire hill to be a solid iceberg. One thing, however, is certain as to this mammoth, viz. that whether it was imbedded in a matrix of pure ice or of frozen earth, it must have been rapidly and totally enveloped in that matrix before its flesh had undergone decay, and that whatever may have been the climate of the coast of Siberia in antecedent periods, not only was it intensely cold within a few days after the mammoth perished, but it has also continued cold from that time to the present hour.

Remains of the rhinoceros also appear to be nearly co-extensive with those of the elephant in these northern regions. Pallas mentions the head of a rhinoceros which was found beyond Lake Baikal, near Tshikoi, and four heads and five horns of this animal from various parts of Siberia on the Irtis, the Alei, the Obi, and the Lena. These horns in the frozen districts are so well preserved, that splices of them are used by the natives to strengthen their bows.

Pallas conceived that these remains are not derived from animals that ever inhabited Siberia, but from carcasses drifted northward from the southern regions by some violent aqueous catastrophe, and that there is proof both of the violence and suddenness of this catastrophe in the phenomenon of an entire rhinoceros found with its skin, tendons, ligaments, and flesh preserved in the *frozen soil* of the coldest part of Eastern Siberia. On the arrival of Pallas in Ircutia in March, 1772, the head of this animal was laid before him, together with two of its feet, having their skin and flesh hardened like a mummy; it had been found in December, 1771, in the sand banks of the Wiluji, which runs in about 64° of north latitude into the Lena; the head and two feet only were taken care of; the rest of the carcass, though much decayed, was still enclosed in its skin, and was left to perish: the bones were yellow; the foot had on its skin many hairs and roots of hairs. On various parts of the skin were stiff hairs from one to three inches long.

If we compare these phenomena of the arctic regions with those of other countries, and especially with England, we shall find it by no means peculiar to the northern extremities of the world to afford extensive deposits of diluvial mud and gravel, containing the remains of extinct species of the elephant and rhinoceros, together with those of horses, oxen, deer, and other land quadrupeds. A large portion of the east coast of England, particularly of Essex, Suffolk, Norfolk, Yorkshire, and Northumberland, is composed of similar deposits of argillaceous diluvium, loaded in many places with bones of the same species of quadrupeds: these deposits are not only on the low grounds and lands of moderate elevation, but also on the summits of the highest hills. *e. g.* on the chalky cliff of Flamborough Head, four hundred and thirty feet above the sea. In the central parts of England, near Rugby, we have similar deposits, containing bones, tusks, and teeth of the same species of animals. In Scotland we have the same argillaceous diluvium on the east coast, near Peterhead, and near the western coast, at Kilmaurs, in Ayrshire, where it contains tusks of elephants and bones.

The analogies which these deposits offer to those in the arctic regions are very striking. In both cases the bones are of the same species of animals. In both cases they are imbedded in superficial deposits of mud and gravel of enormous extent and thickness. In both cases the deposits derive no accession from existing causes, and are suffering only continual loss and destruction by the action of the atmosphere, of rivers, and of the sea. Their chief peculiarity in the polar regions seems to consist in the congelation, to which the diluvium itself, as well as the remains included in it, are subject, from the influence of the present polar climate. Examples might be quoted to show the occurrence of similar remains in diluvial deposits all over Europe, and largely in America. Having then such extensive accumulations of the bones of animals, and the detritus of rocks, all apparently resulting from the simultaneous action of water, but which the operation of existing seas and rivers in the districts occupied by this detritus can never have produced, and are only tending to destroy, we may surely be justified in referring them all to some adequate and common cause, such as the catastrophe of a violent and general inundation alone seems competent to have afforded.

The facts we have been considering are obviously much connected with the still unsettled question respecting the former climate and temperature of that part of the earth in which they occur. Too much stress has, I think, been laid on the circumstance of the mammoth in Siberia being covered with hair. We have living examples of animals in warm latitudes which are not less abundantly covered with hair and wool in proportion to the size, than the elephant at the mouth of the Lena. Such is the hyæna villosa lately noticed at the Cape by Dr. Smith, and described (vol. xv. plate 2, page 463, Linn. Trans.) as having the hair on the neck and body very long and shaggy, measuring in many places, but particularly about the sides and back, at least six inches; again, the thick shaggy covering on the anterior part on the body of the male lion, and the hairy coat of the camel (both of them inhabitants of the warmest climates), present analogies which show that no conclusive argument in proof that the Siberian elephant was the inhabitant of a cold climate can be drawn from the fact of the skin of the frozen carcass at the mouth of the Lena having been covered with coarse hair and wool, but even if it were proved that the climate of the arctic regions was the same both before and after the extirpation of these animals, still must we refer to some great catastrophe to account for the fact of their universal extirpation; and from those who deny the occurrence of such catastrophe, it may fairly be demanded why these extinct animals have not continued to live on to the present hour. It is vain to contend that they have been subdued and extirpated by man, since whatever may be conceded as possible with respect to Europe, it is in the highest degree improbable that he could have exercised such influence over the whole vast wilderness of Northern Asia, and almost impossible that he could have done so in the boundless forests of North

America. The analogy of the non extirpation of the elephant and rhinoceros on the continent and islands of India, where man has long been at least as far advanced in civilization, and much more populous than he can ever have been in the frozen wilds of Siberia, shows that he does not extirpate the living species of these genera in places where they are his fellow-tenants of the present surface of the earth. The same non-extirpation of the elephant and rhinoceros occurs also in the less civilized regions of Africa; still further, it may be contended, that if man had invaded the territories of the mammoth and its associates, until he became the instrument of their extirpation, we should have found, ere now, some of the usual indications which man, even in his wildest state, must leave behind him; some few traces of savage utensils, arrows, knives and other instruments of stone and bone, and the rudest pottery; or, at all events, some bones of man himself would, ere this, have been discovered among the numberless remains of the lost species which he had extirpated. It follows, therefore, from the absence of human bones and of works of art in the same deposits with the remains of mammoths, that man did not exist in these northern regions of the earth at or before the time in which the mammoths were destroyed; and the enormous accumulation of the wreck of mountains that has been mixed up with their remains points to some great aqueous revolution as the cause by which their sudden and total extirpation was effected. It cannot be contended, that like small and feeble species, they may have been destroyed by wild animals more powerful than themselves. The bulk and strength of the mammoth and rhinoceros, the two largest quadrupeds in the creation, render such an hypothesis utterly untenable.

The state of the argument then respecting the former climate of the polar regions is nearly as follows:—It is probable that in remote periods, when the earliest strata were deposited, the temperature of a great part of the northern hemisphere equalled or exceeded that of our modern tropics, and that it has been reduced to its present state by a series of successive changes. The evidence of this high temperature and of these changes consists in the regular and successive variations in the character of extinct plants and animals which we find buried one above another in the successive strata that compose the crust of the globe. These have in modern times been investigated with sufficient care and knowledge of the subject to render it almost certain that successive changes, from extreme to moderate heat, have taken place in those parts of the northern hemisphere which constitute central and southern Europe; and although we are not yet enough acquainted with the details of the geology of the arctic regions to apply this argument to them with the same precision and to the same extent as to lower latitudes, still we have detached examples of organic remains in high latitudes sufficient to show the former existence of heat in the regions where they are found—a few detached spots within the arctic circle, that can be shown to have been once the site of extensive coral reefs, are as decisive in proof that the climate in these spots was warm at the time when these corals lived and grew into a reef, as, on the other hand, the carcass of a single elephant preserved in ice is decisive of the existence of continual and intense cold ever since the period at which it perished. We have for some time known that in and near Melville Island, and it has been ascertained by Captain Beechey's expedition, that at Cape Thompson, near Beering's Strait, there occur within the arctic circle extensive rocks of lime-stone containing many of the same fossil shells and fossil corals that abound in the carboniferous lime-stone of Derbyshire: the remains of fossil marine turtles also (chelonia radiata) have been ascertained by Professor Fischer to exist in Siberia. These are enough to show that the climate could not have been cold at the time and place when they were deposited; and the analogy of adjacent European latitudes renders it probable that the same cooling processes that were going on in them extended their influence to the polar regions also, producing successive reductions of temperature, accompanied by corresponding changes in the animal and vegetable creation, until the period arrived in which the elephant and rhinoceros inhabited nearly the entire surface of what are now the temperate and frigid zones of the northern hemisphere.

Assuming then on such evidence as I have alluded to, the former high temperature of the arctic circle, and knowing from the investment in ice and preservation of the carcass of the mammoth, that this region was intensely cold at the time immediately succeeding its death, and has so continued to the present hour; the point on which we are most in want of decisive evidence is the temperature of the climate in which the mammoth lived. It is in violation of existing analogies to suppose that any extinct elephant or rhinoceros was more tolerant of cold than extinct corallines or turtles; and as this northern region of the earth seems to have undergone successive changes from heat to cold, so it is probable that the last of these changes was coincident with the extirpation of the mammoth. That this last change was sudden is shown by the preservation of the carcass in ice; had it been gradual, it might have caused the extinction of the mammoth in the polar regions, but would afford no reason for its equal extirpation in lower latitudes: but if sudden and violent, and attended by a general inundation, the temperature preceding this catastrophe may have been warm, and that immediately succeeding it intensely cold; and the cause producing this change of climate may also have produced an inundation, sufficient to destroy and bury in its ruins the animals which then inhabited the surface of the earth.

I shall conclude these observations with quoting in his own words the opinions of Cuvier, which have always appeared to me the most correct and most philosophical that have been yet advanced upon this subject.*

* Tout rend donc extrèmement probable que les elèphans qui ont fourni les os fossiles habitoient et vivoient dans les pays où l'on trouve aujourd'hui leurs ossemens.

Ils n'ont pu y disparoître que par une révolution qui a fait périr tous les individus existans alors, ou par un changement de climat qui les a empêché de s'y propager.

Mais quelle qu'ait été cette cause, elle a dû être subite les os et l'ivoire si parfaitement conservés dans les plaines de la Sibérie, ne le sont que par le froid qui les y congèle, ou qui en général arrête l'action des élémens sur eux. Si ce froid n'étoit arrivé que par degrés et avec lenteur, ces ossemens, et à plus forte raison les parties molles dont ils sont encore

MEXICAN BEES.

SOME ACCOUNT OF THE HABITS OF A MEXICAN BEE,

PARTLY FROM THE NOTES OF CAPTAIN BEECHEY: WITH A DESCRIPTION OF THE INSECT AND OF ITS HIVE, BY E. T. BENNETT, ESQ., F.L.S., &C.

In the hives of the domesticated bees of Mexico we meet with a structure altogether peculiar. They exhibit little of the regularity of construction which characterizes the hives of the bees of the old continent, and are far inferior in this respect to the habitations of wasps. In one particular they approximate to the nests of the European humble bees; the honey which they contain is deposited in large bags distinct from the common cells. It is somewhat singular that so interesting a point of natural history has never been particularly noticed; our previous knowledge scarcely extending beyond the facts, that some of the bees of America form nests, like those of wasps, attached to, or suspended from trees, and covered by an outer case constructed by themselves; while

quelquefois enveloppés auroient eu le temps de se décomposer comme ceux que l'on trouve dans les pays chauds et tempérés.

Il auroit été surtout bien impossible qu'un cadavre tout entier, tel que celui que M. Adams à découvert, eût conservé ses chairs et sa peau sans corruption, s'il n'avoit été enveloppé immédiatement par les glaces qui nous l'ont conservé.

Ainsi toutes les hypothèses d'un refroidissement graduel de la terre, ou d'une variation lente, soit dans l'inclinaison, soit dans la position de l'axe du globe, tombent d'elles-mêmes.

Cuvier, Ossemens Fossiles, 1821, *tom. i. p.* 203.

others, incapable apparently of forming this outer crust for their hives, seek cavities ready formed for their reception, and in them construct their habitations. Instances of each of these kinds of hives are mentioned by Piso in his Natural History of both the Indies (page 112); and Hernandez, in his history of Mexico (Lib. ix. p. 133), states, that the Indians keep bees analogous to ours, which deposit their honey in the hollows of trees. Little information beyond that furnished by these older writers is contained in more modern works; and even the Baron Von Humboldt, to whose acute observation science is indebted for so many discoveries respecting the New World, appears not have noticed, with his usual care, the peculiarities of its bees. Had that distinguished traveller directed his attention to the habits of the species which he collected during his memorable journey, M. Latreille would doubtless have given to us the necessary details in his excellent Monograph of the American Bees, included in the Observations Zoologiques of M. Humboldt. In the valuable essay prefixed to this Monograph, M. Latreille has collected from authors numerous statements relating to the habitations of bees, and especially of those of America; but has added to them no new facts as regards the hives of the New World. The subject may, therefore, be regarded as altogether novel, and as requiring some little detail in its explanation.

In the domestication of the bees of Mexico but little violence is done to their natural habits. Inhabitants, in their wild state, of cavities in trees, a hollow tree is selected to form their hive. A portion of it, of between two and three feet in length, is cut off, and a hole is bored through the sides into the hollow, at about its middle. The ends of the hollow are then stopped up with clay, and the future hive is suspended on a tree, in a horizontal position, with the hole opening to the cavity directed also horizontally. Of the hive, thus prepared, a swarm of bees speedily take possession, and commence their operations by forming cells for the reception of their larvæ, and sacs to contain the superabundant honey collected by them in their excursions. Two such hives, completely formed and occupied, were brought to England, safely packed in re-

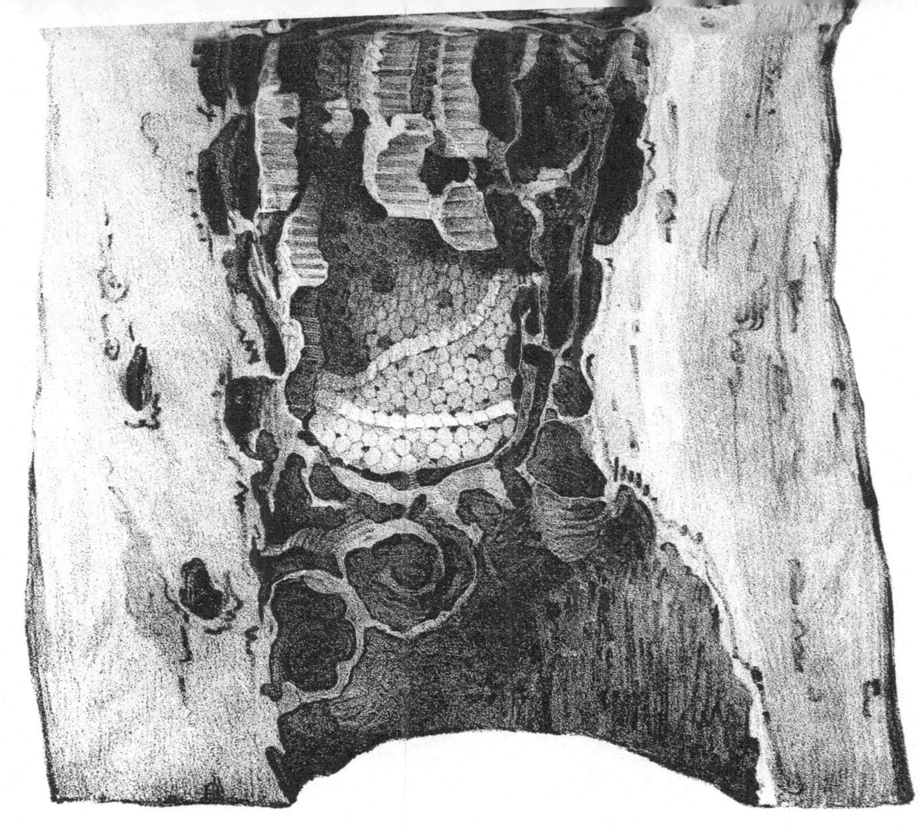

INTERIOR OF A BEE HIVE OF MEXICO

cent hides. Of these one has been forwarded to M. Huber, eminently distinguished for his highly interesting observations on the manners of bees; the other has been presented to the Linnæan Society. The latter has been carefully divided longitudinally, so as to expose its interior; a representation of which is given on the opposite plate, one half of the natural size. In this view nearly the whole of the interior is visible; scarcely a score of the cells, and very few of the honey sacs, having been removed with the upper portion of the trunk. It represents the comb as it would be seen in its natural horizontal position, by an observer looking upon it from above.

The eye of an observer, accustomed to the regular disposition of the comb in the hive of the European bee, is at once struck with the opposite directions assumed by it in different parts of that of the Mexican. Instead of the parallel vertical layers of comb, we have here layers, some of which assume a vertical, while others are placed in a horizontal direction; the cells of the latter being the most numerous. The cells, of course, vary in their direction, in the same manner as the comb which they form: those of the horizontal layers of comb being vertical, with their openings upwards, while the cells of the vertical comb are placed in a horizontal direction. In the horizontal cells the mouths are partly directed away from the entrance to the hive, and partly towards it; the former direction being given to those cells which occupy the middle layers of comb, and the latter to the cells which are placed on the side of the hive opposed to the opening. All the combs, both vertical and horizontal, are composed of a single series of cells applied laterally to each other, and not, as in the European hive bee, of two series, the one applied against the extremities of the other. The horizontal combs are much more regularly formed than the vertical, the latter being broken, and placed at uncertain distances, while the horizontal are perfectly parallel with each other, forming uniform layers, and placed at equal distances. Between these parallel combs are processes of wax, partly supporting them, and passing from the base of one cell to the junction of others in the next layer.

These columns are considerably stronger and thicker than the sides of the cells which they support.

The cells appear to be destined solely for the habitation of the young bees; for in all that have been examined bees have been found. The bee is placed in the cell with its hinder parts directed towards the mouth of the cell, which is covered by a granular mass, probably composed of the pollen of plants. The form of the cells is hexangular, but the angles are not sharply defined, and the mouth is scarcely, if at all, thicker than the sides. In their dimensions and relative proportions they differ materially from those of the European, and still more from those of the Indian bees, as may be seen by the subjoined table:

	Mexican.	European.	Indian.
Diameter of cell .	$2\frac{2}{3}$	$2\frac{2}{3}$. $3\frac{1}{3}$	$1\frac{1}{2}$. $2\frac{2}{3}$
Depth of ditto .	4	5 . 6	$4\frac{1}{3}$. 6

All of those which are visible appear to be uniform in size; nor without the destruction of the specimen can it be ascertained whether there are any larger cells for the larvæ of the males of the queen.

The combs are placed together at some distance from the opening of the hive, and form a group of an oval shape, consisting of five horizontal and parallel layers, occupying the part most remote from the opening; of an interrupted vertical layer applied to the side opposed to the opening; and of two principal, and two or three smaller, vertical layers in the middle. The whole of these are supported by wax, spread out into layers borne on processes of the same material, resting either on the wood of the cavity, or on other parts of the fabric of the comb. In these processes and layers of wax are numerous openings of various sizes, at once admitting of ready access for the inhabitants to every part of the hive, and economizing the use of the material of which they are constructed. Some of the openings are of large dimensions. The entrance into the hive is continued into a long gallery, which, to judge from the direction taken by a flexible substance introduced into it, leads beneath the combs to their very extremity. It is therefore probable that at the extremity the work of building commenced.

Surrounding the combs are several layers of wax, as thin as paper, irregular in their form, and placed at some little distance from each other, the instertices varying from a quarter to half an inch. One of these supports a vertical comb; the others are connected to the combs only at their edges, or by processes or layers of wax. Externally to these are placed the sacs for containing honey, which are generally large, and rounded in form. They vary in size, some of them exceeding an inch and a half in diameter. They are supported by processes of wax from the wood of the cavity, or from each other, and are frequently applied side by side, so as mutually to afford strength to each other, and to allow of one side serving equally for two sacs. Their disposition is altogether irregular, and bears some resemblance to that of a portion of a bunch of grapes, rendering it probable that Nieremberg was acquainted with a similar nest, if not actually that of the same bee, although he denies to the one described by him the power of constructing combs. Some of the honey-sacs are placed apart from the others, forming a distinct cluster of the same general appearance as those immediately adjoining the comb.

From this singular position of the honey-sacs a most important advantage is obtained by the cultivators of the Mexican hive bee. To possess themselves of its honey it is unnecessary to have recourse to the means adopted in Europe for stupefying, or even destroying, the inhabitants of the hive. All that is necessary is, to remove the plug from the end of the cavity employed as a hive, to introduce the hand, and withdraw the honey-sacs. The store of the laborious bee is thus transferred to the proprietor of the hive without injuring, and almost without disturbing its inhabitants. The end of the hive is then again stopped up; and the bees hasten to lay up a fresh store of honey in lieu of that of which they have been despoiled, again to be robbed of their precious deposit. A hive treated in this way affords, during the summer, at least two harvests.

The honey is usually pressed from the sacs by the hand. Its consistence is thin, but its flavour is good, although inferior to that of the whiter honey furnished by the Spanish bee (probably our *Apis mellifica*, L.). It does not readily ferment, some of that contained in the hive being perfectly sweet and grateful, even after its arrival in England.

The wax is coarse in quality; its colour is a dark yellowish-brown. The whole of it appears to be similar in texture and properties, as well that used in the construction of the cells, as that which is applied to the coarser work of forming honey-sacs and supports; the only remarkable difference being, that in the former it is apparently paler, probably owing to the layers employed being considerably thinner and more delicate.

Of the varnish-like substance known by the name of propolis, and used by the European bees to cover the foreign substances with which they frequently come in contact, scarcely any vestige is exhibited, although we have evidence of its existence. The wood of the inside of the hive, except where wax is applied to it, is perfectly naked.

The hollow of the trunk forming the hive now before us is irregular in its outline, and varies in its breadth in different parts. Its average diameter, however, is about five inches. The length occupied by the cells is more than seven inches; and the total length between the extremities of the honey-sacs is fifteen inches. The number of its inhabitants, assuming that of the cells as a guide, must have been considerably under a thousand; a number trifling in comparison with that contained in the hives of the European bee, which commonly amounts to as many as twenty-four thousand.

The bee by which this nest is constructed is smaller than the European hive bee; its abdomen, especially, being much shorter than that of our common species. Like all those American bees which approach in their habits to our European race, it is readily distinguished from that, and from all other hive bees yet discovered in the Old World, by the form of the first joint of its hinder tarsi, which is that of a triangle, with the apex applied to the tibia. On account of this variation in the form of a part so important to the economy of bees, modern entomologists have universally agreed in the propriety of regarding the American races as constituting a distinct group from the bees of the Old World. M. Latreille has gone further, by subdividing the American bees into two genera; *Melipona*, in which the mandibles are not toothed; and *Trigona*, in which these organs are dentate. Of the propriety of this subdivision, which hitherto seemed to be supported by the general appearance of the insects referred to each group, the examination of the bee whose nest has been just described has given rise to considerable doubts. In it one of the mandibles is toothed, and the other is nearly entire. Its technical characters, therefore, are intermediate between the two genera, with a leaning toward *Trigona;* but its general appearance is entirely that of a *Melipona*, approaching very closely to that of *Melipona favosa*, Latr., *Apis favosa*, Fab. That it cannot be that species, or any of the nearly related ones described by M. Latreille in the Observations Zoologiques, is evident from the dentation of its mandible, and it may, therefore, be regarded as new to science. It is described in a note *. The

* MELIPONA BEECHEII.—*Mel. nigrescens, margine postico segmentorum abdominis quinque anteriorum flavo: mandibulá sinistrá apice bi- vel tridentatá.*

DESCR.—*Corpus* totum nigrescens, præter abdominis segmentorum margines.

Mandibula sinistra apice bi- vel tri-dentata, dextera submutica: ambæ pallidè refescentibrunneæ, basi apiceque tantum brunneo-nigris.

Clypeus albido-villosus, maculis tribus nigris: duæ laterales elongatæ, unica apicalis rotundata.

Antennæ articulo primo brunneo-fusco, pallidiori: reliquis saturatioribus.

Facies infernè albido-, supernè fusco-, villosa.

Thorax totus rufescenti-tomentosus.

Abdomen rufescenti-pubescens: segmentorum quinque anteriorum margines postici flavi.

Venter albido-villosus: segmentorum quinque anteriorum margines postici albido-flavescentes.

Pectus albido-tomentosum.

Femora tibiæque nigræ, albido-villosæ, tibiæ posticæ maculâ mediâ rufescenti-brunneâ.

Tarsi fulvi, anticè albido-, posticè et ad apices, rufo-villosi.

Alæ dilutè rufescentes, nervis rufescentibus.

name which is there proposed for it is a just tribute to the observer, to whom we owe the first opportunity possessed in Europe of becoming acquainted with its habits and economy.

Some curious stories are related by the possessors as to the manners of these bees, one of which deserves to be recorded. They assert that at the entrance of each hive a sentinel is placed to watch the outgoings and incomings of his fellows, and that this sentinel is relieved at the expiration of twenty-four hours, when another assumes his post and duties for the same period. On the duration of this guard some doubts may seasonably be entertained, but of its existence ample evidence was obtained by repeated observation. At all times a single bee was seen occupying the hole leading to the nest, who, on the approach of another, withdrew himself within a small cavity, apparently made for this purpose on the left-hand side of the aperture, and thus allowed the passage of the individual entering or quitting the hive; the sentinel constantly resuming his station immediately after the passage had been effected. That it was the same bee which had withdrawn that again took his station in the opening, could not be mistaken; for his withdrawal was only into the cavity on the side of the hole, in which his head was generally in view during the brief interval while the other was passing; and that head again immediately started forwards into the passage. During how long a time the same individual remained on duty could not be ascertained; for although many attempts were made to mark him, by introducing a pencil tipped with paint, he constantly eluded the aim taken at him, and it was therefore impossible to determine with certainty whether the current reports concerning him were or were not founded in fact. With the paint thus attempted to be applied to the bee, the margin of the opening was soiled; and the sentinel, as soon as he was free from the annoyance he suffered from the thrusts repeatedly made at his body, approached the foreign substance to taste it, and evidently disliking the material he withdrew into his hive. The hole was watched to see what would be the result of this investigation of the substance, and a troop of bees was soon observed to advance towards the place, each individual bearing a small particle of wax or of propolis in his mandibles, which he deposited in his turn upon the soiled part of the wood. The little labourers then returned to the hive, and repeated the operation until a small pile rose above the blemished part, and completely relieved the inhabitants from its annoyance.

If the existence of such a sentinel as has just been described can safely be admitted, his utility would be unquestionable, as being at all times prepared to encounter a straggling stranger, or to give warning of the approach of a more numerous body of foes. Such foes actually exist in moderately sized black ants, which sometimes in small, and occasionally in large, bands attack the hive, and between which and the industrious bees desperate conflicts often take place. In these struggles the bees generally obtain the victory; but they are occasionally mastered by the overpowering numbers of their opponents.

VOCABULARY

OF

WORDS OF THE WESTERN ESQUIMAUX.

This vocabulary contains a collection of words made by Mr. Collie, Mr. Osmer, and myself, from straggling parties of Esquimaux, whom we met principally in Kotzebue Sound. It agrees in many respects with that given by Captain Parry in his second voyage, particularly in the numerals, the elements, and celestial bodies, and the names of animals which are common to both places, and leaves no doubt of the two languages being radically the same; though, as might be expected, the idioms are somewhat different.

Captain Parry's remarks upon the language of the Eastern Esquimaux seem to apply equally to that of the Western nation, of which the very few words beginning with *b*, *d*, *l*, *g*, *r*, *u*, and the absence of the letters, *f*, *q*, *v*, *x*, may be adduced as instances. The Western Esquimaux also appear to have the peculiarity of varying their pronunciation, but without materially softening the words. On the whole, the Western Esquimaux language has more gutturals, and the words in general have a harsher sound than those of the Eastern tribe.

My knowledge of the language is too limited to enable me to offer any further remarks on this subject, and I can only submit the vocabulary as it is, persuaded, from the care that has been bestowed upon it, that it will prove useful to persons who may visit the N. W. coast of America. At the same time I cannot vouch for a perfect understanding always subsisting between the inquirer and the respondent, though I have reason to think it was not often otherwise. The most ridiculous mistakes have occasionally been made by collectors of words of barbarous languages, and I was early warned to be careful, by being innocently enrolled in the number of persons who had been deceived. I one day showed an Esquimaux an engraving of a musk bull, to which he immediately applied the name of Mignune, and I wrote in my vocabulary accordingly, but I soon afterwards discovered that the word applied only to the material with which the bull appeared to be drawn, that is, *plumbago*, of which the Esquimaux have a great deal, and that the proper name for the animal was a very different word.

The initials in the small column denote the collector; those words which have a C affixed to them are to be pronounced according to the following directions:

" *A* is either a^{2} (fat), or a^{4} (far), as in Walker's English Pronouncing Dictionary. This letter is generally marked, and when it is not, its association with the consonants will point out the particular sound to be used.* *E* is generally the e^{2} of the above-mentioned Dictionary, but when marked é, it is to be pronounced as in *me*. *Ei* is to be pronounced as in German, *ein*, *eisel*, *gieser*. *Eu* as in French. *I* as in pi^{2}n. *O* as in no^{1}t, or otherwise, as marked. *Oo* as in *moon*. *U* as in lu^{2}b.

Ll is according to the Spanish pronunciation, and *gl*, where mentioned, according to the Italian; where not, as in English. *Qu* as in English. *R* as in French, and $\overset{+}{R}$ still more roughly guttural as in the Northumbrian dialect. This mark (+) always points out a guttural pronunciation, and is very common in the language; some individuals, however, using it much more than others. *S* is pronounced as in *side*, and *ss* as in *glass*. *Y* as in *yet*, and never used as a vowel. *Z* is sounded as in *lizard*. *Ch* is the Spanish *x*, and the Greek χ of the Scotch Universities. *Gh* has nearly the same sound.

* *Au* is to be sounded as *aout* in French—very nearly as *ou* in the English world *sound*. *Ai* is heard in *wild*. *G* is hard, unless otherwise mentioned.

English Names.	Esquimaux Names.	
Adze, an	Oo-lĕe mā, Oo lim-ma	C.
	Oo-lee mow (axe)	
	Tschik-luk	B.
Anchor	Kee-sock	B.
Arm, the	At tsik	C.
below the elbow ..	Tad-leek	B.
Armlet, a thin piece of ivory or leather formed so as to cover part of the wrist and defend it from the bow-string, &c.	Man-yēra (yconst)	C.
Arrow	Kak-a-rōok	B.
head of bone, sharp ..	Ka-wĕe ruk	C.
ditto, blunt ..	Koo-koo-gwait	ib.
ditto, stone ..	Ko-kick-chltēvik	ib.
ditto, ditto ..	A-kal-look-see-goo-tat (see Stone	ib.
dumb-head	Knoo-e-ak	B.
Awl	Poo-toon	O.
Axe	Atti-ghim-nuk	B.
Back, the	Ko-lé-ka[2] ?	C.
Bag, a (of salmon-skin) ..	Poo-tshik (R)	ib.
	Pee a ruk 1	
ditto (of canvas) ..	Porúss ăk ?	ib.
Baidar	Oo-me-ak	B.
	Oō-mēe-ák	C.
Ball, a cannon	O-whak	ib.
Bark, (of a dog)	Ky-muk	O.
Beach, the	Tsinnar	B.
Bead (of any colour) and size	Tshung-au-ník	C.
	Thung-au-rĕt	ib.
ditto	Tshung-au-rā-wik	ib.
	Tshung-au-runnik	ib.
Beads	Chu-nōw-răh	B.
blue	Chū-ou-rēnnĕk	ib.
Bear, a	Tsu-nărr, or Tsu-năck	ib.
skin of	Ib-neĕ ák	C.
Beard and whiskers ..	Oo-mach-oomit	ib.
	Oomich (P)	ib.

English Names.	Esquimaux Names.	
Beaver (the animal) ..	{ Kee-yee-ak . . .	C.
	{ Wai-luk-tuk . .	ib.
Beaver (etched upon bone) ..	Kee-yee-ak . .	ib.
skin	ditto . . .	ib.
Belly	Nai-yak . . .	ib.
of a woman	Nad-djigga . . .	ib.
of a man	Dirdŭck . . .	B.
Bill, of a bird	Ee-ed-djook . .	C.
Bird	{ + Tin-me-a-rit . .	ib.
	{ Ting-mĕ-loŏ-rak	
swimming	Ti-mai-rik . .	ib.
Bite, to	{ Nig-ge-rung-ă . .	C.
	{ Kai-ook-toon ?	
Black colour	{ Kang-no-ak . }	ib.
	{ To-ring-mātik . }	
Blood	{ A-ook . . .	C.
	{ Ka-ōop-e . . .	B.
Blue, it is	Renneck . . .	ib.
Blue and azure colour ..	{ Ka-oo-gli-ak . .	C.
	{ A-gli-oo-ik . .	ib.
Blubber	Tsed-lu-ou-rok . .	B.
Body	A-seet . . .	ib.
Bone	Oa-ee-yak . . .	C.
Boots, native	{ Kum-muk . . }	ib.
	{ Kummugga . . }	
Book	A-gluc i-wick . .	O.
Bore, to	Nee-ook-toon . .	C.
Bottle, (a glass)	Ee-moon . . .	ib.
Bottle	Im-wō-en . . .	B.
Bow, a	Petik, sik, Pitik-shi-a .	ib.
Bow-string	Oo-kwak-ta . .	ib.
Bow (a broken one) ..	Na wik-túk Petiksik .	C.
Bow to shoot with (as carved)	Pe-teik-ta-rik . .	ib.
Bow, for shooting	Pee-tik-seek, or Setka .	B.
Bowl, a	Kalloon-goo-reak .	ib.
Bowl, of wood	Nanna-uck . .	ib.
(a large wooden) ..	Poo-gōō-tuck . .	B.
Box	Chōō-lōō-dit . .	ib.
Box, a small ivory	Aul-toon . . .	C.

English Names.	Esquimaux Names.	
Boy	Ein-yook	C.
Brass, a large hook of	Tsha-wa-kacht	ib.
Branch of a shrub	Ok-pwit	ib.
Bring it	Koki-ghe-e-wha	B.
Bring it here	Kok-ee-gee-ga	C.
Breast, a person's	A-tig-git, Tsha-guga (P)	ib.
Breasts, a woman's	Ei-ing-gek-ku	ib.
Broken, a stick	Na-wik-tuk	ib.
Broken	A-yŭk-se-mărt	B.
Breeches	Koo-now-ita	ib.
	Koke-lek	O.
Bull, a	Moong-mack	B.
Button	Nuck-too-ou	ib.
Butterfly	Dtar-dle-ē-utsik	ib.
	Tak-kull-loo-kwī-tak	C.
Button	Nak-to-ik	ib.
Buttocks, the	Ek-kook (Pd)	ib.
Canoe	Ki-yack	B.
Canoe of skins	Kai-yak	C.
Cap, or hood	Naza-oūn	B.
Cap, an European	Nad-dsaun	C.
Cap, native, of birch bark	Poo-tak	ib.
Catch, to (when throwing)	A-kok-shō	ib.
Chain	Knoo-oo-lok	O.
Chamisso Island	+ Ee-a-roo-ik	ib.
	Eow-ick	ib.
Cheek, the	Oo-loo-ruk-ka, O-u-lu-at (W)	C.
Child, a	Ee gee-lu-gu-ga-ga	ib.
Chin	Ta-bloo-a, Ta-boo-loo / Tub-du-ah (Pd)	ib.
Clouds	Noo-oo-eè-a	ib.
Codlings (small fish)	Mōng-a ?	ib.
Cold, (shivering)	Kai-rung-a	ib.
Comb, a hair	Igli-zucket	B.
	Igli-oo-tik	O.
	I-gli-a-oo-tik	C.
Cork, a	Tshee-mee-at	ib.
	Chim-ēya	B.

English Names.	Esquimaux Names.	
Cough	Koak-tshee-nar	O.
Cow or calf	Nai-mik-toŏk-too	C.
Crab, a	Poo-ōo-i-ak	ib.
	Edloo-az-rey-uk	B.
Creep, to, on hands and feet	Pa-mok-tok	C.
Crow-berries	Azee-ret, A-zee-ak	ib.
Crow-berry bush	Pa-oo-mau-tit	C.
Cry	Ky-rook	O.
Cup, ivory	Kĭ-oon-na	C.
Curlew, a	Shee-ak-too-ok	ib.
Dance, native	Kal-lau-rok-tok	ib.
Dance, or jumping up and down	Ang-a-yoo-rok, In-noo-ret (w)	ib.
Dart, a small	Ninee-uk-puk	B.
Dart for birds	Ni-nask-puk	ib.
Dart with three prongs in the middle	Noo-yak-kwa	C.
Deck, the ship's	Muk-ti-hik	ib.
Deer-skin frock	E-ee-rah, or Atti guy	B.
Deer-skin	Itch-sek	ib.
Dodo (a bird)	Ne-ak-tshuk	C.
Dog, skin of	Kim-muck	B.
Dog, a	Kenma	C.
	Koo-ne-ack	B.
Drill (a native)	E-dill'-leem	C.
Drill-bow	Too-koo-ra	ib.
used also for procuring a light	Too-wachk	ib.
Drill, to	Pee-tak-toon	ib.
Drill-socket	Keng-me-ak	B.
	Omee-yāk	ib.
Drink, to	Ee-moon	C.
Drink	E-mug	B.
Drum, or tambourine	Chowg-suk	ib.
Duck, a	E-wŭck	ib.
	Ee-wāk[26] (I)	C.
	Ee-wark (K)	ib.
Ear, the	Tchee-u-tik	O.
	See-teek	B.
	Tsĕĕ-tuk, Tsee-lig-ga	C.
	Tsheé-dik (w)	

English Names.	Esquimaux Names.	
Earthen-jar	Ig-hu-nee . . .	B.
Eat (or chew) to	A-shad-loo-ik . .	C.
	Ishad-loo-weet . .	B.
Egg	Man-nik . . .	O.
Eight	Penayua . . .	B.
Elbow, the	Ce-koot-sik, or Eedeeootsik	C.
Eye, the	Erick . . .	B.
	Eer ruk-ka, Ee-rik, Erruk	C.
Eyebrow, the	Ka-bloo-ce-a, Ka-blo-otka	ib.
	Ka-bloo-ai . .	B.
Face, the	Kee-na, or Kinna .	C.
	Kenuck . . .	B.
Falcon	Kje-goo-at . .	ib.
Far off	Mūnna . . .	ib.
Fawn (as carved) ..	Eum-nak . . .	C.
Feather, a	Tshoo-lak . . .	ib.
Finger, the first	Teg-heya . . .	B.
middle	Ko-duk-luk . .	ib.
third	Mak-la-e-rah . .	ib.
little	Ekick-koke . .	ib.
Fingers, the	Ta-maridreh . .	C.
nail, the ..	Koo-kwit-ka, Koo-kwik-ka	ib.
	Koo-kwik-kur tamar-drah	ib.
	Koo-gwek (pd) . .	ib.
Finger, the first ..	Tee-ge-ra, Tee-ke-ra .	ib.
second ..	Kei-tik-kluk-a . .	ib.
	Kei-tik-kluk, Kei-tik-klo-a	ib.
third	Muk-gle-rad . .	ib.
	Meu-gigg-le-ra . .	ib.
little	E-rit-ka-mak . .	ib.
Fire, a	Ig-nik . . .	ib.
	Ignuck . . .	B.
Fire, to strike	Ig-ne-dit . . .	C.
Fish	Khallo-ight, Khalloo .	B.
Fish, small	Too-mo-e . .	C.
long	Tyoong-me . .	ib.
Flounder, (or flat-fish) ..	Ik-hanni-luk . .	B.
Fly, a	Ko-kwel-lock . .	ib.
Foot of a man, or animal ..	Iddi-guy . . .	B.

English Names.	Esquimaux Names.	
Foot or feet	Il-te-ga-ra . .	C.
Fork, ivory, used also as a comb	Ni-yik-kik . .	ib.
Fox	Kiok-toot . . .	B.
Fox (as carved)	Ka-ee-yok . .	C.
Fox-skin, brown	Ka-ee-yok-tok[26] . .	ib.
Friend, or term of friendship ..	Il-lipo-lee . .	ib.
Frock, (skin)	Oo-kwăk, Oo-kwa .	ib.
made of gut ..	Nyel-look, also Ka-pee-tuk, Ka-be-took . .	ib.
of hare-skin ..	Oŏ-quad-lik . .	ib.
Funnel, a (copper stove) ..	Kan-mu-yuk . .	ib.
Fur	Mit-koot . . .	ib.
Garnets	Nalloo-na-vitka ? .	ib.
Gaiters	Kammuck . .	B.
Geese, rising	Tattee-ree-gak . .	C.
(as carved) ..	Tattee-regu, Tut-tee-lee-a ruk, and nalloo-yik-ka .	ib.
Give it me	Pee-gle-gi-woong-a Wung-ee-gla-gu .	C.
Glove (of natives) ..	A-dré-get, At-ka-li-ga, A-dre-ret . .	ib.
Gloves	Adj-guy-redt . .	B.
Go away	Illip-se . . .	ib.
Go, to	Il-ti-wal-luk . .	C.
Going away	Pēē-art . . .	B.
Goat	Koo-nē-āk . .	C.
	Ip-na-uck . .	B.
Good, very	Nee-ok-muk . .	C.
Good, I am	Na-koo-roo-oh . .	ib.
Good, it is	Na-koo-rit . .	C.
it, or he is	Ma-may-poke . .	B.
Good, not	Na-koo-rit-nau . .	C.
Grass, engravings of ..	O-kwait ([27]) . .	ib.
Grass	Ee-boo-wit ([27]) . .	ib.
Green colour	O kok . . .	ib.
Grouse	Ar-hay-ghi-uk . .	B.
Gull, a white	Alla-wa, Naw-yet ? .	C.
Gull (parasitic) ..	Ike-muk . . .	ib.
small (L. Sabini) ..	E-ga-goo-i-ak (I) . Kai-ki-ge-gai-ak (K) .	ib.

English Names.	Esquimaux Names.	
Gun, a	Kee-suk . . .	B.
Gun, a great	Tshoo-poon, On-youna .	C.
Guard-fish	Tz near-ōōk . . .	B.
Hair, of the head	Noot-tset, Noot-zatka, Noot-tset-ka . Dtoo-tset (P^{d}) .	C.
Hair, human	Nuchet . . .	B.
Hammer, an iron	Ka-wook, Kè-kek-toon .	C.
Hammer, to	Karroo-o-tuck . .	B.
Hammer, to, or strike with ..	Kar-roo-tok . . . Ktai-roo-ik	ib.
Hand, the	Arge-gei, Ardge-gei .	C.
Hare, a	O-good-logh . . .	ib.
	Quel-luk (K) . .	C.
Harpoon (as thrown) ..	Oo-nāk . . .	ib.
Harpoon, as carried when walking	Oo-nee-yak . . .	ib.
Harpoon, to	Naul-lik-kwa . .	ib.
Harpoon-line (coil of) ..	Allara . . .	ib.
Harpoon	Allik . . .	ib.
	Nenak-pluk . . .	B.
Head, the	Nea-koa . . .	ib.
	Né-ak-kwa . . .	C.
crown of	Ka^{1}b-br^{2}a . . .	ib.
of my	Ka^{4}b-dja^{4}k-ka^{2} . .	ib.
front of	Kă-wă, or Ka^{4}b-wa^{2} .	ib.
Heel	Kite-meek . . .	O.
Hook, fish	Nik-sik . . .	C.
and line	Nicht-siak . . .	ib.
Hoop, for tent	Sow-soro-uk . . .	B.
Husband	Qua-ōōg . . .	C.
I, or me	Wōōng-a . . .	ib.
Image, an	Inné-moo-rok . .	B.
Imber goose, a young ..	Mul-le-kāk . . .	C.
Inlet	Ro-ōōk . . .	ib.
Instrument (musical) made of a bunch of cords and the tips of birds' bills	Ni-mik-taut^{x}ak . .	ib.
Instruments for cutting ivory	Kaigne-noo-străk . .	ib.
Iron	Tsha-wek . . .	ib.

English Names.	Esquimaux Names.	
Instrument for cutting stone arrow heads	Keg-lee-chea . . .	B.
Island, an	Tudra	ib.
Jar, a	Koo-loot-za . . .	ib.
earthen	Ig-hu-nee . . .	ib.
Ivory head carved	Nee-wach-toon . .	C.
Kettle, a	Im-mi-ruk . . .	ib.
Kid, to contain water ..	Mizo-ghau . . .	B.
Kid, to contain oil	Kottoo-ack . . .	B.
Knife, Esquimaux	Seque-tat . . .	ib.
other	Pe-yar-tuk . . .	ib.
of stone	See-goo-tăt . . .	C.
native, of iron ..	Sha-mang-me . Tshau-ung-mun . Tsha-moon . .	ib.
carved for scooping ..	Mid-del-lik . . .	ib.
European	Tshawek . . .	ib.
pen	Pen-ne-yok-ta . .	ib.
Know (I don't)	Ny-loo-gah . . .	O.
Labret (lip ornament, and hole for ditto)	Too-tuk, Poo-tuk . . Poo-tauk (w)	C.
Ladle, a	Imōō-onee . . .	B.
Lake (colour)	Ang-a-ook, Keg-mung-nak . . .	C.
Land, or earth opposed to sea	Tee-drak . . .	ib.
Laugh, to	I-glak-tok . . .	ib.
Lead	Koo-ou-sow-tik . .	B.
Leap, to	Ach-rak-ty[x]ak-took .	C.
Leg	Ka-nuck . . .	O.
Lichen	Ee-buch au-rit . .	B.
Lip, upper	Kok-luk . . .	B.
ornament	Too-otucka . . .	ib.
lower	Kak-ker-luk . . .	C.
Lobster	Poo-tchu-o-tuk . .	B.
Look, to	Te-eg-loo-gook . .	C.
at a thing	Teed-la-book . .	B.
Looking-glass	Tak-a-toon . . .	ib.
Lost, something	Oo-mai-toon . . .	C.

English Names.	Esquimaux Names.	
Maul (a wooden)	Kad-roo-tock . .	B.
Mallet, wooden	Ka-di-oo-tak, and ka-oon	C.
Man, a	Tuăk . . .	ib.
Many (a great many) (a general superlative) .. .. A long way off	Minna minna . .	ib.
Many in number	Ko-lug-na, Ta-maum .	ib.
Marline-spike, small .. of ivory, forlacing bows ..	Ke-poot-tak . .	ib.
Martin (as carved)	Ama-rok . . .	ib.
Match, a, of a cottony or woolly nature	Ee-goo-rit . . .	C.
Mast, ship's	Nake-puk-tuk . .	ib.
Mast, boat's	Doo-bak-ti . . .	B.
Mirror, a	Kaing-nee-gaun . .	C.
Moon, the	Tak-kuk . . .	ib.
his name	Tad-kuck . . .	B.
More	Tshau-loc . . .	C.
Mountain	Mug-wee . . .	ib.
	Magoo-Magoo . .	C.
Mouse and skin	Au-ing-nyak . .	C.
Mouth, the	Kuck-a-luk, ka-klook / Kai-nee-ak (w) .	ib.
	Kan-nuck . . .	O.
Mouse	Kŏŏblă-ŏŏk . . .	B.
Musk-ox	Moong-mak . . .	ib.
ditto	Oo-ming-mi . . .	C.
Musk-rat	Paoona ? . . .	B.
	Kee-boo-gal-lok . / Kee-boo-wal-luk .	C.
Musket, a	Tshoo-poou . . .	ib.
ditto	Tsou-kodt . . .	B.
Nail, an iron	Ke[2]-ke[1]-ak . . .	C.
Narwhal	Tse-dōō-ăck . . .	B.
Neck	Kang-oot-tsitka, koom-oot-tsia . . .	C.
Near	I-muckt . . .	B.
Needle sewing, of wire ..	Mik-koon [26] and [27] mek-koon (w) . .	C.

English Names.	Esquimaux Names.	
Needle-case ivory	Mik-kun-mik	C.
for making nets	Lew-wi-law-tik	ib.
Net, a	Nee-gal-lik	C.
a different sort	Korak, aka-loo-na	ib.
large, for seals	Koo-brak	ib.
No	Naga, Nau-me Tuum, Na-u	C.
	Nāgā, Nā-o, Aūnga	B.
None, I have	Peed-lark	ib.
None, he has	Peed-lo	ib.
No more, none	Pied-lak [26] and [27] (I)	C.
Nose	Nognuck or Kingar	B.
	King-na-ga, king-a-na King-nuk (Pd)	C.
Ochre, red	Eeta	ib.
Oil	O'k-tsho'k	C.
Otter (as carved)	Améo [26]	ib.
Otter-skin	Te-ghĕ-āk-bŏŏk	B.
	Améok-tok [26] and [27] (I) Ami-nak-tok [27] (K)	C.
Owl	Ignă-zēĕ-wyūck	B.
Paddle	Par-hud-duc	ib.
	Par-hua-uk	C.
Paddle, to	Aan nuch[26], Aug-noon[27]	ib.
	Hang-noon (W)	ib.
	Ang-oo-tik	ib.
Pelican (print of a)	Pe-bli-ark-took	B.
Pig	At-kah	O.
Platt	Peez-liar-uk	B.
Plover, the golden	Tood-glict	ib.
ditto	Too-lik [26] and [27] Tood-lik[27]	C.
Plumbago, black lead	Mign-noōn	ib.
Porcupine, a	I-gla-koo-suk	B.
Porpoise, a	Aghi-bee-zee-ak	ib.
Posts over yourts supporting sledges	Ai-ye	C.
Pot, earthen, of natives	Eg[x]gun	ib.

English Names.	Esquimaux Names.	
Pour, to	Eu-koo-i-ruk . . .	C.
Prince of Wales, Cape ..	King-a-gee . . .	ib.
Prongs for small darts ..	A-goot-say . . .	B.
Ptarmigan, a	Kau-wik . . .	C.
Puffin, a	At-pak[27], Ke[1]-lu[2]ng-n[2]a .	ib.
	Kŏŏali-nŏckt . .	B.
Puffin (red feet of)	Itti-guy-it . . .	ib.
Pull out (to turn inside out)	O-li-dju-nauk . .	C.
Pyrites, iron	Ick-nay-ăck . . .	B.
Quiver and bow case ..	Pe-tik-sik-tak . .	C.
Rabbits	Noo-poo-i-tak-tuk .	ib.
Racoon (skin of)	Tsĭch-rĕe-buck . .	B.
Rat, gray-spotted	Tshuch-a-rik, Tshee-ge-rik	C.
Raven	Too-loo-ak . . .	ib.
Rein-deer	Took-too, Took-too .	ib.
Rein-deer (as engraved) ..	Too-too-ak . . .	ib.
Rein-deer	Tootōōt . . .	B.
Right, you are	Ta-mar-dra . . .	C.
Ring (for finger)	Nal-loo-i-a . . .	ib.
River, or stream	Koo-ŭck . . .	B.
River, a large	Koo-rook . . .	C.
River in the Bay of Good Hope	Ma-de-ok . . .	ib.
Root	I-koo-tshook . .	ib.
Rope, a	Lich-loo-nat . . .	ib.
Run, to	Ak-pa-ruk-tuk . .	ib.
Rurick Rock, or Island ..	A-hgho-le-a . . .	ib.
Sail, a ship's	Kaign-nil-bratup ? . Ten-yet-raw-te-taka ?	ib.
Salmon, fresh	Tee-lang-uk . . .	ib.
ditto	I-shalloo-ok . . .	B.
Salmon, dried	A-kol-loo ruk [25] and [27] A-ral-la-roo-ak . A-dal-gunuk-roo-ak	C.
ditto	I-shalloo-roo-ok . .	B.
Salmon-skins, dried ..	Ka-look-peoit, Ka-loo-kwit[25] . . .	C.
Salmon-skin bag	Ick-pai-ruck . . .	B.

English Names.	Esquimaux Names.	
Sand	Koo-wee-a	ib.
Scar	Kee-lee-ak	O.
Scrape, to	Kee-lee-ak-tok-tok	ib.
Scraper of stone for hides	Waing-nee-a	C.
Scraper of bone	Tsal-loo-ee-ga	ib.
Scratch, to	Ko-mee-ak-tok	O.
Sea, or water generally	Ee-muk-ka, Ee-mik I wa[4]k	C.
Sea, the	Tarri-ooke	B.
Sea-horse	I-week	O.
Seal, large	Kasi-guak	B.
Seal	Kasi-gōō-ăk	ib.
ditto	Nik-tsuk, Nik-zak[25]?	C.
Seal (a different sort)	Too-wut-ka-roo-a	ib.
Seal (long and short)	Oo-grook	ib.
Sew, to	Keydli-ark-too-uk	B.
Shake (with cold)	Tchoo-look-tak-tok	O.
Ship or boat	Oo-mee-ak	B.
Ship, a	Oo-mi-ak	C.
Sheep, a	La-loo-iga	ib.
Show it	Tush-e-tush	B.
Sheep	Olk-sŭk	ib.
Shell (murex)	Na-goo-uk	ib.
Shell (of fish)	Yeu-wul-luk	C.
Ship, go on board	Oo-mi-ak-puk	ib.
Shoe	Pin-e-yuk	O.
Shoulder, the	Too-ee-dee-a, Too-eek	C.
Shrew, a	Au-ru-nak	ib.
Shovel or spade	Noo-oun	B.
Sing, to	Poo-doo-a-gar	O.
Skin	A-tuk-tok	C.
Skin for tambourine	E-red-lark	B.
Skin of rein-deer used for tents	I-tshik	C.
Skin (covering of tents)	Kan-nig-it	B.
Skin of brown squirrels	It-re-ak-pook	C.
Skin of birds	O-kor-ree	C.
Skin, shirt of	Iman-nickt	B.
Skinning an animal (as carved)	Tail-lo, Ach-lak-talli	C.
Skull of porpoise	See-shuak	ib.
Sky, the	Keil-yak, Pung-na[26]	ib.

English Names.	Esquimaux Names.	
Slate, sharping	Seed-lin . . .	ib.
Slap, to	Tee-glu-a-gar . .	O.
Sleep	Chenek-tunga . .	B.
Sleep, to	{ Tshin-ik-tuk-ka[2] . Tshung-ek-lunga .	} C.
Sleep, first	Tsinnya--karbeeta .	B.
Sledge, a	{ Oo-nyak . . . Ai-yăk . . .	C. B.
Sling, a	Igli-ok-took . . .	O.
Smoke	Ee-shak . . .	C.
Snake, a (as carved) ..	Malli-goo-i-ak . .	ib.
Snare for birds	Tshe[2]-run-nun . .	ib.
Snipe	Nuck-too-o-lit . .	B.
Son, or my son	{ Oo-wing-ee-laka . Oo-wing-e[2]-loo-eek .	} C.
Shrimp	Nowd-len-nok . .	B.
Skins of deer made into a blanket	Oo-ghe-od-luk . .	ib.
Spear for whales	Ka-poo-ak . . .	C.
Spear, or lance	Tank-pook[26] . .	ib.
Spectacles, native	Eee-gee-yak ? . .	ib.
ditto	Ish-gack . . .	B.
Spoon or ladle	{ Obo-wik, Imoom, or Imoon, ali-oo-tack	. C. . ib.
Spoon	Ou-levo-book . .	B.
Star, a	{ O-blo-a-ret . . Og-bloo-ret . .	C. B.
Stamp with the foot ..	Kee-meak-tok . .	O.
Stick, a forked one ..	Kai-week-loo-ek . .	C.
Squirrel, a	Tsēy-kĕ-rĕck . .	B.
Squirrel, skin-frock ..	Oo-gōō-ar . .	ib.
Steel for striking fire ..	Iknew-igning . .	ib.
Stone	Kallook-row-rok . .	ib.
in general	Ang-mak . . .	C.
bluish	Illi-a-rik . . .	ib.
rounded on beach ..	Och-roo-rak . . .	ib.
for killing of seals ..	Oo-run-nee . . .	ib.
Straightener (a native instrument)	Nalla-ro-ik, A-louik .	ib.
Strike, to, with a mallet ..	Ka-rok-tok (see to hammer)	
Sun, the	{ Bait-tsäach, Maisak, Nei-ya . . .	ib.

English Names.	Esquimaux Names.	
Sun	Bidsuk, or Bizuck .	B.
Swan	Tadi-drōkt . . .	ib.
Swim, to	Kalee-ak-shook . .	O.
Swim, rein-deer swimming	Nallook-look . .	C.
Table	O-goo-luck . . .	O.
Tail of an animal worn by some of the men	Pop-tit . . .	C.
Take it	Mik-ki-krin . . .	ib.
Talons of a bird	Ee-gee-geit . . .	ib.
Tambourine	Kol-laun, Killaun . .	ib.
Tattooing on chin of women ..	Took-nauk . . .	ib.
	Tabloo ó-tay . . .	O.
Tent (as of skins)	Tie-poŏ-eet, Topak Tōō-pek	C.
Tent	Too-pōte . . .	B.
Tooth, a	Kōōtay . . .	ib.
Teeth, the	Kau-tit-ka, Kee-wee-dit-ka, Kewk-teet (pd)	C.
This, and here take it ..	Oona, oona-oona-oona .	ib.
Thong of thick hide ..	Au zoo-nak . .	ib.
Thumb	Kooble-doóa . .	B.
Thumb, the	Tamar-doot-ka (pd) Koo-boo-lo, koo-bloo-a	C.
Thumb, nail of	Koo-gay . . .	B.
Tobacco	Tau-wāk . . .	ib.
Tobacco	Tau-wap . . .	C.
culling for	Tau-wak-i-rim-mik .	ib.
Tobacco-pipe	Nuk-kak[27], Och-whait[26] .	ib.
Wife, old (a fish)	Neet-ar-muck . .	B.
Toe, great	Woo-doo-ah (w) . .	C.
little	In-mee-ga[2] . .	ib.
Tool for sharpening stones, arrow-heads, &c.	Ké-gla . . .	ib.
To-morrow	Ar-hāgo . . .	B.
Tongue, the	Oo-wār . . .	ib.
Tongue, the	Oo-kwak-ka, Oo-kwāā .	C.
Tree, or rather shrub (carved)	A-ning-onung-a . .	ib.
Trousers	Nellikāk-nellikak-kin .	ib.
of different sorts ..	Moo-gwa . . .	ib.

English Names.	Esquimaux Names.	
Trowsers, of a particular sort	Kak-a-leek . . .	C.
Tusk of walrus	Tuak . . .	ib.
Venison	Too-toot . . .	O.
Volcano (from a drawing of one)	Ar-wōu-ŭk . . .	B.
Vulture	Keegle-oght . . .	ib.
Walk, to	Pee-shook-tuk . .	C.
Walrus, the	Ai-wik, Ai-wa . .	C.
	I-bwuck . . .	B.
Wash, to (the hands) ..	E-wick-tok . . .	O.
Water or sea	Ee-muk-ka, Ee-mik .	C.
fresh	Ee-mik-kook . .	ib.
ditto	E-mik . . .	B.
ditto	Imung-yak-toke . .	ib.
	Tschu-dooat . .	ib.
Water, salt	Tarre-oke . . .	ib.
Wave, a	Ky-ōd-sŏ-root . .	ib.
Whale, the	Ah-hōw-loo . .	ib.
	A-ru-ak, A-whee-beek .	C.
Whale-bone	Tsock-kōyt . .	B.
Whale-line	Unga-shark . .	ib.
What is it, or its name ..	Sooua-goona . .	C.
White cloth	Kow-look . . .	ib.
Whistle, to	Oo-wing-nak-tok . .	ib.
Wing, a bird's	Ee-sa-gweh . .	ib.
Wolf, the (engraved) ..	A-ma-ok . . .	ib.
Woman, a young	Kang-neen . . .	ib.
(generally) ..	Oo-leĕ-a . . .	ib.
Woman, or female (generally)	Oōng-na . . .	B.
Wrist	Taor-nōw-tik . .	ib.
Wind	Anoog-way . . .	B.
Wood	Oo-māk-se-lăk . .	ib.
(drift on beach) ..	Oo-nak-sih . . .	C.
Wood, log of	Kai-doo-ik . . .	ib.
(general term) ..	Ta-gnit, Kei-yu . .	ib.
Wound, a small	Killi-ak-toch . .	ib.
	Killi-ak-toch-pep-pin .	ib.
Whiskers	Oomg-yăy . . .	B.

English Names.	Esquimaux Names.	
Yes	A²	C.
Yourt (as carved)	Shi-rak	ib.
Yellow colour	Tshong-ak	ib.
Yellow (bird?)	Pook-taun	ib.

NUMERALS.

One	A-dow-wēet-sesūng-neek	B.
	Te¹-ga²-ra², a-dai-tsuk	C.
Two	Ma-loy-sesungnek	B.
	Mil-lei-tsung-net Ee-pāk? Adri-gak?	C.
Three	Ping-hĕt-see sūngnek	B.
	Pin-get-tsook? Pin-ge-yook Pin-get-tsa-tsung-net	C.
Four	Setūmní-sūngnak	B.
	Tse-tum-mat Sé-tum e²t	C.
Five	Ta-leĕma	B.
	Tad-glé-mat, Adreyeet	C.
Six	Ark-būnna	B.
	Agh-win-nak Ak-ka-oo in-el-get	C.
Seven	Aīt-pă	B.
	Ach-win-nigh-i-pagh-a Mulla-roo-nik, Bo³l-ruk	C.
Eight	Pena-yūa	B.
	Pen-ni-yoo-ik Pé-ge²s-se³t	C.
Nine	See-tūmna	B.
	Tee-i-dim-mik?	C.
Ten	Tād-leĕma	B.
	Kó-lit² (R)	C.

NAUTICAL REMARKS.

PASSAGE FROM TENERIFFE TO RIO JANEIRO.

June 5 *to July* 11.

In June, 1825, His Majesty's ships Wellesley and Bramble sailed from Santa Cruz for Rio Janeiro, and three days afterwards the Blossom departed for the same place. About the same time the packet, the Hellespont, and another merchant vessel made the passage from England. The Bramble crossed the equator in 18° W., the Wellesley in 25° W., the packet in 29½° W., the Blossom in 30° W., the Hellespont in 32° W., and the merchant brig, of which I shall speak presently, in 39° W. The Hellespont, which sailed indifferently, was forty-six days, the packet forty-six days, the Blossom thirty-six, the Wellesley forty-five, and the Bramble forty-eight days. Thus, making a reasonable allowance for the difference between England and Teneriffe, the Hellespont made the best passage, the packet and Blossom next, the Wellesley next, and the Bramble the worst; by which it appears that in proportion as the vessels were to the westward the passages were shortened. The merchant brig, however, was too far to the westward, as she could not weather Cape St. Roque, and, like the King George, Indiaman, she was obliged to stand back to the variable winds to regain her easting, so that her passage occupied a hundred and ten days!

This passage is so frequently made, that remarks upon it might be thought almost superfluous; but I am not disposed to undervalue this sort of information, which is in general too much neglected. There is no doubt that the route from England to Rio Janeiro ought to be varied according to the time of the year; for even in the Atlantic the trade-winds are affected by monsoons, and it is only by a long series of observations that we can ascertain at what time of the year it is advisable to cross the equator in any particular longitude. The journals of the packets for one year would afford valuable information on the subject. In the passage of the Blossom we carried the N.E. trade from Teneriffe to 8° N., and met the S.E. wind in 5° 30′ N. and 25° 50′ W., which carried us to Cape Frio. The trades were steady, and in the northern hemisphere fresh.

From the time of leaving Teneriffe until we lost the N.E. trade, the current set S. 54° W. 115 miles in ten days, or at the rate of 11½ miles per day. With the change of wind occurred an immediate alteration in the direction of the current, and the next twenty-four hours we were set N. 86° E. twenty-three miles. The meeting of the currents was marked by a rippling of the water, which could be seen at a considerable distance. The four succeeding days the current ran between S. 45° E. and S. 89° E. at the average rate of thirteen miles per day. During this time we changed our position from 7° 21′ N. latitude, and longitude 26° 58′ W. to 3° 56′ N., and 26° 44′ W., and had had the S.E. trade one day. We now got into a strong N.W. current, which ran between N. 58° W. and N. 72° W. at an average rate of twenty-two and a half miles per day, until we made Fernando Noronha.

From Fernando Noronha the current changed its direction, and ran between S. 78° W. and S. 21° W. at an average of twenty-seven miles per day, until a hundred miles due E. of Cape Ledo. We stood on the southward; and as we neared the land about Cape Augustine the velocity of the current abated, and our daily error was reduced to seven miles S. 52° W.; but as we drew off the land, still continuing to the southward, the current again increased, and became variable. The first hundred miles from Cape Augustine it ran S. 87° W. twenty six miles; the next due S. twenty-seven miles; the following S. 76° W. twenty-one miles, and then S. 80° W. eleven miles, until our arrival off Cape Frio, when the whole amount of current from Teneriffe was two hundred and seventy-four miles S. 57° W.

From this it appears that the N.E. trades propelled the waters in a S.W. by W. direction, at the rate of eleven and a half miles per diem*; and the S.E. trades to the W.N.W.,

* All the rates are averages.

with double the velocity, or twenty-two and a half miles per day*; and that in the intermediate space, where light variable winds prevailed, there obtained a strong current, which ran in a contrary direction to both these, at the rate of thirteen miles per day.

It appears from numerous observations that in both hemispheres the rate of the current is accelerated on approaching the Gulf of Mexico; and as my route was rather more to the westward than that usually pursued, the above-mentioned average rates are greater, probably, than will be experienced under ordinary circumstances.

REMARKS ON THE PASSAGE FROM RIO DE JANEIRO ROUND CAPE HORN TO CONCEPTION.

August 15 *to October* 8.

This passage was unusually long, owing to the prevalence of contrary winds, particularly in the vicinity of the River Plate. We sailed from Rio de Janeiro on the night of the 15th August, with a westerly wind, the Corcovado and Sugar Loaf capped with clouds. On the 16th, the wind shifted to the eastward; and towards night a gale suddenly arose, accompanied with thunder and lightning. The flashes of lightning passed frequently between the masts; and latterly the electrical fluid settled upon the mast-heads and topsail-yard-arms, and remained there for fifteen minutes. We had been warned of the approach of this storm by the appearance of the sky and a few flashes of lightning, and reduced our sail in time, otherwise it might have done much mischief from the suddenness and violence with which it commenced. This breeze went round to N. and N.W. to W. by S., then to S.E., S.S.W., N.W., southerly again, and S.E., E., and S. by W., until the 25th, the weather being gloomy, and the winds light or of moderate strength.

On the 25th, in latitude 36° and longitude 48° W., we encountered the first pampero, which came on with a heavy squall from S.S.W. attended with rain. For nine days we had these winds; during which time we could seldom carry

* All the rates are averages.

more than the main topsail, in consequence of the violence of the squalls. At the commencement of this bad weather, the squalls were harder and more frequent than towards its termination, and were accompanied with rain, hail, and sleet. Towards the close of it the general strength of the wind was increased, but the violence of the squalls was comparatively moderate, and the intervals longer. Still these gusts of wind gave no warning, and indeed during the whole period, excepting in the squalls, there was a clear blue sky, and apparently fine weather. From the commencement of these pamperos to their termination we had a reduction of nineteen degrees in the temperature of the air, and of fifteen in the surface of the sea. The remarks of Captain Heywood in Captain Horsburg's "Directory," a valuable book, and well known in the navy, will be found very useful in anticipating these squalls.

To these pamperos succeeded a calm, then light and moderate breezes from N.W., E.N.E., N., E.N.E., S.W., S.E. with cloudy weather, until in latitude 48° S. and longitude 54° W., when we fell in with a W.N.W. wind, which the next day carried us into soundings off the Falkland Islands. As we neared the land, the wind died away. The barometer was low, standing at 28·6, and the weather was misty, with drizzling rain at times. About one o'clock P.M. on the 9th September, the mist began to disperse, and a bright yellow sky was seen under anarch to the S.W.; the wind at the same time inclined that way, and in less than an hour we were under close-reefed topsails and storm staysails. This gale lasted about eighteen hours, and then veered to W. by N. and W., with which we advanced to the parallel of Cape St. John. Here we encountered strong S.W. winds with long heavy seas, and stretched to the southward to 58° 02′ S., regretting that we had not passed inside the Falkland Islands, as in that case we should have been nearly a day's run further to the westward before we encountered these adverse winds. After two days the wind veered to S.S.W. and blew hard, but the sea was not high. We now stood to the N.W , and on the 17th in latitude 56° 21′ S. and longitude 61° 51 W., we had a few hours' calm. This was succeeded by a breeze from the southward, which continued moderate with fine weather and a smooth sea; and the next day, having carried us one hundred and twenty-three miles, we made Cape Horn, fourteen miles distant on the lee-beam, bearing N. 2° W., true; the wind still from the southward.

Between Cape Horn and Diego Ramirez we had soundings with forty-five fathoms rock, and sixty fathoms sand; and afterwards from eighty-four to sixty fathoms gravel, coarse and fine sand, and some coral. That night we passed to the northward of Diego Ramirez at nine miles distant, not having less than sixty-six fathoms on a bottom of coarse sand. The following morning the island of Ildefonso bore N. 5° W., true, nine miles, and we had seventy-three fathoms fine sand; and at noon Yorkminster, at the entrance of Christmas Sound, bore N. 37° E., true, nineteen miles, eighty-two fathoms coral and stones. Not liking to range the shore of Terra del Fuego so close during the night with a southerly wind, we tacked; and with the wind still at S.S.W. stood for thirty-six hours to the S.E. into the meridian of Diego Ramirez; and when thirty-six miles S. of it, we again kept W. by S., with the wind at S. by W. We stood on, and had light winds, fine weather, and a smooth sea until the 24th, when there was a calm for twelve hours, with a little swell from N.E. On the 25th early, we got a north-easterly wind, which commenced with fine weather and smooth water; and at noon, on the 26th, carried us to the 79th meridian and 53d parallel of latitude, when we considered ourselves round the Horn. In this situation we were one hundred and forty-three miles due west of Cape Pillar; having numbered exactly fourteen days from the time at which we were a hundred miles due east of Staten Land. We passed Cape Horn on one Sunday, and on the following crossed the meridian of Cape Pillar. Our greatest south latitude in the whole passage was 58° 02′ S. The gales of wind which we experienced were attended with a long swell, that by no means strained the ship, and we did not see a particle of floating ice.

Having reached the meridian of 82° W., there appears to be no difficulty in making the remainder of the passage to

Conception or Valparaiso. In high latitudes the prevailing winds are from W.N.W. to S.W., which are, at worst, leading winds. In latitude 44° 16′ S. and longitude 78° 36′ W. we got S.E. winds, which, with a few hours' intermission of wind from N.E. by E., brought us to Conception on the tenth day from that on which we considered ourselves fairly round the Horn. Some officers are of opinion that near the coast of Chiloe moderate weather and southerly winds are more prevalent than in the offing, which I think highly probable; and if, after reaching the 81st meridian, the winds came from N.W., I should certainly prefer the in-shore track to stretching again to the S.W.

With regard to the best time of the year for rounding Cape Horn, there is a great difference of opinion, as in the same months both good and bad passages have been made; but I should certainly not select the winter time if I had my choice. Independent of the cold, which, during gales of wind, is severely felt by a ship's company necessarily wet and exposed, and the probability of meeting with floating islands of ice, surely the long nights, as Captain Hall has justly observed, must augment in a serious degree the difficulties of the navigation.

From the passage of the Blossom, a preference might be given to the month of September; but in the very same month Captain Falcon in the Tyne had a very long and boisterous passage. I concur in opinion with Cook, Perouse, Krusenstern, and others, in thinking there is no necessity whatever for going far to the Southward, and I should always recommend standing on that tack which gained most longitude, without paying any regard to latitude, further than taking care to keep south (say a degree) of Cape Horn. With a N.W. wind I would stand S.W., and with a S.W. wind N.W., and so on. If there was a doubt, I should give the preference to the southern tack, unless far advanced in that direction. We did not find the strongest winds near the land, but on the contrary; and I am of opinion that here, as is the case in many other places, they do not blow home, and that within thirty miles of the land the sea is partly broken by the inequality of the bottom. There is, however, great objection to nearing the land eastward of Cape Horn, in consequence of the velocity with which the current sets through Strait Le Maire, particularly with a southerly wind. This does not obtain to the westward of Diego Ramirez, in which direction I see no objection to approaching the coast within forty or sixty miles. Cook ranged this shore very close in December, and on more than one occasion found the current setting off shore, and at other times slowly along it to the S.E.

In the first part of this passage the currents ran to N.W., but after passing the latitude of 40° S. they set to the eastward; and when we arrived off Cape Horn the ship was S. 40° E. 116 miles of her reckoning.

While we were in the neighbourhood of Diego Ramirez there was little or no current, but to the westward it ran to the W.N.W. It however, soon after changed, and on our arrival off Conception the whole amount of current was N. 49° E. 147 miles. In rounding Terra del Fuego with a southerly wind full four points must be allowed for variation and current. For in this high latitude there will, in most ships, be found ten or twelve degrees more variation with the head west than east; and though the true variation be but 24° E., at least 29° or 30° must be allowed going westward.

We found the barometer in this passage an invaluable instrument: upon no occasion did it deceive us. In passing these latitudes my attention was drawn to the changes in the temperature of the water, which I usually found to precede a shift of wind from south to north, and vice versa, even before that of the temperature of the air. I subjoin a short statement of these changes, for the satisfaction of such as may feel interested in them.

On the 29th of August, at eight A.M. the temperature of the surface was 58°, the weather moderate and cloudy, and the wind W. N.W.; from this time to midnight it gradually fell until it stood at 48°. The wind now increased, and the next morning shifted to S.W. and S.S.W., and blew fresh gales: the breeze continuing, on the 31st the temperature of the surface underwent a further fall of 3½°; and we had hard squalls, with hail and sleet. It afterwards fluctuated four days between 46° and 49½°, during which time the winds were variable from S.S.W. to N.W. by N., and E.N.E.—the weather for the most part moderate and cloudy; but on the 5th (Sept.) the temperature (always alluding to that of the surface of the sea) rose to 53°, and the wind came from N.E. by N. and N., but light. The next day it shifted to S.E. by S., and the temperature rather decreased, but the breezes were light. On the afternoon of the 7th, after a calm, during which it remained at 50°, there was a decrease of 8°; and thirty-six hours after a gale from S. by E. suddenly arose. During the five following days it was nearly stationary, at the temperature of 39½°, and the wind was variable from W.N.W., S.W., and W. blowing hard. From noon on the 12th to four A.M. on the 13th it fell to 36°, and that night we had a gale at W. by S.; which continued all the next day. At night there was a further decrease of 4°, when the wind veered to S. by W., and blew strong gales. The temperature kept down at 35° until midnight of the 15th, when it rose 5°: and the 16th, at four A.M., the wind changed to W.N.W. and N.W. by W. The temperature, however, soon decreased again 4°, and at nine A.M., the following day, the wind came from S.W. by S. and S.S.E. where it continued, and the temperature remained nearly stationary until we made Cape Horn, when it rose to 42°.

It would, perhaps, be too hazardous to assert upon such short experience that these changes are the forerunners of shifts of wind, though I found similar variations attend the southerly gales off Spitzbergen, where we had always indication of their approach by the increase of the temperature of the sea.* I am, however, persuaded that, like the barometer, it speaks a language which, though at times not the most intelligible, may nevertheless often prove useful.

* See also vol. I. p. 324 of this work.

HOME PASSAGE FROM COQUIMBO TO RIO JANEIRO.

June 3d to July 21st, 1828.

This passage was considerably lengthened by not getting to the westward in low latitudes. From the time of leaving Coquimbo there was a difficulty in making progress in that direction, and we could scarcely weather Massa Fuera. From here the weather became boisterous, the breeze generally beginning at W.N.W., and ending in a moderate gale at S.W.: then backing again, and in the course of the twenty-four hours finishing at S. W., blowing hard, as usual; so that what distance was gained to the westward in the early part of the day was lost toward the close of it. In this manner we were driven down upon the coast, and obliged to stand to the westward, when, had we been a hundred miles further off shore, we should have had a fair wind. On the 22d June we had an easterly wind, which veered to S.E., and drove us away to the latitude 56° 18′ S. and longitude 75° W., when we encountered S. by E. winds, which carried us past Cape Horn on the 30th. Our winds were now fair; but off the Falkland Islands they were variable, until they settled in the E.S.E. quarter. With this we advanced to 35° N., when we encountered N.E. and N.W. gales, with heavy cross seas, and then several pamperos, which were attended by vivid lightning. We afterwards made progress to the northward, and arrived at Rio Janeiro on the 21st July.

In this passage, which was made in the depth of winter, the greatest cold was 21°. From Cape Horn to the Falkland Islands we had thick showers of snow, and had we been bound the opposite way, I have no doubt we should have felt the weather severely. The barometer, as on the former occasion, proved an invaluable monitor. From the time we quitted Massa Fuera until we were off Staten Land, the winds were advantageous for making the passage to the Pacific, and so far they favour the opinion of the winter time being the most desirable for this purpose. The current in this passage ran to the south-east to the latitude 46° S., then north two days, and from 48° to 57° S., between N.W. and S.W., at the rate of thirteen miles a day. From 57° S. and long. 68° W., they ran to the N.E., until we had passed the Cape, and then westerly and north-westerly to the Falkland Islands. Off the River Plate they ran to the S.W. and S. On our arrival at Rio Janeiro the whole effect of the current from Coquimbo was S. 62° W. eighty-two miles.

From the experience of these two passages round the Horn, I am of opinion that a ship bound to the Pacific should pass inside the Falkland Islands, and round Staten Land, as closely as possible; as she will most likely encounter S.W. winds directly the Pacific is open. A north-west wind off the Falklands will, I think, generally veer to W. and S.W. on approaching Staten Land. With S.W. winds off Staten Land, nothing is left of course but to stand to the southward. I should not, however, recommend keeping this board longer than to get an offing, except westing was to be made by it; and if not, I would go about directly a mile of longitude was to be gained on the in-shore tack; avoiding, however, a near approach to Terra del Fuego, eastward of Cape Horn, on account of the north-east set through Strait Le Maire, with southerly winds. I see no good reason for going to a high southern latitude, if it can be avoided without loss of longitude. With regard to the fact, that gales of wind are stronger near the land, I own I cannot concur in such an opinion. On a comparison of the Blossom's passage out with that of a brig commanded by a Lieutenant Parker, which rounded the Horn at the same time, it appeared that whilst she was experiencing strong winds and heavy seas, which washed away some of her boats, the Blossom, close in with the land, had fine moderate weather, and no other indication of the gales the Hellespont was encountering than by a long southerly swell setting upon the shore; and that the Blossom had the advantage of a westerly current, while the brig was put back twenty miles daily by one in the opposite direction.

When clear of Terra del Fuego, I should recommend stretching to the westward as far as the meridian of 82° or 83°, about the parallel of Cape Pillar, before shaping a course along the coast of Chili.

From Chili to the Atlantic ships should pass outside Massa Fuera, and if opportunity offered, get as far west as 85° or 90°, in order that the south-westerly winds, which they will afterwards be certain to meet, and generally blowing strong, may be turned to advantage. I would even recommend keeping to the westward of 83° until past the parallel of 53° S. This precaution appears to be the only one necessary, as the remainder of the passage from that situation is in general very easily performed. With regard to passing inside or outside the Falkland Islands, I think the latter preferable, especially in winter, as the winds sometimes hang in the eastern quarter at that period, and are apt to run a ship in with the River Plate.

From the Falkland Islands to Rio Janeiro the winds are very uncertain. Ships may, however, generally reckon upon encountering at least one pampero between 33° and 37° N.,* and on meeting with northerly or north-north-easterly winds, when within two hundred or three hundred miles of Cape Frio. It is better, in the latter instance, to stand out to the eastward in preference to the other tack, as it will almost always happen that they will there meet an easterly wind to carry them up to the Cape. It has been found very difficult to get up near the shore from Ila Grande and St. Catherine's.

* These winds appear to be of frequent occurrence off the River Plate; they are generally preceded by strong N.W. winds, and a low altitude of the barometer. Care is necessary to avoid being taken aback by the wind shifting suddenly to the S.W., which it sometimes does after a heavy squall. In deep laden ships it would be prudent to lie to with the head to the N.E., as they would then bow the sea, which often runs very high on the shift of wind; whereas on the other tack they would have their stern exposed to it.

PASSAGE FROM VALPARAISO TO OTAHEITE.

November 4th to March 18th.

This was made in the summer, when the trade-wind extends further south than at other times of the year; otherwise it would be advisable to get into a lower latitude than that in which our course was directed. The winds with us were very variable, but always fair. I know nothing worthy of remark here except the current, which, on our arrival at Elizabeth Island, was found to have set the ship three hundred and forty-nine miles to the westward in thirty-nine days, or at the rate of 8·95 miles per day.

Our route from Elizabeth Island was directed to each of the islands lying between it and Otaheite, and we afterwards met with too many interruptions to estimate the rate of the currents; but at this time of the year in particular, there does not appear to be much in any direction. At other times, however, I am told that there is great difficulty in getting to windward. In this sea the westerly monsoon, which sometimes extends as far as these islands, checks the regularity of the trade-wind, and it is not uncommon at such times to meet a westerly wind with heavy rain. This is liable to occur from December to February or March. The trade-wind in this route in general hangs more to the eastward than the S.E. trade in the Atlantic.

REMARKS ON THE PASSAGE FROM THE SOCIETY GROUP TO THE SANDWICH ISLANDS.

April 26th to May 18th.

In making the passage from the Society Group to the Sandwich Islands, the time of the year should be considered. Between the months of April and October the trade-wind is said to hang more to the eastward than at other times, and is consequently favourable to the passage; but it is advisable even at that season to cross the line well to windward, if possible between 145° and 148°, as all that is gained in that direction will ultimately be of use. Between December and April a more northerly trade may be expected, and consequently easting is of more consequence. The S.E. trade is not as regular as that to the northward of the Equator. It generally blows at E. or E.N.E., and when the sun is to the southward of the equator it is sometimes interrupted by N. and N. W. winds. These should be taken advantage of in order to get to the eastward, even at the expense of a few miles of latitude, until well advanced to the northward, and until the N.E. trade is fallen in with.

The Blossom left Otaheite on the 26th of April, 1826, and crossed the equator on the 9th of May in long. 150° 01′ W. From the time of sailing the winds were light from the E. and E. N. E., but sometimes veered to N. E. and N.; with these we tacked and endeavoured to gain easting, but did not succeed as we wished. We kept the easterly wind to the lat. of 4° N. and long. 149° 47′ W., when the N.E. trade met us; it commenced with hard squalls and rain at N.E. by E., at which point it continued with scarcely any variation; and we had as much wind as would allow us to carry, conveniently, courses and double-reefed topsails, and latterly topgallant sails, until we made Owyhee on the 18th, about forty miles due west of us. We now felt the advantage of being well to windward, and keeping the same distance in order to ensure the sea breeze throughout the night, made Mowee the following morning, and the same night arrived close off Diamond Point (Woahoo).

The current from Otaheite to the equator set to the W.N.W. from ten to thirty miles per day, at an average rate of sixteen and a half miles per day. From the equator to the fourth degree N., when we met the N.E. trade, it ran N.N.E. fifteen to twenty-three miles a day, averaging eighteen miles a day, after which it ceased entirely. On our arrival off Owyhee the current from leaving Otaheite had set N. 54° W. 164 miles, or 7.1 miles per day.

REMARKS ON THE PASSAGE FROM THE SANDWICH ISLANDS TO AWATSKA BAY, KAMSCHATKA.

June 1st to 28th.

This passage was very favourable, both in regard to wind and weather, and occupied only twenty-seven days. On quitting Oneehow, instead of keeping within the tropics for the advantage of a fresh trade-wind, I endeavoured to pursue the 30th or 31st parallel down to 191° or 192° W.; and then to avail myself of the westerly winds, said to prevail there, in order to get to the northward.

Quitting Oneehow, I passed to the north-eastward of Bird Island, and the chain of reefs situated near the French Frigate's Bank, and then bore away west. We kept the trade-wind with but one interruption, until in latitude 29° 46′ N. and longitude 185° W., which was on the 10th day of our departure; here the wind veered to the S. and S.S.W., and continued fair three days. On the thirteenth day (June 15th), in lat. 33° N., long. 192° W., it shifted suddenly to N.W. by W. I was now near the situation I had been desirous of reaching, and ready for this wind, but it did not continue; and for five days we were retarded by light winds from all points of the compass, except that quarter. On the 20th June we had a N.E. wind again, which veered to E., S.E., S., and on the 5th day to W. S. W., when it left us in 46° N. and 199° W. An easterly wind succeeded, but, before the twenty-four hours were expired, veered round by S. to W., which, with the exception of a few hours N.N.E. wind, carried us close off the light-house of Awatska Bay on the 28th June.

The weather during this time had been moderate; it had scarcely been necessary to take in top-gallant sails the whole period. It will be seen that, with the exception of three days, we had a leading wind the whole of the way, and that our greatest delay was occasioned by light winds about the parallels of 34° and 35° N. The trade-wind may be said to have attended us as far as 30° N. and 185° W. About the 30th parallel, a S.S.W. wind brought thick rainy weather with squalls, which was cleared away by a northerly breeze in latitude 34° N. We had now light winds and clear weather, but from the 39th parallel to the day of making the land of Kamschatka, with the exception of one day in latitude 50° N., we were attended by a thick fog and drizzling rain.

On comparing the route of the Blossom with those of Captains Clerke and Krusenstern, who quitted the Sandwich Islands for St. Peter and St. Paul's, and both of whom, as well as myself, endeavoured to run down the longitude until sufficiently far to the westward to reach the place of destination without inconvenience from westerly winds, it appears that a preference is due to the course pursued by the Blossom. As the three tracks from the parallel of 33° or 34° N. and long. 166° or 168° E. nearly coincided, I shall divide the passage into two parts; the first from the Sandwich Islands to that situation, and the second from thence to the day of making the land of Awatska.*

Captain Clerke ran down his longitude near the northern tropic, lost the trade-wind in lat. 28° N., and long. 172° E., on the twenty-first day of his departure, and reached the above situation on the twenty-sixth day.

Krusenstern kept to the southward of 20° N., lost the trade in 27° N. and about 176° E. on the seventeenth day, and reached the above place on the twenty-second day.

The Blossom kept to the northward of 30°, lost the trade in 30° N. and 175° E. on the tenth day of her departure, and was in the above-mentioned situation on the thirteenth day.

From this situation to the second point or the day of arrival off Awatska, it is remarkable that the three passages are nearly of the same duration, that of Captain Clerke occupying thirteen days; of Krusenstern thirteen; and of the Blossom thirteen and a half. By which it is evident that the advantage was gained by the Blossom in the first part of the passage, and this was not confined to time alone, but extended

* I limit the passages to the time of making the land, as Captain Clerke was five days off the port.

to personal comfort, as the Blossom escaped the heat of a tropical climate, of which Captain King complains so much, and on the whole had better weather.

The currents on the first part of this passage were very irregular, varying their direction from N.N.E. to W.N.W.; the preponderance being in the latter direction, and in one day amounting to thirty-eight miles. After losing the trade-wind we had no current of consequence, excepting on three days in lat. about 35° N. and long. 194° W. during very light winds. On one of these days it ran S. 45° E. forty miles, on another S six miles, and on the third S. 31° E. nineteen miles. The whole effect of the current between Oneehow and Petrapaulski was N. 25° 30′ W. fifty-two miles.

FROM AWATSKA BAY TO KOTZEBUE SOUND.

July 5*th to* 22*d*, 1826, *and July* 20*th to* 5*th August*, 1827.

After clearing the outer bay, between Cape Gavarea and Chepoonski Noss, in both years we experienced much fog; but it cleared away in the vicinity of the islands of Beering and of St. Lawrence. The weather in both seasons was fine, and we met no impediments from winds until after passing the island of St. Lawrence, and then only for a day. The situation of Beering's Island is now well fixed, and so far it may be approached with safety; but the soundings decrease very fast near the land. Fifty-three miles S.W. by W. from the island we had no bottom with four hundred and twenty fathoms; twenty-seven miles in the same direction no bottom at two hundred fathoms; but at four miles we sounded in sixty fathoms fine dark sand. It is not advisable to stand within two miles of the western shore of this island, as there are breakers and low rocky points projecting from that part of the coast; two miles and a half from these breakers we had only nineteen fathoms dark sand; nor should the southern shore be approached within six miles, on account of Seal Rock, unless the weather be fine. From here I would recommend steering for St. Lawrence Island, in preference to the main land. Ships will come into soundings of fifty-four fathoms' mud in about the latitude 61° 25′ N. and 175° 17′ W. long., which depth will gradually decrease to thirty-one fathoms, when the bottom will almost immediately change from mud to fine dark sand. Two miles and a half S. 73° W. from the S.W. cape, there are fifteen fathoms; but off the N.W. end of the island there is a shoal upon which there are only nine fathoms, stony bottom, four miles' distance from the land. It is narrow, and the water soon deepens again, and the bottom changes to fine sand as before.

From St. Lawrence Island there appears to be a current running to the northward at the rate of about three quarters of a mile an hour, which increases as the sea narrows towards the Strait of Beering. Ships may pass either side of the Diomede Islands, but they should not run between them, as the passage is not yet explored. Cook passed between the Fair Way Rock and Krusenstern Island, and had deep water; but no person has, as yet, I believe, been between Ratmanoff and the next island. Near these islands the water deepens to twenty-seven and thirty fathoms, and the bottom in some places changes to stones. The channel to the eastward of the Diomede Islands is the widest; and the only precaution necessary is to avoid a dangerous shoal to the northward of Cape Prince of Wales, upon which the water shoals almost immediately from twenty fathoms to four and a half. Its outer edge lies about north (true) from Cape Prince of Wales. From here, ships may run along shore in safety in ten fathoms near the land.

It is unnecessary to give any directions for the sea to the northward of Kotzebue Sound, as the lead is the best guide, remembering that off Cape Krusenstern, Point Hope, and Icy Cape, the water shoals fast, as those places appear to be washed by strong currents.

In this passage there was not much current between Awatska and St. Lawrence Island: it amounted to only thirty-one miles S. 54° W. Off the island it ran S.S.E. seven-eighths per hour on one trial, and on another seven hours afterwards,

N. E. five-eighths per hour; but between this island and Beering's Strait it ran to the north-westward at about three quarters of a mile an hour. To the northward of the Strait it takes a more northerly direction, and near the land runs first to the N.E. and then N.W.

KOTZEBUE SOUND TO CALIFORNIA.

October 14*th to* 7*th November*, 1826, *and October* 6*th to* 29*th*, 1827.

These passages were made late in the year, when north-westerly winds prevail, and consequently at a favourable time for getting to the southward. In both years they occupied exactly twenty-three days; and it is further remarkable, that in each, the Aleutian Islands were passed on the ninth day after our departure. The route pursued by the Blossom was to the westward of King's Island, and between St. Lawrence Island and the main-land of America, and thence within sight of St. Paul's and St. George's Islands to the Strait of Oonemak.

To the eastward of King's Island the soundings are very irregular, varying from nine to six fathoms; and as at the season above mentioned the weather appears to be generally bad, it is advisable to go to the westward of the island, where the water is deep. Between St. Lawrence Island and the continent of America there is a bank with eleven fathoms water upon it. If, on approaching it in foggy weather, it be doubtful, from the shoaling of the water, whether it be not the island that is the occasion of the decrease of soundings, haul over to the American shore, and the water will deepen. To the southward of St. Lawrence it is necessary only to mention the islands of St. Paul and St. George, which apparently may be safely approached within four or five miles; but I could not get near them in either year to ascertain what dangers lie close off the shore. In the geographical table I have given the positions of these islands, which were before considered so uncertain that they were not placed on our charts.

I should recommend the passage being always made to the eastward of these islands, as between them and Oonemak there is a strong current from Bristol Bay, which in 1827 drifted the Blossom thirty-five miles to the S.W. in the course of the day. The Strait of Oonemak, lying between the islands of Oonemak and Coogalga, appears at present to be the safest opening to the Pacific from the Kamschatka Sea. The Aleutian Islands in the autumn appear to be enveloped in fog about half-way down, and to have a region of mist lying to windward of the Archipelago, which makes it necessary for a ship to be certain of her position before she attempts any of the channels, as she might be led down so close upon the land in the fog, that she would not have room to rectify a mistake, should she unhappily incur any, which is very likely to happen, from the irregularity and velocity of the currents about the islands. Under these circumstances I should recommend making the north-west end of Oonemak, and afterwards keeping along the coast of that island to the southward. As this island lies forty miles to the northward of the other islands of the chain, Amnak excepted, which is three degrees to the westward, it cannot be mistaken, unless the reckoning of the ship is very incorrect indeed. And by so doing, in the event of not liking to attempt the passage, a vessel will still be far enough to windward, supposing the breeze to be from the northward, to weather the other islands of the chain; and if from the westward, she may reach into Bristol Bay.

We had no opportunity of seeing the summits of either Oonemak or Alaska, which, when clear, are good guides for the strait;* but when the low land of the former can be seen, the south-west point of Oonemak may be known by a pointed rock situated near the base of a remarkable wedge-shaped cliff, conspicuous from the northward and north-westward. The narrowest part of the strait is between this rock and Coogalga Island, and the distance exactly nine miles and a half, in a S. 1° 30′ E. (true) direction. In a line between

* See Cook's Third Voyage, vol. II.

these, at the distance of four miles from the rock, there are soundings in thirty fathoms, and I understand that if necessary there is anchorage close under Oonemak.

Coogalga Island is about four miles in length, and may be known by a remarkable peak near its N. E. extremity, in latitude 54° 16′ 52″ N., and longitude 164° 47′ 06″ W. The variation off it is 20° 50′ E.

From the Aleutian Islands to San Francisco we steered nearly a direct course, with winds generally from the N. W. and W., and made Punta de los Reyes on the 3d November. In this passage the currents were variable. From Beering's Strait to the Aleutian Islands they prevailed to the westward, and near the islands ran strong, but afterwards they continued between S. E. and S. W. On our arrival off California, the whole amount, in 1826, was S. 89° W. sixty-four miles; and in 1827, S. 26° W. forty miles.

REMARKS ON THE PASSAGE FROM MONTEREY (NEW CALIFORNIA) TO WOAHOO, SANDWICH ISLANDS.

January 5th to 25th, 1827.

This passage was begun at a period when the north-west and westerly winds are proverbially prevalent upon the coast of New Albion, and extend a considerable distance to the westward.

We sailed from the Bay of Monterey on the 5th January, and immediately took a northerly wind, which carried us into the trades; and we arrived off Mowee on the twentieth day. Our passage might have been considerably shorter, had we not taken a circuitous route in search of some islands reported to lie to the southward, and had sail been carried throughout the twenty-four hours, instead of hauling to the wind as soon as it was dusk, to maintain our position during the night, that nothing might be passed unseen within the limit of our horizon.

As we left the extra tropical latitudes, the atmosphere gradually became more hazy and humid, the clouds increased, and in 18° N. we had some showers of rain. On the 18th, in latitude 16° 18′ N. and longitude 136° W., we had a very strong trade at N. E., with squally weather, and a long cross sea from the westward, which was afterwards found to be the effect of a gale of wind in the parallel of 21° N.; but which did not reach us.

There was very little current in this passage; this little generally ran to the southward and westward, and averaged 3.6 miles a day. The barometer, though so far entered in the tropical latitudes, was perceptibly affected by the changes of weather, but maintained its horary oscillations.

On my arrival I found that from the 15th to the 21st there had been very strong gales from the westward at Woahoo, and from the S. W. at Owyhee. This was, no doubt, the cause of the high cross sea we experiencd from the 18th to the 23d. I found also that the harbinger, an American brig which quitted Monterey nine days after the Blossom, was obliged to lie to for three days, from the 20th to 23d January, in a strong gale from the S. W. She had steered a direct course for the Sandwich Islands, in which she experienced very variable winds, and, on the whole, had bad weather, and was only one day less performing the passage than ourselves: whence I think it fair to conclude that nothing is lost by running well into the trade. During the winter season, I should recommend ships gaining the 17th parallel before they shaped a direct course for the islands. This seems to me to be the best mode to ensure a good passage and fine weather.

REMARKS ON THE PASSAGE FROM THE SANDWICH ISLANDS TO MACAO (CHINA).

March 1st to April 10th.

This passage was made at a late period of the season; the north-east monsoon had become very faint, and about the Bashee Islands appeared altogether to have finished.

From Woahoo to the Ladrones the passage occupied twenty-six days; thence to the Bashee Islands twelve days; and from the Bashee to Macao three days; in all forty-one days.

The first part of the run was within the limit of the trade-wind; it hung generally in the eastern quarter, and with the exception of a few days' rain, squalls, and very vivid lightning, in latitude 19° N., and longitude 170° W., the weather was very fine.

Off the Ladrones we had a short calm; then a breeze at north; and made the passage to the Bashees with light and variable winds, first from the northward, and latterly from the S. and S. W. The weather during this period was remarkably fine. Off Formosa we took a strong northerly wind, which carried us to Macao.

The currents from Woahoo to the Ladrones ran generally to the *eastward*, and averaged 6.9 miles per day.

I should recommend to ships making this passage to run down the parallel of 18° 30′ N. or 19° N., taking care of Wake's Island, which is said to lie in latitude 19° 18′ N. They should make the Island of Assumption, in latitude 19° 42′ N., and longitude 214° 34′ W., and pass to the southward of it.

Twelve miles to the southward of Assumption, Captain Freycinet has discovered a reef of rocks, which may be avoided by keeping close to the above-mentioned island. Assumption is a small conical island, 2096 feet high, and apparently without any danger. Perouse anchored in thirty fathoms, within three quarters of a mile of its western shore. The Mangs bear from its eastern point N. 27° 07′ W. (true).

In the N.E. monsoon I would steer from here for the North Bashee Island, and thence pass northward of the Prata Shoal; but with the S.W. monsoon a different route is necessary, for which see Captain Horsburgh's India Directory. The Bashees, Vele Rete, and Botel Tobago Xima, are all very well laid down in Horsburgh's chart; but the Cumbrian Shoal has since been found to lie in the situation first assigned it, fifteen miles due S. of Little Tobago Xima, and in latitude 21° 42′ 15″ N. In its vicinity we found very strong ripplings, which, when the winds were light, sounded like breakers; but they did not affect our reckoning much, for on the 10th of April, in the forenoon, we made Pedro

Branco, as we expected. This rock is an excellent landmark; by our observation it lies 1° 33′ 13″ E. of the west end of the Typa. Shortly after noon we got sight of the Great Lemma, and that night anchored between Lantao and Chichow.

FROM THE ARZOBISPO ISLANDS TO KAMSCHATKA.

June 16th to July 3d.

At the commencement of this passage it was my endeavour to get nearly into the meridian of Petropaulski before I shaped a course for that place, in order to escape the inconvenience likely to arise from the prevalence of easterly winds, which we unexpectedly encountered the preceding year.

Between the parallels of 30° N. and 35° N. we had light and variable winds, as in our first passage; and in 39° N. took a southerly wind, which continued with a very thick wetting fog, as before, until within a day's sail of Petropaulski, when it veered to the S.W., and soon after came fresh off the land, precisely as it had done the preceding year. In the summer I recommend making the land a little to the southward of Cape Gavarea, as the wind generally blows off shore, and to the eastward of the promontory veers to the northward; and if a vessel is not well in with the Cape, she will find much difficulty in beating up. Until we were in latitude 34° N. longitude 153° E., the currents ran between N.W. and S.W. twelve miles per day; they then changed to S. five miles per day as far as 40° N., and off the Kurile Islands ran strong to the S.E. The weather throughout this passage, with the exception of the fog, was very fine.

FROM SAN FRANCISCO TO SAN BLAS (MEXICO).

December 6th to 21st.

We found no difficulty in getting to the southward, the prevailing wind at this season being from the N.W. It is advisable, however, to stand about forty or fifty leagues off the coast, to avoid interruptions from variable winds which occur near the land. These winds are in general taken advantage of by vessels bound in the opposite direction to that of our present course.

The weather throughout this passage was remarkably fine. The wind was from W.N.W. to N.N.E. until we made Cape San Lucas, when it veered to E.N.E., and obliged us to pass between the Tres Marias Islands. This route occasioned the loss of a day, and I should advise any vessel making the passage to close the land to the northward of Cape San Lucas, provided the wind were in the north-east quarter; as in addition to the inconvenience which a shift of wind to the E. would occasion, there is another arising from a strong current, which generally sets out of the Gulf of California. From the Cape steer for Isabella Island, and thence for Piedra de Mer.

Between 33° N. and Cape San Lucas we found a current to the westward, and from the Cape to the Tres Marias to the southward. The whole effect of current from San Francisco to these islands was S. 58° W. eighty miles.

SAN BLAS TO ACAPULCO AND VALPARAISO.

March 8th to May 1st, 1828.

At this season north-westerly winds prevail upon the coast between San Blas and Acapulco, inclining toward the land in the day, and to the sea in the night. We passed four miles to westward of Corveteña (a small rock, situated N.W. by N. nineteen miles from Cape Corrientes) without having soundings in eighty fathoms. On the 10th we were within sight of the volcano of Colima, 12,003 feet above the sea, and on the 13th anchored at Acapulco.

At San Blas we heard various opinions upon the best route from Acapulco to Valparaiso, some being in favour of a passage to the eastward of the Gallapagos, by keeping along the land, and carrying the N.W. wind, and others to the westward, by steering at once out to sea. We adopted the latter mode of proceeding; and after light and variable winds, principally from the eastward, crossed the equator in 99° 40 W., on the eleventh day of our passage, about two degrees more to the westward than was intended.

After two days' unsettled weather and hard showers of rain, we got the S.E. trade in 3° S. latitude. It at first held to the southward, but, as we proceeded, veered gradually to the eastward, and obliged us to make a long sweep, in which we went as far to the westward as 108°, and having brought us into 23° S. and 106° W. it left us. We had afterwards variable winds and squally weather, and found some difficulty in approaching our destination. At this season very unsettled weather prevails on the coast of Chili, and storms and heavy rains from the northward are by no means unfrequent. It appears to me to be advisable at this period to steer direct for the port, if possible, and to disregard the chances of winds and of currents near the land. The currents in the first part of this passage ran about seven miles a day to the eastward, but from 8° N. and 98° W. to 19° S. and 108° W. they flowed in a S. 88° W. direction, at the average rate of about twenty-eight miles per day, and on our arrival at Valparaiso they had drifted the ship S. 81° W., four hundred and one miles, or at the average rate of eleven and a half miles a day.

On account of these strong currents it is desirable to cross the equator well to the eastward, in about 96° or 97° W., and to pass the latitudes in which they prevail as quickly as possible, by keeping clean full.

RIO JANEIRO TO ENGLAND.

August 5th to September 25th.

This passage was remarkable for strong S.W. winds between the trades. Upon leaving Rio, N.E. winds obliged us to stand to the S.E. to the lat. 27° S. and long. 36° W., where we met the S.E. trade-wind, which carried us across the equator in 24° 20′ W., and left us in 5° N. latitude. It was there succeeded by strong south-west winds, attended by a long swell from the same quarter. This continued to 15° N., and was succeeded by the N.E. trade, which prevailed as far as 27° N. and 35° W. We had here six days calm, and then variable winds, with much bad weather and long seas from the northward, and did not arrive in England until fifty-one days after our departure from Rio. Had we been farther westward when the N.E. trade failed, the passage would have been shortened, and as at this season N.W. winds prevail on the coast of America, I should endeavour on another occasion to arrive at a more westerly longitude before I outran the trade-wind.

The current in this passage was very different to that which we experienced on the outward voyage, and was no doubt influenced by the strong S.W. winds. From the tropic of Capricorn to the equator it ran N. 88° W. a hundred and fifty-one miles, or ten miles per day, and from that latitude to the termination of the S.E. trade S. 66° W. twenty-five miles a day. Here we encountered the winds from the westward, which, while they lasted, occasioned a current to the eastward at the rate of twenty-six miles a day. With the N.E. trade there was very little in any direction.

OBSERVATIONS ON THE COAST OF CHILI.

CONCEPTION.

During the summer months southerly winds prevail along this coast, and occasion a strong current to the northward. It is advisable, therefore, to make the land well to the southward of the port, unless certain of reaching it before night. Punta Rumena appears to me to be a preferable land-fall to that of Saint Mary's Island, which has been recommended, as it may be seen considerably further, and has no danger lying off it. But should the latter be preferred, it may be known by its contrast to the mainland, in having a flat surface and perpendicular cliffs, as well as by a remarkable peaked rock off its N.W. extremity*. If the port cannot be reached before dark, it would be advisable to bring to the wind, between Saint Mary's and the Paps of Bio Bio, as there will almost always be found a southerly wind in the morning to proceed with. In doing this, take care of the Dormido Bank, lying off the

* This rock bears S. 53° 08′ W. true, from the Look-out Hill, Talcahuana, and is 24′ 48″ W. of it. Its latitude is 32° 58′ 10″ S., as found by Mr. Forster.

N.W. end of Saint Mary's. Having daylight to proceed by, close the land near the Paps of Bio Bio, and, keeping one and a half mile from the shore, stand along the coast of Talcahuana Peninsula.

Should the Paps of the Bio Bio be clouded, the land about them may still be known by the opening into Saint Vincent's Bay, and by the land receding in the direction of the Bio Bio river, as well as by high rocks lying off the points. The capes of Saint Vincent's Bay on both sides are high and terminate abruptly, and the south one has a large rock lying some distance off it. The northern cape is tabled, and has a small tuft of trees near its edge. Table land extends from here to Quebra Ollas. The Paps viewed from the westward appear like an island; the wide opening of the Bio Bio being seen to the southward, and Saint Vincent's Bay to the northward. The high rocks off the capes, at the foot of the Paps, are an additional distinguishing mark; and when near enough the rock of Quebra Ollas will be seen lying off the N.W. end of the peninsula. About one third of the way between Quebra Ollas and Saint Vincent's Bay, there is a large rock called the Sugar Loaf. All this coast is bold, and may be sailed along at a mile and half distance. Quebra Ollas rock lies the farthest off shore, and is distant exactly one mile and a quarter from the cliff; it may be rounded at a quarter of a mile distance, if necessary, but nothing can go within it.

Having passed Quebra Ollas, steer to the eastward, in order to round Pajaros Ninos as closely as possible, and immediately haul to the wind (supposing it from the southward), for a long beat up to the anchorage. There are two passages into Conception, but the eastern is the only one in use. On the eastern shore of this channel there is no hidden danger, until near Punta Para and Lirquen, when care must be taken of the Para Reef, the Penco Shoal, and the flat of Roguan. When near the two latter the southern head of Saint Vincent's Bay comes open with Talcahuana Head, it will be time to go round; and it is not advisable *at any time* to open the *northern* cape of Saint Vincent's Bay, distinguished by a tuft of trees upon it, with Talcahuana Head. These two landmarks a little open, and the pointed rock at the south extremity of Quiriquina a little open with Point Garzos, the N.E. extremity of the peninsula, will put you on a two and a half fathom shoal. There is a safe channel all round this shoal: but ships can have no necessity for going to the southward or eastward of it.

On the Quiriquina side of the channel avoid the Aloe shoal (situated one-sixth of a mile off the first bluff to the northward of the low sandy point), by keeping the north-west bluff of Espinosa ridge open a sail's-breadth (5°) with Talcahuana Head *, and do not stand into the bay between the Aloe shoal and the sandy point. The low sandy point, Punta Arena, may be approached within three hundred yards, after which it is advisable not to shut in Espinosa Bluff with Talcahuana Head, both mentioned before: for although there is a wide channel between the Belen Bank and Fronton Reef (off the south end of Quiriquina), yet, as there are no good cross marks for the shoal, a stranger had better not run the risk, particularly as there will be found ample space to work between this line and the Para Reef. When the hut on Look-out Hill is over the N.W. extremity of Talcahuana village, and the Fort S. Joa bears W. by S. ½ S. the Belen is past †, and the anchorage may be safely approached by a proper attention to the lead. Be careful to avoid drifting down upon the Belen, either in bringing up in squally weather, or in casting; and remember that on approaching it the soundings are no guide, as it has eight fathoms close to it. There is no passage inside the shoal for ships, except in case of urgent necessity. There is no good land-mark for the channel.

Men of war anchor in six or eight fathoms; Fort St. Augustine S. 45° W., true; Fort Galvez, N. 57° W., true; Talcahuana Head, S. 7° 30 W., true. Merchant vessels usually

* These are two remarkable bluffs situated to the left of Talcahuana, Espinosa being the furthest inland.

† This mark, it must be remembered, carries you well clear of the Belen, and in bringing them on, take care not to shoot too far over toward Talcahuana Head, or to shoal the water on that side to less than five fathoms.

go quite close in shore, between the Shag Rock, a flat rock near the anchorage, and Fort Galvez, and anchor in three or four fathoms; in doing this, until the Shag Rock is passed, keep a *red mark*, which will be seen upon a hill south of Espinosa Ridge, open with Talcahuana Head. A good berth will be found in three fathoms' mud, close off the town; the *eastern* slope of Espinosa Hill in one with Talcahuana Head. At Talcahuana moor open hawse to the north-eastward; but many think this unnecessary, as the holding ground is so excellent, that it is sufficient to steady the ship with a stream.

Should it happen by any accident that ships, after having passed Quebra Ollas, should not be able to weather Pajaros Ninos (supposing the wind to be from the northward), or should be set upon the northern shore of Talcahuana Peninsula, off which lie scattered rocks, they may run through the channel between Quiriquina and the peninsula. In doing this it is safest to keep close over on the island side, but not in less than seven fathoms water. On the opposite shore a reef extends, eastward from the Buey Rock, to the distance of seven or eight hundred yards from the foot of the cliffs; the mark for clearing it is Fort St. Augustine, *open* with all the capes of Talcahuana Peninsula: but this danger will generally show itself, except the water be particularly smooth, as there is a small rock near its outer edge which dries at half tide *.

Having passed the Buey Rock, haul a little to the westward to avoid a reef off the S.W. extremity of Quiriquina, and be careful not to stand into either of the sandy bays of Quiriquina, between this point and the range of cliffs to the northward of it, or towards the peninsula, so as to bring the Buey Rock to bear to the eastward of N., true, until you have advanced full half a mile to southward, when the lead will serve as a guide. If it be found necessary to anchor, haul into Tombez Bay in the peninsula, and bring up in seven or eight fathoms' mud. This is in the northernmost bay, and may be known by several huts and a large storehouse. When through,

* The narrowest distance between this rock and the reef on Quiriquina sides, is exactly half a mile.

give S. and S.W. points of Quiriquina a berth of half a mile, and having passed them, steer over towards Lirquen, until the two heads (Espinosa and Talcahuana) are open; then pursue the directions before given.

If vessels put into Conception for supplies, the anchorage of Talcahuanha is unquestionably the best, on account of being near the town; but if wood and water only be required, or if it be for the purpose of avoiding bad weather from the northward, &c., the anchorage under the sandy point of Quiriquina will be found very convenient: it is in many respects better sheltered than Talcahuana, particularly from the northerly, north-westerly, and north-easterly winds. The depth is twelve fathoms, the bottom a blue clay, and the marks for the anchorage south point of Fronton S. 76° 20′ W., true; Punta Arena N. 45° E., true; one-sixth of a mile off shore; the sandy point being shut in with Point Darca, and the south end of Quiriquina in one with a hut which will be seen in a sandy bay in the peninsula. On rounding the sandy point (Punta de Arena), which may be done quite close, clew all up, and the ship will shoot into a good berth. Wood may be procured at the island at a cheaper rate than at Talcahuana, and several streams of water empty themselves into the bay to the northward of the point.

The common supplies of Talcahuana are wood, water, fresh beef, live stock, flour, and a bad sort of coal. We found stock of all kinds dear, and paid the following prices: for a bullock, twenty-nine dollars; sheep, three dollars; fowls, three reals each, or four and a half dollars a dozen; nine dollars per ton for coal, although we dug it ourselves.

It is high water, full and change, at Talcahuana at 3h. 20m.; and the tide rises six feet seven inches; but this is influenced by the winds.

GAMBIER ISLANDS.

This group consists of eight high islands, surrounded by coral islands and reefs, enclosing a lagoon, in which there are several secure anchoring places; but the lagoon has many knolls, which render necessary a good look-out from aloft, and even the precaution of keeping a boat ahead. As the islands afford only a supply of water, the anchorage under Mount Duff is the most desirable.

The best channel to enter by lies on the eastern side of the group, to the southward of all the coral islands; and with Mount Duff bearing N. 39° W., true, *in one with the south tangent* of the *easternmost* high island. With these marks steer boldly over the reef, upon which there is in this part six fathoms water, and pass close to the southern extreme of the island, before in one with Mount Duff. Then keeping a boat ahead, proceed under easy sail for the anchorage, about a quarter of a mile south of Mount Duff, the peaks bearing about north, true; but do not attempt to go to the northward, as all that part of the lagoon is full of reefs and knolls. In this situation a ship will be abreast of two streams of good water; but there will be some difficulty in procuring it, on account of the ledges of coral which surround this and all the other islands. As the ground is rocky, it is advisable to use a chain cable. There are several other anchorages, and water may also be had at the north-eastern island, but this appears to me to be, on the whole, the most convenient.

There are also other passages over the reef; and the islands lying to the south-east may be passed on either side, but those which I have recommended are the best and most convenient for navigation with the trade wind. The western channel must not be attempted, and all the south-western part of the group should be avoided as dangerous. The best passage to sail out at bears about S., true, from Mount Duff, the eastern bluffs of Peard Island, upon which Mount Duff is situated, *in one*. This mark will lead over the bar in six and a quarter fathoms. Though this channel lies to leeward of the group, there is generally a very heavy swell upon the reef; and it would not be advisable to attempt it in light winds, as there is no anchoring ground outside; and the swell and the currents, which sometimes run strong, might drift a vessel upon a shallow part of the bar, either to the eastward or westward of the channel, upon which the sea breaks heavily in four fathoms, and outside which there is no bottom at eighty fathoms, within forty yards of the breakers.

The plan which I have given of these islands must not be considered complete, as such a survey required more examination than I could bestow; and there are, no doubt, many knolls of coral in the lagoon which we did not discover. A careful look-out from aloft is therefore absolutely necessary.

It is high water here at 1h. 50m. full and change; but a current generally sets to the westward in the day-time, and runs strong in the western channel.

OTAHEITE.

In clear weather the mountains of Otaheite may be seen ninety miles from the deck. The ports most frequented are situated on the north side of the island, and may be approached without difficulty when the trade wind is blowing. It, however, sometimes happens in the winter months that the trade is interrupted by breezes from the N.W. and W., and at others that calms and unsettled weather prevail. At such times avoid getting into the bay between Otaheite and Tyraboo, especially on the south-west side of the island, as the swell rolls in heavily upon the shore, and there is no anchorage outside the reefs.

Arrived within a few miles of the north-eastern part of Otaheite, several points covered with cocoa-nut trees will be seen stretching from the foot of the hills. One of these is Point Venus, and may be known by One-tree Hill, which, with the exception of the western extremity of the island, is the last bluff head-land upon this part of the coast.

Matavai Bay, on the south-western side of Point Venus, may be considered a safe anchorage from April to December; but during the remainder of the year the trade is liable to interruptions from westerly winds, which blow directly into Matavai, and occasion a high sea. The protection to the anchorage is afforded by Point Venus and the Dolphin shoal, a coral bank, with only two and a quarter fathoms upon its shallowest part. Between it and Point Venus there is a channel about fifty yards wide, with 17, 15, and 10 fathoms close to the reef; and by anchoring a boat on the edge of the shoal, a vessel may enter with perfect safety, provided the breeze be fair. It is, however, better to pass to the southward of the bank, which may be ascertained by two remarkable cocoa-nut trees in the E.N.E. being seen, to the southward of an European built house on the beach, bearing E. by N., and haul round it towards the anchorage, taking care not to get to leeward, so as to bring the N.E. bluff of One-tree Hill to bear to the southward of S.E., as there are several coral banks in that direction. Anchor in eight and a half or nine fathoms, mud, off old Pomarre's house, taking care of the reef that lies off that part of the shore.

To the westward of Matavai there are three good harbours, Papawa, Toanoa, and Papiete, of which the latter is the largest and the most frequented. The others, however, are the most healthy. The entrances to all are extremely narrow, and a stranger ought to take on board a pilot; but he should bear in mind that some of the persons who act in that capacity, though well acquainted with the channels, understand very little about navigating a vessel.

Toanoa is four miles west of Matavai, and may be known by a remarkable ragged mountain, which will be seen through a deep valley when abreast of it. When near, this ragged mountain is very conspicuous, and at night it is a good guide to the entrance.

The channel into Toanoa is only three hundred and thirty yards wide; off the eastern side of the passage there is a rock upon which the sea sometimes breaks, lying N.W. sixty fathoms from the breakers, and another on the *inner* side of the opposite reef. Neither of these rocks, however, narrow the channel much, and are only dangerous in the event of the wind breaking the ship off, or in rounding the reefs closely. With a fair wind sail boldly in, keeping mid-channel, and, clueing all up, allow the ship to shoot into a berth about two cables' length from the shore in thirteen or fourteen fathoms. Here she must wait until the wind falls, and then tow into the harbour; or if the wind be off the land, set fore and aft sails, and keep the boats ready with the lines in them. There are three channels to the inner harbour: of which the two south ones only are frequented, on account of the currents running strong through that to the northward. Perhaps the centre channel, though scarcely broader than a frigate, had better be used going in, and the south coming out. In the centre channel there are eight or twelve fathoms water; but in the southern one a shoal extends from the shore which renders it necessary to keep close to the rock. Anchor in eight and a half fathoms about midway between the outer reef and the shore, opposite some cottages; and moor head and stern by fastening cables to the trees on shore, and carrying out the small bower close to the outer reef.

To proceed to sea it is necessary to warp into the outer anchorage after the sea breeze has done in the evening, or very early in the morning, before it sets in, and push through the channel before the current makes strong. In all these entrances the current sets out in the daytime, sometimes at the rate of two or three knots, and rather sweeps over the reef to the leeward. There is another entrance to Toanoa from Papete, but that just described is the most convenient.

The harbour of Papawa is not frequented, and as it cannot be entered without a pilot, I shall give no directions for it.

PAPIETE.

Two miles to the westward of Toanoa there is a harbour, called by the natives Papiete, capable of containing at least thirty vessels. The entrance is even narrower than that at Toanoa, being only three hundred and seventy feet in the clear, and has a bar with only four and a quarter fathoms upon it. The current here runs out faster than through the channel to the northward, and in blowing weather the sea breaks quite across. This is also a more intricate and dangerous channel than the other; and the only way for a stranger to ensure safety is to moor a boat in the middle of the channel.

There are no good marks for this spot; but as a general remark keep about forty yards from the western extremity of two rocks, which lie eighty yards off the dry part of the eastern reef. These two rocks have only one and a half fathoms upon them, and generally break. There is another rock about sixty yards north of the eastern reef, but this lies out of the channel. On the western side of the channel there is a shoal with only one and a half fathom water upon it, which extends midway between the dry reefs. From this description it is evident that a pilot is necessary for this port, and that the boats should be in readiness to tow or run out kedges as required, whether the pilot advises it or not.

After the entrance is passed, steer S. by E., true, until the first rock on the inside, bearing S. E. by S. one-eighth of a mile from the eastern dry reef, is passed; then haul towards the missionary church and beat up to the anchorage between that shore, which may be approached within a half cable's length, and the reefs which extend from the Moto, or low island, towards the S.W. These reefs will be seen, and may be approached as close as convenient. Another rock lies S. by W., true, 2000 feet from the entrance; but with the trade wind this will be weathered.

If it be necessary, the Moto may be passed to the eastward; but the channel is very narrow, and can only be safely navigated by a person acquainted with it.

Papiete is a very convenient harbour in many respects, but it is subject to calms and much hot weather, in consequence of its being rather to leeward, and the trade wind being obstructed by woods of cocoa-nut trees.

The tides in all these harbours are very irregular. It is generally high water at half an hour after noon every day, and low water at six in the morning.

AWATSKA BAY.

KAMSCHATKA.

It is desirable to make the coast well to the southward of

Cape Gavarea, and to round it as closely as possible, as the wind will in all probability veer to the northward on passing it. If the weather be clear, two mountains will be seen to the west and north-west of the cape, and three far off to the northward and eastward. The eastern one of the two former, called Villeuchinski, is 7.375 feet high, and peaked like a sugar-loaf, and is in latitude 52° 39′ 43″ N., and long. 49′ 46″ W. of Petropaulski. The highest and most northern of the three latter is the mountain of Awatska, in latitude 53° 20′ 01″ N. and long. 3′ 47″ E. of the before-mentioned town. Its height is 11.500 feet, and in clear weather it may be seen a very considerable distance. The centre hill of the three is the volcano, but it emits very little smoke. These peaks are the best guide to Awatska Bay, until near enough to distinguish the entrance, which will then appear to lie between high perpendicular cliffs. Upon the eastern one of these, the *lighthouse bluff*, there is a hut and a signal-staff, and when any vessel is expected a light is sometimes shown. If the harbour be open, a large rock, called the Baboushka, will be seen on the western side of the channel, and three others, named the Brothers, on the eastern side, off the lighthouse. The channel lies in a N. by W. direction, true, and when the wind is fair it may be sailed through by keeping mid-channel; but it frequently happens that vessels have to beat in, and as the narrowness of the channel renders it necessary to stand as close to the dangers as possible, in order to lessen the number of tacks, it is requisite to attend strictly to the leading marks:

The outer dangers are a reef of rocks lying S.E., about two miles from the lighthouse, and a reef lying off a bank which connects the two capes opposite, *i. e.* Stanitski Point with the cape to the southward. To avoid the light-house reef, do not shut in the land to the northward of the lighthouse bluff, unless certain of being at least two miles and a half off shore, and when within three quarters of a miles only, tack when the lighthouse bluff bears N. or N. ½ E. The Brothers Rocks in one with the lighthouse is close upon the edge of the reef. The first western danger has a rock above water upon it, and may be avoided by not opening the Baboushka with the cape beyond, with a flag-staff upon it, or by keeping Stanitski Point well open with the said signal bluff. In standing towards this rock, take care the ebb tide in particular does not set you upon it. A good working mark for all this western shore is the Baboushka, open with *Direction bluff*, the *last* cape or hill on *the left upon the low land* at the head of Awatska Bay. The bay south of Stanitski Point is filled with rocks and foul ground. The lighthouse reef is connected with the Brothers, and the cape must not be approached in any part within half a mile, nor the Brothers within a full cable's length. There are no good marks for the exact limit of this reef off the Brothers, and consequently ships must estimate that short distance. They must also here, and once for all, in beating through this channel, allow for shooting in stays, and for the tides, which, ebb and flood, sweep over toward these rocks, running S.E. and N.E. They should also keep good way on the vessel, as the eddy currents may otherwise prevent her coming about.

To the northward of the Brothers, two-thirds of the way between them and a ragged cape (Pinnacle Point) at the south extreme of a large sandy bay (Ismenai Bay), there are some rocks nearly awash; and off *Pinnacle* Point, (N.N.W. one mile and three quarters from the lighthouse) there is a small reef, one of the outer rocks of which dries at half tide. These dangers can almost always be seen: their outer edges lie nearly in a line, and they may be approached within a cable's length. If they are not seen, do not shut in the Rakovya signal bluff. Off Pinnacle Point the lead finds deeper water than mid-channel, and very irregular soundings.

To the northward of Stanitski Point the Baboushka may be opened to the eastward a little, with the *signal*-staff bluff, but be careful of a shoal which extends about three cables' length south of the Baboushka. Baboushka has no danger to the eastward at a greater distance than a cable's length, and when it is passed there is nothing to apprehend on the western shore, until N.N.W. of the signal-staff, off which there is a long shoal, with only two and two and a half fathoms. The water shoals gradually towards it, and the helm may safely be put down in four fathoms and three quarters; but a certain guide is not to open the *western* tangent of Baboushka with Stanitski Point south of it. There is no other danger on this side of the entrance.

When a cable's length north of Pinnacle Reef, you may stretch into Ismenai Bay, guided by the soundings, which are regular, taking care of a three-fathom knoll which lies half-way between Pinnacle Point and the cape north of it. This bay affords good anchorage, and it may be convenient to anchor there for a tide. There is no other danger than the above-mentioned knoll. The large square rock at the northern part of this bay (Ismenai Rock) may be passed at a cable-length distance. This rock is connected with the land to the northward by a reef, and in standing back toward it the *Pinnacle* Point must be kept *open* with the *lighthouse.* When *in one,* there are but three fathoms and a half. Rakovya signal-staff to the northward in one with the bluff south of it (which has a large green bush over-hanging its brow), will place you in five fathoms close to the rocks.

Off the north bluff of Ismenai Bay there extends a small reef to a full cable-length from the shore; until this is past do not shut in Pinnacle Point with the lighthouse. But to the northward of it you may tack within a cable-length of the bluffs, extending that distance a little off the signal-staff bluff, in consequence of some rocks which lie off there.

Northward of Rakovya signal-staff the only danger is the Rakovya shoal, upon the W. part of which there is a buoy in the summer, and to clear this keep the Brothers *in sight.*

There is no good mark for determining when you are to the northward of this shoal, and as the tides in their course up Rakovya Harbour are apt to set you towards it, it is better to keep the Brothers open until you are certain, by your distance, of having passed it; (its northern edge is seven-eighths of a mile from Rakovya bluff) particularly as you may now stretch to the westward as far as you please, and as there is nothing to obstruct your beat-up to the anchorage. The ground is every where good, and a person may select his own berth.

Rakovya Harbour, on the eastern side of Awatska Bay, will afford good security to a vessel running in from sea with a southerly gale, at which time she might find difficulty in bringing up at the usual anchorage. In this case the Rakovka shoal must be rounded and left to the northward; five and five and a half fathoms will close upon the edge of it, but the water should not be shoaled under nine fathoms.

The little harbour of Petropaulski is a convenient place for a refit of any kind. In entering it is only necessary to guard against a near approach to the signal-staff on the peninsula on the west. The sandy point may be passed within a few yards' distance.

Weighing from the anchorage off the Peninsula flag-staff with light winds and with the beginning of the ebb, it is necessary to guard against being swept down upon the Rakovya shoal, and when past it upon the signal bluff on the same side. There are strong eddies all over this bay; and when the winds are light, ships often become unmanageable. It is better to weigh with the last drain of the flood.

Tareinski Harbour, at the S.W. angle of Awatska Bay, is an excellent port, but it is not frequented. It has no danger, and may safely be entered by a stranger.

It is high water at Petropaulski at 3h. 30m. full and change.
Tide rises . . . 6ft. 7 inch. spring tides.
2.2 neap tides.

SAN FRANCISCO.

CALIFORNIA.

The harbour of San Francisco, for the perfect security it affords to vessels of any burthen, and the supplies of fresh beef and vegetables, wood, and fresh water, may vie with any port on the N.W. coast of America. It is not, however, without its disadvantages, of which the difficulty of landing at low water, and the remoteness of the watering-place from the only anchorage which I could recommend, are the greatest.

Ships bound to San Francisco from the northward and westward should endeavour to make Punta de los Reyes, a bold and conspicuous headland, without any danger lying off it sufficiently far to endanger a ship. In clear weather, when running for the land before the latitude is known, or the Punta can be distinguished, its situation may be known by a table-hill terminating the range that passes at the back of Bodega. This hill in one with the Punta de los Reyes bears E. (mag.). If ships are not too far off, they will see, at the same time, San Bruno, two hills to the southward of San Francisco, having the appearance of islands; and from the mast-head, if the weather be very clear, the South Farallon will in all probability be seen. Punta de los Reyes, when viewed from the W. or S.W., has also the appearance of an island, being connected by low land to the two hills eastward. It is of moderate height, and as it stands at the angle formed by the coast line, cannot be mistaken. Soundings may be had off this coast, in depths varying with the latitude. In the parallel of the Farallones they extend a greater distance from the main land, in consequence of these islands lying beyond the general outline of the coast.

The Farallones are two clusters of rocks, which, in consequence of the shoals about them, are extremely dangerous to vessels approaching San Francisco in foggy weather. The southern cluster, of which in clear weather one of the islands may be seen from the mast-head eight or nine leagues, is the largest and highest, and lies exactly S. 3° E. true, eighteen miles from Punta de las Reyes. The small cluster of rocks lies to the N.W., and still further in that direction there are breakers, but I do not know how far they extend from the rocks above water. In a thick foggy night, we struck soundings in twenty-five fathoms, stiff clay, near them; and on standing off, carried regular soundings to thirty-two fathoms, after which they deepened rapidly.

Coming from the southward, or when inside the Farallones, the position of the entrance to San Francisco may be known by the land receding considerably between the table-hill already mentioned, and San Bruno Hill, which, at a distance, appears to terminate the ridge extending from Santa Cruz to the northward. The land to the northward or southward of these two hills has nothing remarkable about it to a stranger; it is, generally speaking, sufficiently high to be seen thirteen to fifteen leagues, and inland is covered with wood.

About eight miles and a quarter from the fort, at the entrance of San Francisco, there is a bar of sand, extending in a S. by E. direction across the mouth of the harbour. The soundings, on approaching it, gradually decrease to four and a quarter and six fathoms low water, spring tide, depending upon the situation of the ship, and as regularly increase on the opposite side to no bottom with the hand-leads. In crossing the bar, it is well to give the northern shore a good berth, and bring the small white island, Alcatrasses, in one with the fort or south bluff, if it can be conveniently done, as they may then ensure six fathoms; but if ships get to the northward, so as to bring the south bluff in one with the Island of Yerba Buena, they will find but four and a quarter; which is little enough with the heavy sea that sometimes rolls over the bar; besides, the sea will sometimes break heavily in that depth, and endanger small vessels: to the northward of this bearing the water is more shallow. Approaching the entrance, the Island of Alcatrassses may be opened with the fort; and the best directions are to keep mid-channel, or on the *weather side.* On the south shore the dangers are above water, and it is only necessary to avoid being set into the bay between the fort and Point Lobos. If necessary, ships may pass inside, or to the southward of the *One Mile Rock;* but it is advisable to avoid doing so, if possible. On approaching it, guard against the tide, which sets strong from the outer point toward it, and in a line for the fort. Off Punta Boneta there is a dangerous reef, on which the sea breaks very heavy: it lies S.W. from the point, and no ship should approach it nearer than to bring the fort in one with Yerba Buena Island.

In the entrance it is particularly necessary to attend to the sails, in consequence of the eddy tides and the flaws of wind that come off the land. The boats should also be ready for lowering down on the instant, as the entrance is very narrow, and the tides running strong and in eddies, are apt to sweep a ship over upon one side or the other, and the water is in general too deep for anchorage; besides, the wind may fail when most required. The strongest tides and the deepest water lie over on the north shore. Should a ship be swept into the sandy bay west of the fort, she will find good anchorage on a sandy bottom in ten and fifteen fathoms out of the tide; or in the event of meeting the ebb at the entrance, she might haul in, and there await the change. There is no danger off the fort at a greater distance than a hundred yards.

As soon as a ship passes the fort, she enters a large sheet of water, in which are several islands, two rocks above water, and one under, exceedingly dangerous to shipping, of which I shall speak hereafter. One branch of the harbour extends in a S.E. by S. direction exactly thirty miles, between two ridges of hills, one of which extends along the coast towards the Bay of Monterey, and the other from San Pablo, close at the back of San José to San Juan Baptista, where it unites with the former. This arm terminates in several little winding creeks, leading up to the Missions of Santa Clara and San José. The other great branch takes a northerly direction, passes the Puntas San Pablo and San Pedro, opens out into a spacious basin ten miles in width, and then converging to a second strait, again expands, and is connected with three rivers, one of which is said to take its rise in the rocky mountains near the source of the Columbia.

As a general rule in San Francisco, the deepest water will be found where the tide is the strongest; and out of the current there is always a difficulty in landing at low water. All the bays, except such as are swept by the tide, have a muddy flat, extending nearly from point to point, great part of which is dry at low water, and occasions the before-mentioned difficulty of landing; and the north-eastern shore, from Punta San Pablo to the Rio Calavaros beyond San José is so flat that light boats only can approach it at high water. In low tides it dries some hundred yards off shore, and has only one fathom water at an average distance of one mile and a half. The northern side of the great basin beyond San Pablo is of the same nature.

After passing the fort a ship may work up for the anchorage without apprehension, attending to the lead and the tides. The only hidden danger is a rock with one fathom on it at low water, spring tides, which lies between Alcatrasses and Yerba Buena islands. It has seven fathoms alongside it: the lead therefore gives no warning. The marks when on it are, the north end of Yerba Buena Island in one with two trees (nearly the last of the straggling ones) south of Palos Colorados, a wood of pines situated on the top of the hill, over San Antonio, too conspicuous to be overlooked; the left hand or S.E. corner of the Presidio just open with the first cape to the westward of it; Sausalito Point open ¼ point with the north end of Alcatrasses; and the island of Molate in one with Punta de San Pedro. When to the eastward of Alcatrasses, and working to the S.E., or indeed to the westward, it is better not to stand toward this rock nearer than to bring the Table-peak in one with the north end of Alcatrasses Island, or to shut in Sausalito Point with the south extreme of it. The position of the rock may generally be known by a ripple; but this is not always the case.

There are no other directions necessary in working for Yerba Buena Cove, which I recommend as an anchorage to all vessels intending to remain at San Francisco.

In the navigation of the harbour much advantage may be derived from a knowledge of the tides. It must be remembered that there are two separate extensive branches of water lying nearly at right angles with each other. The ebbs from these unite in the centre of the bay, and occasion ripplings and eddies, and other irregularities of the stream, sometimes dangerous to boats. The anchorage at Yerba Buena Cove is free from these annoyances, and the passage up to it is nearly so after passing the Presidio. The ebb begins to make first from the Santa Clara arm, and runs down the south shore a full hour before the flood has done about Yerba Buena and Angel Island; and the flood, in its return, makes also first along the same shore, forcing the ebb over the Yerba Buena side, where it unites with the ebb from the north arm.

The flood first strikes over from the Lime Rock*, and passing the Island of Alcatrasses, where it diverges, one part goes quietly to Santa Clara: the other sweeping over the sunken rock, and round the east end of Angel Island, unites with a rapid stream through the narrow channel formed by Angel Island and the main, and both rush to the northward through the Estrecho de San Pablo to restore the equilibrium of the basin beyond, the small rocks of Pedro Blanco and the Alcatrasses Island lying in the strength of the stream.

The mean of eighty observations gave the time of high water (full and change) at Yerba Buena anchorage	10h. 52m.			
The tide at the springs rises	7ft. 10in. sometimes 8ft. 3in.			
Neap . . .	1 10			
Average rate of ebb at spring tide	2k. 0f.	at neap .	1k. 0f.	
Flood . .	1 0	——— .	0 6	
Duration of flood . . .	5h. 25m.			
At Sausalito the mean of seventeen observations gave the time of high water (full and change)	9 51			
Rise (full and change) . .	6ft. 0in.			
Neap	2 6			
Duration of flood . . .	4h. 43m.			

* See the Chart.

On quitting San Francisco, the direction of the wind in the offing should be considered. If it blow from the S.W. there would be some difficulty in getting out of the bay to the southward of Punta de los Reyes. The residents assert that an easterly wind in the harbour does not extend far beyond the entrance, and that a ship would, in consequence, be becalmed on the bar and perhaps exposed to a heavy swell, or she might be swept back again, and be obliged to anchor in an exposed situation. Northerly winds appear to be most generally approved, as they are more steady and of longer duration than any others: they may, indeed, be said to be the trade-wind on the coast. With them it is advisable to keep the north shore on board, as the strength of the ebb takes that side, and as on the opposite shore, near the One Mile Rock, the tide sets rather *upon* the land. In case of necessity, a ship can anchor to the eastward of the One Mile Rock; but to the S.W. of the rock the ground is very uneven. The wind generally fails in the entrance, or takes a direction in or out. From the fairway steer S.W.½,W. and you will carry seven fathoms over the bar, ½ ebb, spring tide. This I judge to be a good course in and out with a fair wind. I would avoid, by every endeavour, the chance of falling into the sandy bay to the southward of Lobos Point, and also closing with the shore to the N.W. of the Punta Boneta.

MONTEREY.

CALIFORNIA.

The anchorage at Monterey is at the south extremity of a deep bay, formed between Punta Ano Nuevo and Punta Pinos. This bay is about seven leagues across, and open in every part except that frequented by shipping, where it is shut in by Point Pinos. Ships should not enter this bay in light winds in any other part than that used as an anchorage, as there is generally a heavy swell from the westward, and deep water close to the shore.

It is impossible to mistake Point Pinos if the weather be at all clear, as its aspect is very different to that of any part of the bay to the northward. It is a long sloping rocky projection, surmounted by pine-trees, from which it takes its name; whereas the coast line of the bay is all sandy beach. There is no danger in approaching Point Pinos, except that which may ensue from a heavy swell almost always setting upon the Point, and from light winds near the shore, as the water is too deep for anchorage. With a breeze from the southward, Point Pinos should be passed as closely as possible; a quarter of a mile will not be too near: and that shore should be hugged in order to fetch the anchorage. In case of having to make a tack, take care of a shoal at the S. E. angle of the bay, which may be known by a great quantity of sea-weed upon it: there is no other danger. This shoal has three and a half and four fathoms upon its outer edge, and seven fathoms near it. With a fair wind steer boldly towards the sandy beach at the head of the bay, and anchor about one-sixth of a mile off shore in nine fathoms, the fort upon the hill near the beach bearing W. S. W., and moor with the best bower to the E. N. E.

This anchorage, though apparently unsafe, is said to be very secure, and that the only danger is from violent gusts of wind from the S. E. The north-westerly winds, though they prevail upon the coast, and send a heavy swell into the bay, do not blow home upon the shore: and when they are at all fresh they occasion a strong off-set in the bay. This, I believe, is also the case at Callao and at Valparaiso, to which this anchorage bears a great resemblance.

There is no good water to be had at Monterey, and ships in want of that necessary supply must either proceed to San Francisco, or procure a permit from the governor, and obtain it at Santa Cruz, or some of the missions to the southward.

By the mean of many observations on the tides at this place, it is

High water (full and change) at 9 h. 42 m.
Rise is about . . . . 6 ft. 0 in. at spring-tide,
And . . . . . 1 2 at the neaps.

There is very little current at the anchorage.

HONORURU.

SANDWICH ISLANDS.

The harbour of Honoruru has a bar, with only twenty feet water upon it at low water, and the channel is so narrow and intricate that no stranger should attempt it. The natives understand the signal for a pilot, and will come off if the weather is not too boisterous. In consequence of this difficulty ships anchor outside, in about sixteen fathoms water; the Punch-bowl bearing N. N. E. half E., and the highest part of Diamond Point E. by S. ¼ S.

Should it be necessary to enter the harbour, the morning is the best time, as there are then leading winds through the passage; but after the trade wind has set in it cannot be entered. It is necessary to adopt the precaution of having the boats ready to tow or run out lines to the reefs.

From the outer anchorage run along shore in nothing less than eleven fathoms, and look out for a large grass-hut, which stands conspicuous upon the wharf at the north head of the harbour, on the western side of a new yellow European house. When the north end of this hut is *in one* with the *eastern* chimney of an European built house,* with a ship's figure-head attached to it,† haul directly in for the opening between the breakers, which will now be seen.

The bar is about fifty fathoms in breadth, and consists of smooth coral rock, having ten fathoms close to its outer edge, and seven fathoms on the inner.

When on the bar, the King's residence (an European built house with a slate-coloured pointed roof), situated to the N. E. of the town, will be open to the westward of the north-west hummock of Punch-bowl Hill; the before-mentioned mark of the hut and chimney will also be on, and is to be kept so until the outer cocoa-nut tree in Wytiete Bay comes in one with a small rise on the northern part of Diamond Hill. Then

* The only house that had a chimney in 1827.

† These in one bear N. 20° E. by compass.

bring the eastern tangent of the cluster of cocoa-nut trees nearest the fort, in one with a remarkable saddle on the mountain at the back of the town, until the outer part of the dry ground on the right comes on with Diamond *Point*, or until a large hut standing by itself on the north shore of the harbour is in one with *four* cocoa-nut trees in a cluster. With these marks, steer for the *four* trees, open the trees to the eastward until they are a sail's-breadth apart; and when the fort flag-staff is one with the trees eastward of the fort, anchor in four and a quarter fathoms, mud.

These directions will, I think, be intelligible to a person on the spot; but I must repeat, that no stranger should run for this harbour, except in cases of absolute necessity. Should it be attempted, a good look-out from the jib-boom end, or fore-yard, will be found serviceable.

In consequence of the sea that rolls over the reef, and breaks in four or five fathoms water, it is necessary that boats should follow nearly the directions that have been given for vessels, except that when the eastern point of the dry land on the right of the entrance comes on with Diamond Hill, they may then steer for the south end of a stone wall, which will be seen on the western side of the harbour; and when the before-mentioned yellow house opens, they may steer for the landing place. Unless they adopt these precautions, they will in all probability run upon the reefs, or be upset. And in entering the harbour, it is necessary for boats as well as shipping to keep the marks strictly on.

I shall conclude these remarks, the greater part of which have been furnished by Mr. Elson, the master, by observing, that the water in the wells in the town is unwholesome upon a voyage, and that it is proper to send the casks up the river to be filled.

TYPA.

MACAO.

The depth of water in the Typa has diminished within these last thirty years, as there are now not more than nine and a half or ten feet water, at the lowest spring-tides, and no vessel drawing more than fourteen or at the most fifteen feet, can enter at the top of the tide.

There are no marks required for this channel; but with the last of the flood (say three-quarters), enter between Kaloo and Kai-kong, keeping about mid-channel, and when the *western* point of the *western* Kai-kong opens with the ragged point at the S. W. extremity of the *eastern* Kai-kong, keep a little to the northward, and pass the ragged point at the distance of a quarter of a mile; then steer mid-channel between the islands, remembering not to attempt the channel between the *western* Kai-kong and the island of Makarina The water will now deepen, and when the town of Macao opens with the *west* Kai-kong, and when the ragged point bears *east*, anchor in about eighteen or twenty feet water; in which berth you will have good riding ground over a muddy bottom.

The time of high water is 9 h. 30 m.
The tide at full and change rises 7 ft. 1 in.; rate about 2 k. 4 f.;
at the neap . . 2 1 . . . . . 1 6.

The flood sets to the northward from the anchorage, and branches off on meeting the tide setting westward to the north of Kai-kong.

NAPAKIANG.

LOO CHOO.

Ships bound to Napakiang may pass close round the south extremity of the island, and sail along the western coast at the distance of a mile or a mile and a half. They will then see a sandy island in latitude 26° 05′ 50″ N., and longitude 7′ 40″ W. of Abbey Point, which is the only danger to the westward of Loo Choo that I am acquainted with, until near the Kirrama Islands, or to the northward of the entrance of Napakiang.

Abbey Point, at the south extremity of the port of Napakiang, may be known by its ragged outline, and by a small wooded eminence called Wood Point, situated about a mile and a half to the southward of it. The mainland here falls back, and forms a bay, which is sheltered by coral reefs stretching to the northward from Abbey Point; they are, however, disconnected, and between them and the point there is a channel sufficiently deep for the largest ship. Nearly in the centre of this channel, outside withal, there is a coral bank named Blossom Rock, having a good passage on either side of it. The channel between it and Abbey Point should be adopted with southerly winds and flood tides, and that to the northward with the reverse. A reef extends off Abbey Point, which, for convenience of description, will be called Abbey Reef. When off Abbey Point a rocky headland will be seen, about a mile and a half north of the town; this I shall call Kumi Head, and upon the ridge of high land beyond it three hummocks will be seen to the left of a cluster of trees. In the distance, a little to the left of these, is Mount Onnodake, in latitude 26° 27′ N. A remarkable rock, which, from its form, has been named Capstan Rock, will next appear; and then, to the northward of the town, a rocky head with a house upon its summit, which I shall call False Capstan Head. At the back of Capstan Rock there is a hill, named Sheudi, upon which the upper town is built. The highest southern point of this is one of the landmarks to which I shall have to refer.

Having opened out the Capstan Rock, haul towards Abbey Reef, and bring the right-hand hummock about 4° to the east of Kumi Head, and steering in with these marks on, you will pass through the south channel in about seven fathoms water, over the tail of Blossom Rock. You may now round Abbey Reef tolerably close, and steer in for the anchorage. Should the wind veer to the eastward in the passage between Blossom Rock and Abbey Point, with the above-mentioned marks on, you must not stand to the northward, unless the *outer* cluster of trees near the extremity of Wood Point are in one with, or open to the westward of, Table Hill, a square rocky headland to the southward of it. This mark clears also the tongue of Oar Reef, which with Blossom Rock forms the other western channel.

It is advisable, with the wind to the north-eastward, to beat through the channel north of Blossom Rock (Oar Channel), in preference to that above-mentioned. To do this, bring the *false* capstan-head in one with a flat cluster of trees on the ridge to the *right of the first gap south* of Sheudi: this will clear the *north* tongue of Blossom Rock; but unless the Table Hill be open to the eastward of Wood Hill, you must not stand to the southward, but tack directly the water shoals to less than twelve fathoms, and endeavour to enter with the marks on. Having passed to the N. E. of Blossom Rock, which you will know by Wood Hill being seen to the *right* of Table Hill, stand towards Abbey Point as close as you please; then tack, and on nearing Oar Reef take care of a tongue which extends to the eastward of it, and be careful to tack immediately the *outer trees* of Wood Point open with Abbey Point. In entering at either of the western channels, remember that the flood-tide sets to the northward over Blossom's Rock, and the ebb to the southward.

The best anchorage is in Barnpool, at the N. E. part of the bay, in seven fathoms water, where a vessel may ride in perfect security. The outer anchorage, I should think, would be dangerous with a hard westerly gale. The Blossom anchored there in fourteen fathoms muddy bottom: Abbey Bluff, S. 43° 20′ W.; Capstan Rock, S. 75° 40′ E.; (mag.); variation 53′ 59″ E.

The entrance to Barnpool lies between Barnhead and the reef off Capstan Rock. In entering, you are not to approach Barnhead nearer than to bring the north tangent of *Hole Rock* (to the northward of Capstan Rock) in one with the before-mentioned *flat clump of trees* on the hill south of Sheudi, until the point of the burying ground (Cemetery Point) is seen just clear of Capstan Head. You may anchor in any part of Barnpool.

As the northern channel into Napakiang is very dangerous, I shall not tempt any person to sail through it, by giving directions for it.

It is high water at Napakiang at 6 h. 28 m., full and change; rise from five to seven and a half feet, but this was very irregular during our stay at the place.

ARZOBISPO ISLANDS.

PORT LLOYD.

This group of islands lies N. by E. and S. by W., and is divided into three clusters, extending from 27° 44′ 35″ N. to 26° 30′ N. and beyond. As I have described these islands in my narrative, I shall here give only the necessary direction for entering Port Lloyd, which is the best harbour in the group, and indeed, the only one that should be frequented.

DIRECTIONS FOR ENTERING PORT LLOYD.

Having ascertained the situation of the port, steer boldly in for the *southern head;* taking care not to bring it to the *northward* of N. 47° E, true, or to shut in with it two paps on the N.E. side of the harbour, which will be seen nearly in one with it on this bearing. *In this position they are a safe leading mark.* To the southward of this line there is broken ground.

If the wind be from the southward, which is generally the case in the summer time, round the south Bluff at the distance of two hundred yards, *close to a sunken rock,* which may be distinctly seen in clear weather. Keep fresh way upon the ship, in order that she may shoot an end through the eddy winds, which baffle under the lee of the head *; and to prevent her coming round against the helm, which would be dangerous. The winds will at first break the ship off, but she will presently come up again: *if she does not,* be ready to go about, as you will be close upon the reefs to the northward, and put the helm down *before the south end of the island off the port to the westward* comes on with *the High Square Rock at the north side of the entrance.*

If she comes up, steer for a high *Castle Rock* at the east end of the harbour, until a pointed rock on the sandy neck to the eastward of the *south* headland comes in one with a high sugar-

* Keep the top-gallant clew-lines in hand.

loaf shaped grassy hill to the southward of it. After which you may bear away for the anchorage, taking care not to open the sugar-loaf again to the westward of the pointed rock †. The best anchorage, Ten-fathom Hole excepted, which it is necessary to warp into, is at the northern part of the harbour where the anchor is marked in the plan.

In bringing up, take care of *a spit which extends off the south end of the small island* near Ten-fathom Hole, and not to shoot so far over to the *western* reef as to bring a rock, at *the outer foot of the south bluff*, in *one* with *some black rocks* which will be seen near you to the south-westward. The depth of water will be from eighteen or twenty fathoms, clay and sand.

If the wind be from the northward, beat between the line of the afore-mentioned *Sugar Loaf and Pointed Rock* westward, and a north and south line from the Castle Rock to the eastward. This rock on the western side, as well as the bluff to the northward of it, may be *shaved* if necessary. The hand-leads are of very little use in beating in here, as the general depth is twenty or twenty-four fathoms.

The best watering-place is in Ten-fathom Hole. It is necessary to be cautious of the sharks, which are very numerous in this harbour. It is high water 6h. 8m., full and change.

TRES MARIAS AND SAN BLAS.

WEST COAST OF MEXICO.

The Tres Marias, situated 1° 15′ west of San Blas, consist of three large islands, steep and rocky, to the westward, and sloping to the eastward with long sandy spits. Off the S. E. extremity of Prince George's Island (the centre of the group) we found that the soundings decreased rapidly from seventy-five fathoms to seventeen, and that after that depth they were more regular. Two miles from the shore we found ten and twelve fathoms, bad holding ground. There is nothing to make it desirable for a vessel to anchor at these

† This rock is white on the top with birds' dung, and looks like an island.

islands. Upon Prince George's Island there is said to be water of a bad description; but the landing is in general very hazardous.

There are passages between each of these islands. The northern channel requires no particular directions: that to the southward of Prince George's Island is the widest and best; but care must be taken of a reef lying one third of a mile off its S.W. point, and of a shoal extending a mile and a half off its south-eastern extremity. I did not stand close to the south Maria, but could perceive that there were breakers extending full three quarters of a mile off its S.E. extremity; and I was informed at San Blas that some reefs also extended from two to four miles off its south-western point. There is an islet off the north-west part of this island, apparently bold on all sides; but I cannot say how closely it may be approached.

From the south channel Piedro de Mer bears N. 76° E. true, about forty-five miles. It is advisable to steer to windward of this course, in order that, as the winds, during the period at which it is proper to frequent this coast, blow from the northward, the ship may be well to windward.

The Piedro de Mer is a white rock, about a hundred and thirty feet high, and a hundred and forty yards in length, with twelve fathoms all round it, and bears from Mount St. Juan N. 77° W. thirty miles.

Having made Piedro de Mer, pass close to the southward of it, and unless the weather is thick, you will see a similarly shaped rock, named Piedro de Tierra, for which you should steer, taking care not to go to the northward of a line of bearing between the two, as there is a shoal which stretches to the southward from the mainland. This course will be S. 79° E. true, and the distance between the two rocks is very nearly ten miles.

To bring up in the road of San Blas, round the Piedro de Tierra, at a cable's length distance, and anchor in five fathoms, with the low rocky point of the harbour bearing N. ½ E., and the two Piedros in one. This road is very much exposed to winds from S.S.W. to N.N.W., and ships should always be prepared for sea, unless it be in the months in which the northerly winds are settled. Should the wind veer to the westward, and a gale from that quarter be apprehended, no time should be lost in slipping and endeavouring to get an offing, as a vessel at anchor is deeply embayed, and the holding ground is very bad. In case of necessity a vessel may cast to the westward, and stand between the Piedro de Tierra and the Fort Bluff, in order to make a tack to the westward of the rock, after which it will not be necessary again to stand to the northward of a line connecting the two Piedros.

The road of San Blas should not be frequented between the months of May and December, as during that period the coast is visited by storms from the southward and westward, attended by heavy rains, and thunder and lightning. It is besides the sickly season, and the inhabitants having all migrated to Tepic, no business whatever is transacted at the port.

It is high water at San Blas at 9h. 41m., full and change; rise between six and seven feet spring tide.

MAZATLAN.

The anchorage at Mazatlan, at the mouth of the Gulf of California, in the event of a gale from the south-westward, is more unsafe than that at San Blas, as it is necessary to anchor so close to the shore, that there is not room to cast and make a tack. Merchant vessels moor here with the determination of riding out the weather, and for this purpose go well into the bay. Very few accidents, however, have occurred, either here or at San Blas, as it scarcely ever blows from the quarter to which these roads are open between May and December.

There is no danger whatever on the coast between Piedro de Mer and Mazatlan; the lead is a sure guide. The island of Isabella is steep, and has no danger at the distance of a quarter of a mile. It is a small island, about a mile in length, with two remarkable needle rocks lying near the shore to the eastward of it.

Beating up along the coast of Sonora, some low hills, of which two or three are shaped like cones, will be seen upon the sea-shore. The first of these is about nine leagues south of Mazatlan, and within view of the island of Creston, which forms the port of Mazatlan. A current sets to the southward along this coast, at the rate of eighteen or twenty miles a day.

Having approached the coast about the latitude of 23° 11′ N., Creston and some other steep rocky islands will be seen. Creston is the highest of these, and may be further known by two small islands to the northward of it, having a white chalky appearance. Steer for Creston, and pass between it and a small rock to the southward, and when inside the bluff, luff up, and anchor immediately in about seven and a half fathoms, the small rock about S., 17° E., and the bluff W. by S. Both this bluff and the rock may be passed within a quarter of a cable's length; the rock has from twelve to fifteen fathoms within thirty yards of it in every direction It is, however, advisable to keep at a little distance from the bluff, to escape the eddy winds. After having passed it be careful not to shoot much to the northward of the before-mentioned bearing (W. by S.), as the water shoals suddenly, or to reach so far to the eastward as to open the *west* tangent of the *peninsula* with the *eastern* point of a low rocky island S. W. of it, as that will be near a dangerous rock, nearly in the centre of the anchorage, with only eleven feet water upon it at low spring-tides, and with deep water all round it. I moored a buoy upon it; but should this be washed away, its situation may be known by the eastern extreme of the before-mentioned low rocky island, between which and Battery Peak there is a channel for small vessels, being in one with a *wedge-shaped protuberance* on the *western* hillock of the *northern island* (about three miles north of Creston), and the N.W. extremity of the high rocky island to the *eastward* of the anchorage being a little open with *a rock off the mouth of the river* in the N. E. The south tangent of this island will also be open a little (4°), with a dark *tabled hill* on the second range of mountains in the east. These directions will, I think, be quite intelligible on the spot.

The winds at Mazatlan generally blow fresh from the N.W. in the evening; the sea-breeze springs up about ten in the forenoon, and lasts until two o'clock in the morning.

It is high water at this place at 9 h. 50 m., full and change rise seven feet spring tide.

These are all the directions which I think it necessary to give in this place, as the ports of Coquimbo and Valparaiso, at which the Blossom touched, are so easy of access, and so well known, as to require none; and Port Clarence and Kotzebue Sound, near Beering's Strait, so little likely to be frequented, and so free from danger, that it would be extending the limits of this work unnecessarily to add any thing on the subject. Besides, the charts of those places which have been published since our return contain all that a vessel can require for her guidance.

GEOGRAPHICAL POSITIONS.

The words in italics in the first column are native names.

NAMES OF PLACES.	Latitude. North.	Longitude. From Meridian of	Longitude. West from Greenwich.	Remarks.
	° ′ ″	*Greenwich.*	° ′ ″	
EIGHT STONES . .	34 48 20	. . . .	16 47 59	Do not exist within the horizon of this spot.
SANTA CRUZ, . . TENERIFFE, . .	28 27 51	. . . .	16 14 23	Saluting Battery.
	South.			
FERNANDO NORONHA,	3 52 55	. . . .	32 15 35	The Church Peak.
RIO JANEIRO, . . .	22 54 37	. . . .	43 04 41	Villegagnon Fort.
TALCAHUANA, . . .	36 42 35	. . . .	72 56 59	Fort St. Augustine.
		Talcahuana.	72 56 56	Observatory.
VALPARAISO, . . .	. . .	1° 27′ 36″ E.	71 29 20	From Conception } Landing place.
		0 16 57 W	71 33 34	From Coquimbo
SALAS Y GOMEZ ISL. .	26 27 46	32 23 12	105 20 08	S. E. extreme.
EASTER ISLAND, .	27 08 46	36 27 40	109 24 36	Perouse Point, Cook's Bay.
	27 06 28	36 15 22	109 12 18	Peaked Hill on N. E. extreme.
	27 11 21	36 28 30	109 24 26	Needle Rock.
	27 03 33	. . . .	. . .	Point St. John.
DUCIE'S ISLAND, . .	24 40 20	51 48 42	124 45 38	N. E. extreme.
HENDERSON'S or ELIZABETH ISLAND,	24 21 18	55 21 31	128 18 27	N. E. extreme.
PITCAIRN ISLAND, .	25 03 37	57 11 13	130 08 09	Village from Talcahuana.
		. .	130 08 23	Ditto from Bow Island.
		Pitcairn Isl.	130 08 23	Village.
HERCULES, or OENO I.	24 01 21	0 32 36 W	130 40 59	N. E. extreme of trees.
CRESCENT ISLAND,	23 20 29	4 26 45	134 35 08	South extreme.
	23 17 39	. . . .	. . .	N. W. extreme.
GAMBIER ISLANDS,	23 08 23	4 46 58	134 55 21	Watering Valley.
	23 07 58	4 46 31	134 54 54	Eastern Peak of Mount Duff.
	23 01 17	. . . .	. . .	N. extreme of Low Island.
	23 15 15	. . . .	. . .	South extreme of breakers.
		Gambier.	134 55 21	
HOOD'S ISLAND, . .	21 30 50	0 37 58	135 33 19	West point.
MINERVA, or CLERMONT DE TONNERE, I	18 33 42	1 06 11	136 01 32	S. E. extreme.
	18 28 48	. . . .	. . .	North point.
SERLE'S ISLAND,	18 16 01	2 05 24	137 00 45	Northern big tree.
	18 22 39	1 59 42	136 55 03	S. E. extreme.
WHITSUNDAY ISLAND,	19 23 38	3 41 27	138 36 48	Large tree near N.W. extreme.
QUEEN CHARLOTTE ISL.	19 17 40	3 47 07	138 42 28	Large tree near E. extreme.
LAGOON (Cook's) ISL. native name *Teay*,	18 43 19	3 51 52	138 47 13	West clump cocoa-nuts.
	18 42 26	. . . .	. . .	North extreme.
	. .	3 47 51	138 43 12	East extreme.
THRUM CAP ISLAND,	18 30 08	4 12 39	139 08 00	Cluster trees on N. W. extreme.
EGMONT ISLAND, .	19 22 59	4 16 42	139 12 03	Cluster near north extreme.
	19 24 26	4 19 13	139 14 34	S. W. extreme.
BARROW ISLAND, .	20 45 07	4 07 48	139 03 09	North extreme.
		4 08 48	139 04 09	West extreme.

NAMES OF PLACES.	Latitude. South.	Longitude. From Meridian of	Longitude. West from Greenwich.	Remarks.
	° ′ ″	*Gambier.*	° ′ ″	
CARYSFORT ISLAND,	20 44 53	3° 27′ 23″ W	138 22 44	Cocoa-nut trees, N.E. extreme.
		3 24 07	138 19 28	East extreme.
OSNABURGH ISLAND, or MATILDA REEF,	21 47 00	. . . .	. . .	North extreme.
	21 53 42	4 04 13	138 59 34	S. W. extreme.
	21 50 32	3 49 07	138 44 28	East extreme.
	21 50 00	3 58 33	138 53 54	Sandy Island on the Bar.
COCKBURN ISLAND,	22 12 25	3 44 32	138 39 53	Hillock at N. E. extreme.
	22 17 09	. . . .	. . .	S. W. extreme.
BLIGH'S LAGOON ISLAND,	21 37 41	5 42 37	140 37 58	North extreme.
BYAM MARTIN ISLAND,	19 40 22	5 27 07	140 22 28	N. W. extreme.
GLOUCESTER ISLAND, *Tooe Tooe,* . . .	19 07 38	5 42 28	140 37 49	N. E. extreme.
	19 08 44	5 45 30	140 40 51	S. W. extreme.
BOW ISLAND, called *Heyou*	18 04 31	. . . .	140 56 58	Morai at entrance.
	. .	*Bow Island.*	140 51 35	Observatory.
	18 04 00	. . . .	. . .	North extreme.
	18 08 31	0 09 24 W.	141 00 59	Cluster cocoa-nuts W. extreme.
	18 26 06	0 13 09 E.	140 38 26	S. E. extreme.
MOLLER ISLAND, . *Amannoo,* . . .	17 44 18	0 16 21	140 35 14	N. E. cocoa-nuts at extreme.
	17 52 51	0 03 09	140 48 26	S. W. extreme.
RESOLUTION *(Towerey),*	17 22 20	0 32 15 W.	141 23 50	Cocoa-nuts S. E. extreme.
CUMBERLAND ISLAND,	19 10 19	0 19 08	141 10 43	North Stony point.
	19 12 20	0 17 31	141 19 06	S. E. extreme.
PRINCE WILLIAM, HENRY ISLAND, or LOSTANGE, . .	18 49 02	. . . .	. . .	South extreme of reef.
	18 45 53	0 51 03	141 42 38	S. W. extreme.
	18 42 54	0 47 50	141 39 25	N. E. extreme.
Two Groups. *Dawahaidy,* .	18 18 10	1 15 08	142 06 43	South extreme.
	18 15 36	1 12 22	142 03 57	S. E. do.
	18 10 08	1 15 08	142 06 43	Two cocoa-nuts near N. extreme.
Maracau, . .	17 58 24	1 16 20	142 08 15	North extreme.
	18 09 58	. . . .	. . .	South extreme.
DOUBTFUL ISLAND, .	17 19 46	1 29 36	142 22 11	East extreme.
MELVILLE ISLAND,	17 34 59	1 47 37	142 39 12	N. W. extreme.
	. .	1 40 11	142 31 46	S. E. extreme.
BIRD ISLAND, . . .	17 48 00	2 13 17	143 04 52	North extreme.
CROKER ISLAND, . .	17 26 30	2 32 07	143 23 42	Two cocoa-nut trees E. extreme.
MAITEA ISLAND, . .	17 53 39	7 9 12	148 00 47	The Peak.
Otaheite. *Tiarraboo,* . . .	17 54 12	. . . .	. . .	South tangent.
—	. .	8 14 07	149 05 42	S. E. extreme.
POINT VENUS, .	. .	8 37 25	149 29 00	
		Toanoa.	149 30 42	Observatory.
EIMEO ISLAND, . .	17 29 51	0 16 22 W	149 47 04	Peak with hole through it.
TETHEROA ISLAND, .	17 02 23	0 00 16	149 30 58	S. E. extreme.
	North.	*Woahoo.*	158 00 00	
ONEEHOW ISLAND, .	21 52 15	2 23 20	160 23 20	Yam Bay S. W. extreme.
PETROPAULSKI, . . .	53 00 58	*Petropaulski.*	201 16 30	Church.
VILLEUCHINSKY MOUNT,	52 40 43	0 22 51 W	201 39 21	
CAPE GAVARIA, . .	52 21 43	0 4 22	201 20 52	
HIGH NORTHERN PEAK,	53 19 30	0 04 05 E.	201 12 25	
BEERING'S ISLAND,	55 22 14	7 16 21 W	194 00 09	North low points.
	55 17 02	7 06 09	194 10 21	West point, or Point Kytroff.
BEERING'S SEAL ROCK,	55 13 35	7 00 51	194 15 39	N. W. end.
CLARKE'S ISLAND,	63 24 40	29 37 00	171 39 30	S. W. cape.
	63 51 10	29 47 00	171 29 30	N. W. cape.
		Chamisso.	161 46 00	
ST. PAUL'S ISLAND, .	57 10 33	8 31 48	170 17 48	The Western peak.

NAMES OF PLACES.	Latitude. North.	Longitude. From Meridian of	Longitude. West from Greenwich.	Remarks.
	° ′ ″	*Petropaulski.*	° ′ ″	
St. George's Island,	56 37 30	7° 46′ 49″ w	169 32 49	The south peak.
Diomede Isl. { N. W. or Ratmanoff Island,	65 51 12	7 17 45	169 03 45	N. W. extremity.
Diomede Isl. { Krusenstern Isl.	65 46 17	7 09 10	168 55 10	South extremity.
Diomede Isl. { S. E. or Fairway Rock, . .	65 38 40	6 57 45	168 43 45	Centre.
East Cape, . . .	66 03 10	7 57 50	169 43 50	South-east extremity.
Cape Prince of Wales,	65 33 30	6 13 10	167 59 10	Bluff under the peak.
Cape Espenberg,	66 34 56	1 50 38	163 36 38	East extreme.
Cape Krusenstern, {	67 08 00	2 00 00	163 46 00	Low cape not defined.
	67 11 05	1 50 45	163 36 45	Western bluff over Cape K.
Cape Deceit,	66 06 20	0 54 32	162 40 32	At S. E. extreme of Kotzebue [Sound.
Point Rodney, . .	64 42 10	4 31 50	166 17 50	
King's Island, . .	64 58 49	6 11 47	167 57 47	Northern peak.
Cape York, ...	65 24 10	5 33 40	167 19 40	
Port Clarence, . .	65 16 40	5 01 50	166 47 50	Poiut Spencer.
Chamisso Island, .	66 13 11	. . . .	161 46 00	The summit by obs.
Cape Mulgrave, .	. . .	2 11 41	163 57 41	Badly defined.
Cape Thomson, . .	68 07 39	4 06 26	165 52 26	
Sharp Peak over Cape Seppings,	67 57 20	2 55 21	164 41 21	
Hope Point, . . {	68 19 50	5 00 24	166 46 24	Sandy point.
	68 19 15	. . . .	. . .	Lieutenant Belcher.
Cape Dyer, . . .	68 37 52	4 22 19	166 08 19	
Cape Lisburne, . {	68 52 09	4 19 39	166 05 39	Flint Station.
	68 52 03	. . . .	. . .	Lieutenant B.
Cape Sabine, . . .	68 56 40	2 49 08	164 35 08	
Cape Beaufort, . .	69 06 47	1 52 28	163 38 28	Coal Station.
Lake Station, . .	69 34 23	1 20 40	163 06 40	Village.
Icy Cape, . . . {	70 20 01	0 00 08	161 46 08	Village.
	70 19 08	. . . .	. . .	Lieutenant Belcher.
Cape Collie, . . .	70 37 24	1 50 36 E.	159 55 24	
Point Barrow, . .	71 23 31	. . . .	156 21 30	Boat expedition.
San Francisco, . .	37 47 50	*San Francisco.*	122 23 07	Observatory.
Punta de los Reyes,	37 59 40	0 36 52 w	122 59 59	The extremity of the cliff.
Great Farallon, .	37 41 55	0 35 51	122 58 58	The Peak.
Table Hill, . . .	37 55 40	0 10 27	122 33 37	
Bolbones Mountain,	37 52 55	0 29 26 E.	121 53 44	Height 3765 feet.
San Francisco, . .	37 48 30	0 04 16 w	122 27 23	The fort.
Notch Hill, . . .	37 30 58	0 00 00	122 23 07	A small peak on the coast.
Monterey, ...	36 36 24	0 31 21	121 51 46	The fort.
Point Pinas, . . .	36 37 15	. . . .	. . .	
Hononoru Fort, (*Woahoo.*) . .	21 18 12	. . . .	158 00 25	
		Woahoo.		
Macao, ...	22 12 00	88 31 18 w	246 31 18	Saluting battery.
		Typa.	246 28 00	
Assumption Island,	19 40 53	31 55 18 E.	214 32 42	W. end Kaikong.
Mangs, ...	19 57 02	31 47 48	214 40 12	The peak.
North Bashee, . .	. . .	8 29 00	238 01 04	{ Peak on Centre Island, The Mangs from east point
Vela Rete, . . .	. . .	7 19 32	239 08 28	Assumption, true N. 27°
Formosa, ...	. . .	7 23 21	239 04 38	Highest rock. [07½′ W.
Pedra Branca, . .	. . .	1 33 13	244 54 47	S.E. tangent.

NAMES OF PLACES.	Latitude.	Longitude.		Remarks.
	North.	From Meridian of	West from Greenwich.	
	° ′ ″	° ′ ″	° ′ ″	
Little Botel Tobago,	21 57 30	8 08 30 E.	238 19 30	N. E. extreme.
Xima, ...	21 57 00	8 07 50	238 20 10	S. W. extreme.
Great Tobago, Xima, .	22 01 40	8 07 45	238 20 15	S. W. extreme.
Great Tobago, Xima, .	22 06 10	8 00 50	238 27 10	N. W. extreme.
Samsanne Island, .	22 41 15	8 00 30	238 27 30	The centre.
Loo Choo, ...	26 12 25	14 10 20	232 17 40	Abbey Point station.
Ditto, ...	26 04 05	. . . .	. . .	South extreme.
		Abbey Point.	232 17 40	
Sandy Island, . .	26 05 50	0 07 40 W	232 25 20	The centre.
Kirrama Island, .	26 09 00	0 25 30	232 43 10	The high wedge-shaped island.
Abp.'s Group. Port Lloyd, .	27 05 35	14 29 30 E	217 48 29	No. 1 station, N. ex. harbour
		From No. 1 sta	217 48 30	
Abp.'s Group. N.W. Island of Parry's Group	27 43 30	0 03 49 W	217 52 19	The N. W. tangent.
Abp.'s Group. Kater Island,	27 29 40	0 00 42 E.	217 47 48	The north extreme.
San Blas, ...	. . .	17 08 35	105 15 30	At Arsenal.
		San Blas.	105 14 43	Town 47″ east of Arsenal.
San Juan Mount, .	21 27 00	0 18 10 E	104 56 33	Southern pap.
Tonalisco Mount, .	21 46 48	0 29 48	104 44 55	
Tepic	21 30 42	. . . .	. . .	Consulate.
Piedra de Mer, . .	21 34 45	0 13 30 W	105 28 13	
Isabella Island, .	21 51 15	0 37 20	105 52 03	The peak.
Tres Marias. Northern Island,	21 32 53	1 13 20	106 28 03	The south bluff.
Tres Marias. San Juanito,	21 45 00	1 23 50	106 38 35	Flat Island, N. W. part.
Tres Marias. San Juanito,	21 44 05	1 24 37	106 39 20	High rock.
Tres Marias. Prince George,	21 28 12	1 09 58	106 24 41	The northern peak.
Tres Marias. Southern Island,	21 19 22	0 57 20	106 12 03	The eastern peak.
Mazatlan, ...	23 11 40	1 07 41	106 22 24	High bluff at extreme.
Corvetena, . . .	. . .	0 33 06	105 47 49	Small rock off C. Corrientes.
Cape Corrientes, .	. . . .	0 24 30	105 39 13	This cape, in one S. Juan Mt bears N. 32° 24′ E., true.
Colima Mountain, .	19 24 42	1 41 42 E.	103 33 01	12,003 feet high.
Acapulco, . . .	16 50 32	5 23 59	99 50 44	Fort San Carlos.
	South.			
Coquimbo, . . .	29 56 57	. . . .	71 16 41	The copper foundery.

AURORA BOREALIS.

We had frequent opportunities of observing the Aurora Borealis in the autumns of 1826 and of 1827. From the 25th of August until the 9th October, about the time of the departure of the Blossom from the northern regions in both years, this beautiful meteor was visible on every night that was clear, or when the clouds were thin and elevated.* It is remarkable that, in both years, its first appearance was on the 25th August. The season of 1826 was distinguished by an almost uninterrupted succession of fine weather and easterly winds, and that of the following year by continued boisterous weather and winds from the westward. In the former year, the weather being fine, the Aurora was more frequently seen than in the latter; but in 1827 the displays were brighter, and the light more frequently passed to the southward of the zenith. It never appeared in wet weather.

In 1826, when, as before mentioned, the weather was settled, the Aurora generally began in the W.N.W. and passed over to the N.E., until a certain period, after which it as regularly commenced in the N.E. and passed to the N.W.; whilst in 1827 the appearance of the meteor was as uncertain as the season was boisterous and changeable. The period when this change in the course of the light took place coincided very nearly with that of the equinox; and as the Aurora Borealis has been supposed to be affected by that occurrence, we imagined that the change might be in some way owing thereto, but the irregularity of the meteor in this respect in 1827 gave a contradiction to this hypothesis. It was, however, uniform in making its appearance always in the northern hemisphere, and generally in the form of elliptical arches from 3° to 7° of altitude, nearly parallel with the magnetic equator. These

* In 1826 it was visible on twenty-one nights; in 1827 only eleven.

arches were formed by short perpendicular rays passing from one quarter to the other with a lateral motion, or by their being met by similar rays from the opposite direction. The arches, when formed, in general remained nearly stationary, and gave out coruscations, which streamed toward the zenith. When at rest the light was colourless, but when any movement took place it exhibited prismatic colours, which increased in strength as the motion became rapid. The coruscations seldom reached our zenith, and more rarely passed to the southward of it, but when that occurred the display was always brilliant: on one occasion only they extended to the southern horizon.

We remarked, that when any material change was about to occur one extremity of the arch became illuminated, and that this light passed along the belt with a tremulous hesitating movement toward the opposite end, exhibiting the colours of the rainbow. An idea may be formed of this appearance from the examination of the rays of some moluscous animals in motion, such as the nereis, but more particularly the beroes. Captain Parry has compared its motion to the waving of a ribbon. See Second Voyage, p. 144. As the light proceeded along the arch, coruscations emanated from it; and as the motion became violent the curve was often deflected and sometimes broken into segments, which were brightest at their extremities, and in general highly coloured. When one ray of the Aurora crossed another, the point of intersection was sometimes marked by a prismatic spot, very similar to that which occurs in the intersections of coronæ about the moon, but far more brilliant; and when the segments, which generally *crooked* toward the zenith, were much curved, colours were perceptible in the bend. Generally speaking, after any brilliant display, the sky became overcast with a dense haze, or with light fleecy clouds.

The Aurora has been frequently observed to rest upon a dark nebulous substance, which some persons have supposed to be merely an optical deception, occasioned by the lustre of the arch; but this appearance never occurs *above* the arch, which would be the case, I think, if these surmises were well founded. We sometimes saw this cloud before any light was visible, and observed it afterwards become illuminated at its upper surface, and exhibit all the appearances above mentioned. It was the general opinion that the lustre of all the stars was diminished by the Aurora, but particularly by this part of it. Captain Parry, however, observes in his Journal, p. 142, that the stars in this dark cloud were unobscured, except by the light of the Aurora. He, however, agrees with us in the lower part of the arch being always well defined, and in the upper being softened off, and gradually mingled with the azure of the sky. It is worthy of notice, that we never observed any rays shoot downwards from this arch, and I believe the remark will apply equally to the observations of Captains Parry and Franklin. We frequently observed the Aurora attended by a thin fleecy-cloud like substance, which, if not part of the meteor, furnishes a proof of the displays having taken place within the region of our atmosphere, as the light was decidedly seen between it and the earth. This was particularly noticed on the 28th of September, 1827. The Aurora on that night began by forming two arches from W. by N. northward to E. by N., and about eleven o'clock threw out brilliant coruscations. Shortly after the zenith was obscured by a lucid haze, which soon condensed into a canopy of light clouds. We could detect the Aurora above this canopy by several bright arches being refracted, and by brilliant colours being apparent in the interstices. Shortly afterwards the meteor descended, and exhibited a splendid appearance, without any interruption from clouds, and then retired, leaving the fleecy stratum only visible as at first. This occurred several times, and left no doubt in my own mind of the Aurora being at one time above and at another below the canopy formed about our zenith. I must not omit to observe here that, on several occasions, when the light thus intervened between the earth and the cloud, brilliant meteors were precipitated obliquely toward the south and south-west horizons.

This supposition of the light being at no great elevation is strengthened by the different appearances exhibited by the

Aurora at the same times to observers not more than from ten to thirty miles apart, and also by its being visible to persons on board the ship at Chamisso Island, after it had vanished in Escholtz Bay, only ten miles distant, as well as by the Aurora being seen by the barge detached from the Blossom several days before it was visible to persons on board the ship, about two hundred miles to the southward of her.

Captain Franklin has mentioned a similar circumstance in his notices on the Aurora Borealis in his first expedition, when Dr. Richardson and Mr. Kendall were watching for the appearance of the meteor by agreement, and when it was seen by the former actively sweeping across the heavens and exhibiting prismatic colours, without any appearance of the kind being witnessed by the latter, then only twenty miles distant from his companion. Captain Parry also, in his Third Voyage, describes the Aurora as being seen even between the hills and the ship anchored at Port Bowen.

Dr. Halley and other philosophers have supposed that the coruscations of the Aurora proceeded always in radii perpendicular to the surface of the earth, in the direction of the magnetic meridian from the poles towards the equator, and the former has ingeniously accounted for the apparent deviations occasionally witnessed on the principles of perspective; but this explanation is not quite satisfactory, as Captains Parry, Franklin, and ourselves, in Kotzebue Sound, have seen these rays emanate from almost all parts of the horizon, and actually pass the zenith. At the same time I am disposed to believe, from my own observation, that the radii in general take the perpendicular direction above alluded to, probably on account of the less resistance they meet in the higher regions of the atmosphere than in such as near the surface of the earth; and this will partly account for the appearance of the cone formed at the zenith of the ships at Melville Peninsula, described in Captain Parry's Second Voyage, page 146, and of another very similar, witnessed by ourselves in Kotzebue Sound on the 26th August, 1827, on which occasion the rays shot up from all directions, and formed over our zenith the perfect appearance of a tent stretched upon a number of poles united at their ends; but even here the rays could not have been quite parallel unless their extremities were infinitely high.

In Kotzebue Sound the Aurora was seldom visible before ten o'clock at night, or after two o'clock in the morning. We never heard any noise, nor detected any disturbance of the magnetic needle: but here I must observe that Kater's compass was the only instrument employed for this purpose, and then on board the ship only, the exposed situation in which we were anchored not admitting of any establishment on shore, either for this purpose or for astronomical observations.

Mr. Collie, the surgeon of the Blossom, whose attention to meteorological phenomena was unwearied, has given an ingenious hypothesis on the subject of the Aurora. After expressing his opinion that this meteor occurs in the region of the thin and higher clouds of the earth's atmosphere, he observes, that "it is highly probably that the two strata of atmospheric fluid proceeding in opposite directions—the one from the equinoctial toward the polar regions, and the other in the reverse direction—are charged with opposite electricities, and that they are in different degrees of temperature and of humidity: the upper stratum, flowing from the equator toward the poles, being of a higher temperature and more charged with vapour than the lower, proceeding from the pole to the equator. They might thus be charged with opposite electricities, which would communicate and neutralize each other.

"The opposite temperatures would be reduced to their mean, and under certain circumstances these changes might be attended with the evolution of electrical light, and with the condensation of transparent vapour into thin clouds (stratus-cirrus, or cirro-stratus). As the watery particles of these clouds form, a certain degree of electric conductibility would be established, by which this subtle fluid might be propagated to short distances; but the greater dryness of the air, both above and below this region of thin mist, would oppose an unconducting barrier to its escape. As soon as one thin cloud, a thin stripe of cirrus, or fleecy portion of cirro-stratus or cirro-cumulus, became charged with electricity, it would occasion, by the laws of electric phenomena, an opposite electrical state in that portion nearest it; and these opposite electricities would instantly attract each other, fly together, burst forth in fire, and become neutralized. If there should be a plane in which such thin clouds are formed, the subversion and re-establishment of the balance of electricity being thus begun would be rapidly propagated throughout the whole of this space, and produce that rapid, undulatory motion which we observe in the Aurora Borealis."

In considering the subject of the Aurora Borealis, my attention was drawn to a fact which does not appear to me to have been hitherto noticed. I allude to the direction in which the Aurora generally makes its first appearance, or, which is the same thing, the quarter in which the arch formed by this meteor is usually seen. It is remarkable, that in this country the Aurora has always been seen to the northward; by the expeditions which have wintered in the ice it was almost always seen to the southward; and by the Blossom, in Kotzebue Sound, 250 miles to the southward of the ice, it was, as in England, always observed in a northern direction. Coupling this with the relative positions of the margins of the packed ice, and with the fact of the Aurora appearing more brilliantly to vessels passing near the situation of that body, than by others entered far within it, as would seem to be the case from the reports of the Greenland ships, and from my observations at Melville Island and at Kotzebue Sound, it does appear, at first sight, that that region is most favourable to the production of the meteor. I do not, however, presume to offer any hypothesis on the subject; but having witnessed the extraordinary change that takes place in the atmosphere, along the whole line of ice covering the Polar Sea, I should be remiss if I omitted to direct the attention of the natural philosopher to the circumstance. There is perhaps no part of the globe where the atmosphere undergoes a greater or more sudden change than over this line of the ice. A diminution of 10 or 15 degrees of temperature constantly occurs within the space of a few miles: the humid atmosphere over the ocean may sometimes be seen laden with heavy clouds, which disperse as they arrive at this line of reduced temperature, and leave the region over the ice exposed to a bright sunshine. Indeed the extraordinary effect of this large body of ice upon the atmosphere, particularly when the sea is deep and the temperature of the ocean and its superstratum of air high, as between Spitzbergen and Greenland, will scarcely be credited by persons who have not witnessed it. Mr. Scoresby has given some extraordinary instances of this in his Arctic Voyages; and to these I will add one of many which fell under my own observation. The ships of the first polar expedition were beset in the ice about nine miles from the open sea. It was blowing a hard gale upon the ice, and we could perceive a ship carrying off under storm stay-sails only. There was nothing between us and the ship to intercept the gale, and yet we were becalmed during the whole of the day. The atmosphere over the open sea was loaded with clouds *(nimbi)*, while that over the ice enjoyed a bright sunshine throughout. The limits of these opposite states of the atmosphere, by seamen called the *ice-blink*, were marked by a well-defined line, nearly perpendicular over the margin of the ice. As the heavy clouds reached this spot they were gradually condensed, the effect of which was precisely similar to that which sometimes occurs about the summits of high mountains, against which the clouds are successively driven, without any being seen to depart, and without any apparent increase.

This remarkable disturbance of the equilibrium of the atmosphere being admitted, I would here merely suggest whether, under certain dispositions of the atmosphere, electricity might not be induced and communicated to the surrounding region, so as to occasion the Aurora Borealis, and to account for its appearance in the before-mentioned directions in preference to others.

I am not aware what would be the effect of the meeting of two atmospheres, one influenced by a large body of ice, the other by an extensive continent, such as that of America, and particularly when the circumstances might be modified by large frozen lakes. But it appears from Captain Franklin's observations at Great Bear Lake, that the Aurora arose in almost all quarters of the horizon, and more frequently illuminated his zenith than the Auroras appear to have done either of those at the before-mentioned places.

Our observations were too limited to justify any remark on the observation of Captain Franklin, that the appearance of the Aurora occurs more frequently in the last quarters of the moon than in others.

THE END.

J. B. NICHOLS AND SON, 25, PARLIAMENT STREET.

Printed in the United States
By Bookmasters